Artificial Intelligence for Wireless Communication Systems

The text provides a comprehensive study of the application of advanced artificial intelligence (AI) in next-generation wireless communications with a focus on theory, standardization, and core development. It further highlights AI-enabled intelligent architecture for sixth-generation (6G) networks to realize smart resource management, automatic network adjustment, and intelligent service layers. The book covers artificially assisted non-orthogonal multiple access schemes for 6G communication.

This book:

- Discusses the use of AI in various aspects of wireless communications, including channel modeling, signal detection, channel coding design, and resource management.
- Explores technical challenges in the ubiquitous fifth-generation (5G) wireless networks and the prospects of introducing AI-based techniques in the envisioned 6G wireless networks.
- Presents potential issues in AI-enabled approaches in wireless communications.
- Covers AI-enabled energy efficiency optimization and cross-layer optimization in next-generation wireless networks.
- Explains artificially empowered security and privacy schemes in next-generation wireless networks and next-generation mobile management.

It is primarily written for senior undergraduates, graduate students, and academic researchers in the fields of electrical engineering, electronics and communication engineering, and computer engineering.

Artificial Intelligence for Wireless Communication Systems
Technology and Applications

Edited by
Samarendra Nath Sur, Agbotiname Lucky Imoize,
Ankan Bhattacharya, Debdatta Kandar and
Jyoti Sekhar Banerjee

CRC Press
Taylor & Francis Group
Boca Raton London New York

CRC Press is an imprint of the
Taylor & Francis Group, an **informa** business

Front cover image: metamorworks/Shutterstock

First edition published 2025
by CRC Press
2385 NW Executive Center Drive, Suite 320, Boca Raton FL 33431

and by CRC Press
4 Park Square, Milton Park, Abingdon, Oxon, OX14 4RN

CRC Press is an imprint of Taylor & Francis Group, LLC

ISBN: 978-1-032-57667-1 (hbk)
ISBN: 978-1-032-85334-5 (pbk)
ISBN: 978-1-003-51768-9 (ebk)

DOI: 10.1201/9781003517689

Typeset in Sabon
by codeMantra

Contents

Preface

Next-generation wireless networks demand intelligent algorithms capable of adapting to well-defined network protocols and achieving efficient resource management to satisfy billions of data-hungry applications. However, traditional techniques for analyzing wireless communication systems are no longer optimal due to increasing network complexity, precise signal processing requirements, and substantial computational costs. Moreover, designing and optimizing 5G and beyond 5G wireless networks pose challenges due to critical requirements for network efficiency and high-quality user experience. To address these problems, artificial intelligence (AI) approaches, including deep learning (DL), deep reinforcement learning (DRL), transfer learning (TL), federated learning (FL), and distributed learning, are currently being introduced to wireless communication systems. AI is gaining widespread popularity as a candidate technology to tackle the design complexity issues in wireless systems. Specifically, AI has demonstrated great potential for providing solutions to network issues that seemed intractable in the past. Additionally, AI has delivered remarkable results in speech synthesis, image detection, and natural language processing. Furthermore, AI finds applications in resource management, channel modeling, and wireless signal processing. Thus, the research focus has shifted to harnessing the vast potential of AI-based technologies in the design and development of next-generation wireless systems.

This book proposes a comprehensive overview of the application of advanced AI in next-generation wireless communications, with a primary focus on theory, standardization, and core development. The book delves into various aspects of wireless communications where AI can play a pivotal role, such as channel modeling, channel estimation, signal detection, channel coding design, resource and mobility management, and localization. Furthermore, the book explores the technical challenges inherent in the pervasive adoption of 5G wireless networks and anticipates the integration of AI-based techniques in the envisioned 6G wireless networks. The book also addresses potential issues that may arise wit AI-enabled approaches in wireless communications and provides recommendations for further research exploration in this dynamic field.

Finally, the book extensively discusses the prospects and societal benefits anticipated with the introduction of AI in future wireless communications, highlighting the transformative potential of this emerging technology.

About the editors

Samarendra Nath Sur (senior member, Institute of Electrical and Electronics Engineers (IEEE)) received an M.Tech degree in Digital Electronics and Advanced Communication from Sikkim Manipal University in 2012 and Ph.D degree in MIMO Signal Processing from the National Institute of Technology (NIT), Durgapur, in 2019. Currently, he is working as an assistant professor in the Department of Electronics & Communication Engineering. His current research interests include broadband wireless communication (MIMO and spread spectrum technology), advanced digital signal processing, remote sensing, and radar image/signal processing (soft computing). He is a senior member of IEEE and a member of IEEE-IoT, the IEEE Signal Processing Society, and the Institution of Engineers (India) (IEI). He has published more than 120 SCI/ Scopus-indexed international journal and conference papers. He is also a regular reviewer of reputed journals, namely IEEE, Springer, Elsevier, Taylor & Francis, IET, and Wiley. He has co-edited several books with Springer Nature, Elsevier, Routledge, and CRC Press. He is also serving as a guest editor for topical collections/special issues of journals like Springer Nature, MDPI, and Hindawi. He is currently engaged with the International Journal on Smart Sensing and Intelligent Systems as an associate editor.

Agbotiname Lucky Imoize (senior member, IEEE) received the B. Eng. degree (Hons.) in Electrical and Electronics Engineering from Ambrose Alli University, Nigeria, in 2008 and the M. Sc. degree in Electrical and Electronics Engineering from the University of Lagos, Nigeria, in 2012. He is a lecturer in the Department of Electrical and Electronics Engineering at the University of Lagos, Nigeria. Before joining the University of Lagos, he was a lecturer at Bells University of Technology, Nigeria. He was, until recently, a research scholar at the Ruhr University Bochum, Germany, under the sponsorship of the Nigerian Petroleum Technology Development Fund (PTDF) and the German Academic Exchange Service (DAAD) through the Nigerian-German Postgraduate Program. He was awarded the Fulbright fellowship as a visiting research scholar at the Wireless@VT Laboratory, Bradley Department of Electrical and Computer Engineering, Virginia Tech., USA, where he worked under the supervision of Prof. R. Michael Buehrer from 2017 to 2018. He worked as a core network product manager at ZTE, Nigeria, and as a network switching subsystem engineer at Globacom, Nigeria. His research interests cover the fields of 6G wireless communication, wireless security systems, and artificial intelligence. He has co-edited seven books and co-authored over 200 articles in peer-reviewed journals and conferences. He is an active reviewer and editor for over 50 international journals and conferences. He is the vice chair of the IEEE Communication Society, Nigeria chapter, a registered engineer with the Council for the Regulation of Engineering in Nigeria, and a member of the Nigerian Society of Engineers.

Ankan Bhattacharya obtained his Ph.D. from the National Institute of Technology, Durgapur, India. He is the author of several research papers, which have been published in many reputed journals and conferences at national and/or international levels. Dr. Bhattacharya is a life member of the Forum of Scientists, Engineers & Technologists (FOSET), a member of the Institution of Engineers India (IEI), and a member of the International Association of Engineers (IAENG). His areas of research are antenna engineering, computational electromagnetics, electronic circuits and systems, signal processing, microwave devices, and wireless communication technologies. He is also an editor/reviewer of many national/international journals of repute. He has organized/participated in many national/international conferences, seminars, and workshops. He is a series editor of the book series, "Modern Aspects of Computing, Devices and Communication Engineering,". Dr. Bhattacharya has been active in delivering Invited Talks and has also been a part of many national/international conferences in the capacity of coordinator, session chair, technical committee member, etc. Dr. Bhattacharya has been appointed as a guest editor of Discover Applied Sciences, a multi-disciplinary, peer-reviewed journal of Springer Nature. Dr. Bhattacharya is the associate editor of the "Journal of Information Processing Systems (JIPS)," an official international journal of the Korea Information Processing Society (indexed in Scopus and ESCI), and also the associate editor of Human-Centric Computing and Information Sciences (HCIS), a SCI-indexed Q1 Journal. Presently, he is associated with Hooghly Engineering & Technology College, Hooghly, West Bengal, India, as an associate professor and head of the Electronics and Communications Engineering Department.

Debdatta Kandar was born in 1977 in Deulia, Kolaghat, Purba Medinipur, West Bengal, India. He has more than 18 years of professional experience. Currently, he is holding the post of Head of the Department and Professor in the Department of Information Technology, North-Eastern

Hill University, Shillong, India. Before joining NEHU, he worked at S.K.P. Engineering College, Tiruvannamalai, Tamil Nadu, and Sikkim Manipal Institute of Technology, Sikkim, India. He worked on a DRDO-sponsored project, and he has several years of industry experience too. He has successfully guided several Ph.D., M.Tech, and B.Tech students. Four of his students have already been awarded Ph.D. degree. He has national and international collaborations for carrying out research work. He has delivered talks in different workshops, seminars, FDP, etc. He successfully organized several national and international seminars, conferences, etc. He also chaired several sessions at different conferences. He has published approximately 90 research papers in different national and international journals, conference proceedings, and book chapters. He also published edited books, etc. He has completed his post-doc from the Department of Electrical and Electronic Engineering Technology, University of Johannesburg, Johannesburg, South Africa. He received his Ph.D. (Engg.) from the Department of Electronics and Telecommunication Engineering, Jadavpur University, Kolkata, West Bengal, India, in 2011. He has been awarded "Young Scientist" award from Union Radio Science International (URSI GA-2005) at Vigyan Bhavan, Delhi, for his research work. The President of India, Dr. A.P.J. Abdul Kalam, invited him to his residence on that occasion. His research interests include wireless mobile communication, artificial intelligence, soft computing, and radar operation.

Jyoti Sekhar Banerjee is currently serving as the Head of the Department in the Computer Science and Engineering (AI and ML) Department at the Bengal Institute of Technology, Kolkata, India. Additionally, he is also the professor in charge of R&D and Consultancy Cell & Nodal Officer of the IPR Cell of BIT. Dr. Banerjee did his postdoctoral fellowship at Nottingham Trent University, UK, in the Department of Computer Science. He also completed the postgraduate diploma in IPR and TBM from MAKAUT, WB. He has teaching and research experience spanning 19 years and has completed one IEI-funded project. He is a member of the CSI, IEEE, ISTE, IEI, ISOC, and IAENG and a fellow of IETE. He is the present honorary secretary-cum-treasurer of the ISTE WB section. He is the current honorary secretary of the Computer Society of India, Kolkata Chapter. He is also the Executive Committee Member of the IETE, Kolkata Centre. He has published over 70 papers in various international journals, conference proceedings, and book chapters. So far, he has published ten edited books and two textbooks. Currently, he is processing a few more edited books for reputed international publishers like Springer and CRC Press. He is also processing four more textbooks; those are now in press. He is serving as editor in chief of the *American Journal of Advanced Computing (AJAC), Smart Society, USA*; Editorial Board Member of the Indian

Journal of Technical Education (IJTE) (UGC-CARE List Journal); and Series Editor for "Advances in Disruptive Technologies and Generative Artificial Intelligence," published by CRC Press, USA. He is the lead author of "A Text Book on Mastering Digital Electronics: Principle, Devices, and Applications." Currently, he is serving as the General Chair of ICHCSC 2022, 2023, and 2024 and General Co-Chair of GCAIA 2021. Dr. Banerjee is the Program Chair of the Human 2023 and AAIDS 2024. Dr. Banerjee is the Organizing Chair of the EAIT 2024, ICNSBT 2023 and 2024, and ICACA 2024. Dr. Banerjee served as a guest editor of ICAUC_ES 2021 and ICPAS-2021 issues in the *IOP Journal of Earth and Environmental Science, Physics, & MAICT 2021* issues in the AIP Conference Proceedings, all of which are Scopus-indexed proceedings. His areas of research interests include computational intelligence, cognitive radio, sensor networks, AI/ML, network security, different computing techniques, IoT, WBAN (e-healthcare), and expert systems.

Contributors

Talatu Adamu
Department of Electrical and
  Electronic Engineering
Air Force Institute of Technology
  (AFIT)
Kaduna, Nigeria

Samuel Adedeji Adeleye
Department of Information Systems
Federal University of Technology
Akure, Ondo State, Nigeria

Christopher Akinyemi Alabi
Telecommunication Engineering
  Department
Air Force Institute of Technology
  (AFIT)
 Kaduna, Nigeria

Saikat Chandra Bakshi
Department of Electronics and
  Communication Engineering
National Institute of Technology
Calicut, Kerala, India

Jyoti Sekhar Banerjee
Department of CSE (AI & ML)
Bengal Institute of Technology
Kolkata, West Bengal, India

Avishek Bhattacharjee
Department of Electronics and
  Communications Engineering,
  Hooghly Engineering &
  Technology College
Hooghly, West Bengal, India

Ankan Bhattacharya
Department of Electronics and
  Communications Engineering
Hooghly Engineering &
  Technology, College
Hooghly, West Bengal, India

Ashim Kumar Biswas
Department of Electronics and
  Communication Engineering
JIS College of Engineering
Kalyani, West Bengal, India

Showkat Ahmad Dar
Public Administration
Annamalai University
Cuddalore, Tamil Nadu, India

Avik Kumar Das
Department of Electronics and
  Telecommunication Engineering
Indian Institute of Engineering
  Science and Technology, Shibpur
Howrah, West Bengal, India

Sayantika Das
Department of Computer Science
   and Engineering
Hooghly Engineering &
   Technology College
Hooghly, West Bengal, India

Subham Das
Department of Electronics and
   Communications Engineering
Hooghly Engineering &
   Technology College
Hooghly, West Bengal, India

Sujoy Das
Department of Computer Science
   and Technology
Narula Institute of Technology
Kolkata, West Bengal, India

Arnab De
School of Computing and
   Informatics
Vignan's Foundation for Science,
   Technology and Research
Guntur, Andhra Pradesh, India

Manoj Kumar Dey
Hooghly Engineering &
   Technology College
Hooghly, West Bengal, India

Arka Gain
Department of Computer Science
   and Technology
Narula Institute of Technology
Kolkata, West Bengal, India

Madhushree Ghosh
Hooghly Engineering &
   Technology College
Hooghly, West Bengal, India

Sumana Hazra
Hooghly Engineering &
   Technology College
Hooghly, West Bengal, India

Monday Abutu Idakwo
Computer Engineering Department
Federal University
Lokoja, Kogi State, Nigeria

Agbotiname Lucky Imoize
Department of Electrical and
   Electronics Engineering
University of Lagos
 Lagos, Nigeria

Ibraheem Temitope Jimoh
Department of Information Systems
Federal University of Technology
Akure, Ondo State, Nigeria

Koushik Karmakar
Department of Computer Science
   and Technology
Narula Institute of Technology
Kolkata, West Bengal, India

Ankit Kumar
Tata Nexarc
Bangalore, Karnataka, India

Krishanu Kundu
Department of Electronics &
   Communication Engineering
GL Bajaj Institute of Technology &
   Management
Greater Noida, Uttar Pradesh,
   India

Koustubh Majumdar
Department of Computer Science
   and Engineering
Hooghly Engineering &
   Technology College
Hooghly, West Bengal, India

Akindeji Ibrahim Makinde
Department of Information Systems
Federal University of Technology
Akure, Ondo State, Nigeria

Subhojit Malik
ECE Department
Hooghly Engineering &
    Technology College
Hooghly, West Bengal, India

Sayantan Mallik
Department of Electronics and
    Communications Engineering
Hooghly Engineering &
    Technology College
Hooghly, West Bengal, India

Sourav Nandi
Hooghly Engineering &
    Technology College
Hooghly, West Bengal, India

Adewale Omotolani Oronti
Department of Information Systems
Federal University of Technology
Akure, Ondo State, Nigeria

Narendra Nath Pathak
Department of Electronics &
    Communication Engineering
Dr. B.C. Roy Engineering College
Durgapur, West Bengal, India

Ankita Pramanik
Department of Electronics and
    Telecommunication Engineering
Indian Institute of Engineering
    Science and Technology, Shibpur
Howrah, West Bengal, India

Shiladitya Pujari
University Institute of Technology
The University of Burdwan
Burdwan, West Bengal, India

Bappadittya Roy
School of Electronics Engineering
Vellore Institute of Technology-AP
    University
Guntur, Andhra Pradesh, India

Manab Kumar Saha
Hooghly Engineering &
    Technology College
Hooghly, West Bengal, India

P. Sakthivel
Department of Political Science and
    Public Administration
Annamalai University
Cuddalore, Tamil Nadu, India

Panagiotis Sarigiannidis
Department of Informatics and
    Telecommunications Engineering
University of Western Macedonia
Kozani, Greece

Rupam Some
Education Specialist
SingleStore India Pvt. Ltd.
Chinsurah, West Bengal, India

Samarendra Nath Sur
Department of Electronics and
    Communication Engineering,
    Sikkim Manipal Institute of
    Technology
Sikkim Manipal University
Rangpo, East Sikkim, India

Tulika Verma
Department of Computer Science
   and Engineering (Artificial
   Intelligence and Machine
   Learning)
Pranveer Singh Institute of
   Technology
Kanpur, Uttar Pradesh, India

Kuldeep Verma
Department of Defence Studies
Hindu College
Moradabad, Uttar Pradesh, India

Pradeep Vishwakarma
Department of Electronics and
   Communication Engineering,
   Sikkim Manipal Institute of
   Technology
Sikkim Manipal University
Rangpo, East Sikkim, India

Chapter 1

Artificial intelligence revolutionizing wireless communication systems

Samarendra Nath Sur, Pradeep Vishwakarma, and Ankan Bhattacharya

1.1 INTRODUCTION

The evolution of advanced computing capabilities, advancements in algorithms, and the revolution in the field of big data are driving a new era of technological revolution in human society, led by artificial intelligence (AI) [1]. AI and machine learning (ML) stand out as the most rapidly advancing and increasingly essential components for the evolution beyond 5G (B5G)/6G communication systems, and it is due to the pressing need to deliver services with heightened system capacity, reduced latency, enhanced reliability, improved spectral efficiency, and the facilitation of massive Internet of Things (IoT) deployments.

With the continuous expansion of ML technologies, wireless communication systems are experiencing rapid advancements. A significant challenge faced by today's wireless networks lies in accommodating large volumes of traffic with diverse quality of service (QoS) and the quality of experience (QoE) for end-user requirements. For instance, 5G networks are engineered to support enhanced mobile broadband (eMBB), ultrareliable low-latency communications (URLLCs), and massive machine-type communications (mMTCs), aiming to deliver high data rates, reliability within strict latency constraints, and massive connectivity for IoT devices, respectively [2]. Similarly, the most promising concept in B5G appears to be cell-free massive multiple-input multiple-output (CF mMIMO) technology. Despite its appealing features, CF mMIMO systems face challenges such as power distribution and channel estimation. Researchers have demonstrated through various studies that deep learning (DL) has proven to be effective in addressing a wide array of these challenges [3]. Incorporating AI technology into 6G networks holds the potential to introduce novel and impactful applications, including intelligent network management, edge computing, and heightened security and privacy measures. Nevertheless, ensuring the responsible development and integration of AI within 6G networks demands technical proficiency and collaborative efforts among stakeholders. Such endeavors are essential to guarantee the ethical and advantageous utilization of this transformative technology [4]. Moreover, effective resource management plays a pivotal role in facilitating information exchange among vehicles, infrastructure, device-to-device (D2D) connections, safety-related information transmission, health updates, and extremely accurate map navigation in 6G networks. Specifically, it presents challenges in concurrently achieving reliability, computational efficacy, fast data rates, reduced latency, and energy-effective network [4]. Hence, there is a stringent requirement to devise comprehensive optimization frameworks that address resource management challenges, including allocation of radio resource, user association, spectrum management, optimal allocation of power, and computational order. These efforts are indispensable for addressing the varied requirements presented by 6G applications in wireless networks.

In regard to wireless communication systems, conventional methods for addressing resource management challenges often rely on suboptimal or heuristic optimization techniques. However, the joint optimization issues concerning different parameters associated with the resource management in forthcoming 6G networks pose NP-hard problems that cannot be effectively tackled using global optimization algorithms or suboptimal approaches. This is mainly attributed to the exponential increase in computational complexity and the challenge of minimizing performance disparities from the perfect approach. For reconcile the trade-off between complexity of computation and inherent performance gaps in addressing NP-hard issues, AI/ML algorithms have emerged as an alternative to traditional methodologies. This shift has spurred research efforts into exploring ML techniques for jointly optimizing the resources of networks through ML-driven intelligent management of resources in 6G [5]. Moreover, the integration of ML with 6G networks holds promise for intelligently optimizing the resources of networks, facilitating learning in real time, and enabling complex choice-making for current autonomous networks. Consequently, it paves the way for extensive network densification while addressing the growing emphasis on global sustainability and equity [6].

This chapter explores the burgeoning synergy between AI and wireless communication, delving into various facets such as channel model, channel decoding and signal detection, resource management, adaptive modulation and coding schemes, and autonomous network management. It also highlights the associated challenges and future direction of AI in wireless communication.

1.2 AI-BASED CHANNEL MODEL

Channel prediction involves mathematically forecasting the natural propagation of signals to aid receivers in approximating the affected signal, particularly crucial in highly dynamic channels. Wireless communication channel modeling typically employs either deterministic or stochastic methodologies. The deterministic approach relies on electromagnetic theories, requiring knowledge of every object within the propagation space, exemplified by ray tracing. In contrast, stochastic modeling utilizes measurements involving statistical distributions of channel parameters. The aim of the channel modeling is to extract and generate channel state information (CSI), encompassing parameters like fading, power delay profile (PDP), delay, Doppler, large-scale fading (LSF), and average the power delay profile (APDP), from the channel impulse response (CIR) [7]. In classical channel modeling, the general expectation–maximization (EM) algorithm is commonly employed to estimate the CSI corresponding to every multipath components. This is achieved by implementing space-alternating generalized expectation–maximization (SAGE) algorithms, which serve as an extension of the EM algorithm. However, with the increasing usage of multiple users, applications, and multiple-input multiple-output (MIMO) systems, SAGE algorithms are becoming more complex, necessitating the exploration of other sophisticated and computationally efficient techniques [8]. Both researchers and academia seek more efficient methods with reduced complexity and enhanced accuracy. The advent of ML technology offers a novel avenue for processing extensive measurement and traffic data pertaining to the wireless channel. Consequently, novel channel learning strategies are being proposed to develop models independent of wireless channel modeling, addressing these challenges.

With the advent of high computational capabilities and abundant data, AL emerges as a transformative force in system design for the new radio 5G. Within AI, various subcategories such as ML and deep learning (DL) techniques, including supervised learning methods, are being leveraged to predict the channel state information (CSI) across diverse environments based on specific datasets. ML plays a crucial role in enabling swift channel modeling for 5G wireless communication systems, utilizing partially relevant channel measurement data and models. A comparative analysis between the classical channel model and ML-based channel approach is presented in Figure 1.1. By employing ML algorithms, channel modeling complexity can be reduced while enhancing accuracy, thereby minimizing the need for extensive measurements. Researchers are also delving into ML methods that establish connections between wireless channel modeling across different systems. Given the extensive operations and measurements involved, researchers increasingly turn to ML to bolster channel modeling predictions. ML techniques can effectively predict and estimate wireless channel parameters, encompassing both large- and small-scale fading phenomena such as path loss, path loss exponent, frequency, Doppler spread, and various random variables explaining large-scale fading effects.

Researchers have dedicated considerable effort to harness AI technologies for acquiring comprehensive deterministic insights into real-world channels, subsequently enabling accurate predictions of these dynamically changing channels. Furthermore, a DL technique can be utilized to train an NN model with the true data. With the acquired knowledge and the

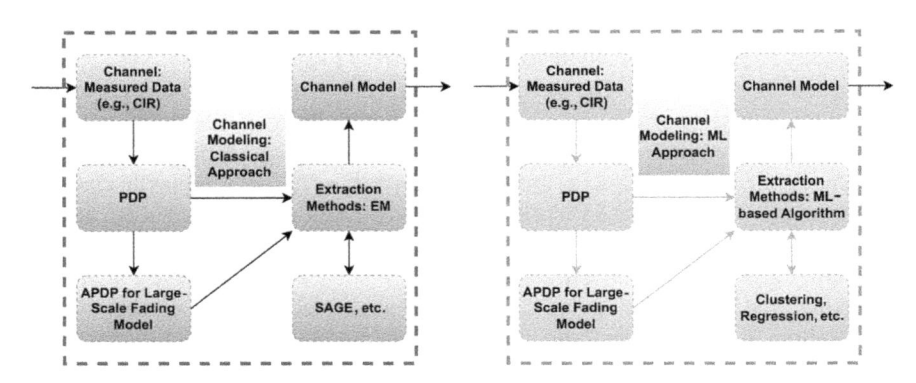

Figure 1.1 Comparative visualization between classical and ML-based channel modeling approach.

utilization of additional information such as channel feedback [9], localization [10], and channel prediction [11] extracted from this trained model, tasks like scheduling, beamforming, and power control can be effectively facilitated. In [12], the authors have employed various ML algorithms, including k-nearest neighbors (KNN) and random forest, to predict the path loss model based on a dataset. Their findings demonstrated high accuracy, with minimal estimation errors observed. The authors in [13] introduced a self-supervised pretraining DL technique aimed at modeling wireless channels for diverse tasks, all while safeguarding personal privacy by abstaining from utilizing labels on users' personal data. The effectiveness of a pretrained channel model significantly relies on the neural network architecture chosen. Given that a realistic channel resembles a natural time series, a recurrent neural network (RNN) emerges as a suitable choice for modeling the channel. The authors in [14] explore bidirectional training by introducing double RNNs for the channel modeling. In [15], the authors have introduced a model for estimating satellite communication channels based on atmospheric data. In their work presented in [16], the authors introduced a framework for channel modeling and simulation based on artificial neural networks (ANNs). This framework aims to address the limitations of traditional geometry-based stochastic modeling (GBSM) and simulation approaches, which struggle to accurately predict a channel that varies with time or position to align with real-world environments. Precise prediction of large-scale channel fading is essential for planning and optimizing 5G millimeter-wave cellular networks. In their study [17], a network called FadeNet was proposed to forecast large-scale fading across the coverage area of base stations. The authors [18] have explored AI-driven fog computing (FC) and proposed a novel AI-based reliable and interference-free mobility management algorithm (RIMMA) for fog computing intra-vehicular networks. Driven by AI, D2D communication stands poised as an allied technology poised to enhance system performance and

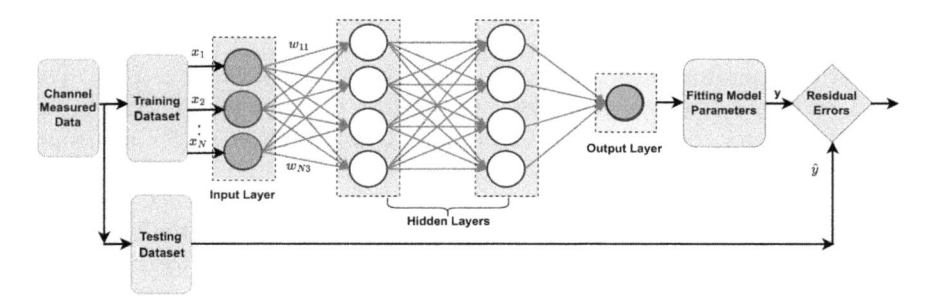

Figure 1.2 Feedforward DNN model.

facilitate novel services in 6G and beyond. To facilitate the channel model approach for D2D network, the authors in [19] have explored four DL models, namely long short-term memory networks (LSTMs), gated recurrent units (GRUs), convolutional neural networks (CNNs), and dense or feedforward networks (FFNs).

In [20], the authors have explored the potential of the convolution neural network (CNN) for the modeling of third generation partnership project (3GPP) channel. Several researchers have explored deep neural networks (DNNs) for the efficient channel model. Figure 1.2 illustrates a feedforward DNN model that is suitable for application in channel modeling for the estimation of channel parameters. The authors in [21] employed DNNs to estimate the channel characteristics of MIMO systems, leveraging the direction of arrival (DOA). The authors trained the DNN using datasets encompassing various channel scenarios, demonstrating superior performance compared to traditional methods. The accuracy and the efficiency of a wireless network panning can be further improved by exploiting channel information. The authors in [22] demonstrated that a wireless channel model based on DNNs can accurately predict the outage probability with minimal computational overhead. They also demonstrated that the proposed model is very effective for highly dynamic condition particularly for the train control systems. Researchers have investigated different approaches to AI-driven channel modeling, and further insights can be explored in related reviews such as [23].

1.3 AI-BASED CHANNEL DECODING AND SIGNAL DETECTION

In contemporary digital communication, spanning mobile networks, and IoT, it becomes imperative to secure efficient and reliable data transmission [24]. However, the capabilities of modern digital communication frameworks are constrained by factors like noise and channel interference, which have the potential to introduce errors during the transmission process.

This give rise to the requirement of the development of sophisticated channel decoding scheme to detect and correct these errors at the receiver end, thereby maintaining the accuracy of transmitted information [25]. The authors in [26,27] have proposed a framework utilizing the DL-based estimation scheme for estimating CSI implicitly and recovering the transmitted symbols directly. The demonstration showed that a DL-based approach can effectively handle channel distortion and accurately detect transmitted symbols, achieving performance comparable to that of the minimum mean-square error (MMSE) estimator. The primary challenge in current mMIMO systems lies in their high computational complexity and intricate spatial configurations, which pose significant obstacles to effectively leveraging the channel characteristics and sparsity inherent in these multi-antenna systems. In [12], a framework is proposed which integrates DL methods into mMIMO systems to tackle DOA estimation and channel estimation challenges [21]. In [28], the DL-based channel estimation scheme has been expanded to tackle a doubly selective channels and the proposed scheme outperforms the conventional estimators in numerous scenarios. Furthermore, in [29], a novel complex-valued neural network architecture called sparse complex-valued neural network (SCNet) is presented for tackling channel estimation within mMIMO systems. Similarly, the authors have utilized DNNs with supervised training to address the joint problem of MIMO detection and channel decoding [30]. Deep neural networks (DNNs) have emerged as a focal point of research in the realm of DL in recent years. The processing superiority associated with DNNs has motivated researchers to investigate their applicability in channel estimation tasks. The utilization of a rectified linear unit (ReLU) DNN for channel estimation has shown that it is not limited to particular signal models and converges asymptotically to MMSE estimation across diverse scenarios, all without the need for prior knowledge of channel statistics [31].

1.4 AI-BASED RESOURCE MANAGEMENT

The future 6G networks will handle high-bandwidth tasks like virtual reality, the Internet of Senses, no-touch cognitive systems, and self-driving cars. To deliver a seamless user experience, these emanating wireless applications demand network services that confront a variety of performance aspects, namely efficient allocation of resource, task off-loading, energy efficiency, latency reduction/minimization, and handover supervision.

1.4.1 Efficient allocation of resource

Conventional allocation of resource hurdles with wireless networks are recreated as well as handled via optimization techniques. Recent works [32–44] employed ML methods to tackle allocation of resource hurdles like

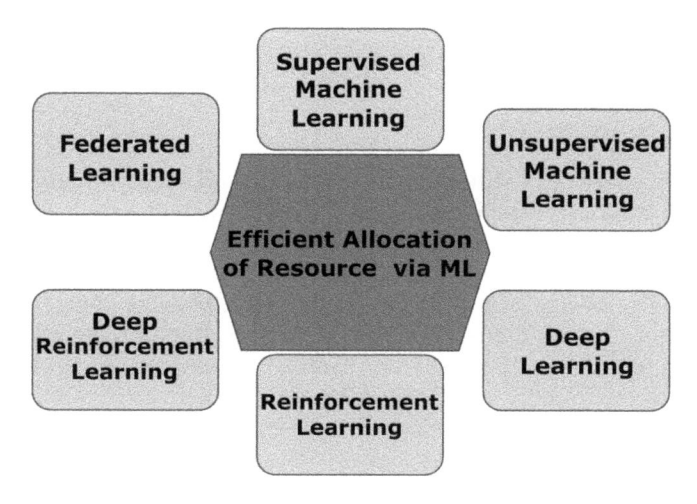

Figure 1.3 Different ML techniques for efficient allocation of resource.

channel distribution, estimation of off-loading, control of power transmit, multi-restriction/constraint QoS optimization, and selection of transmission modes. This section reviews current research on ML-assisted resource allocation. The latest studies are grouped relative to the types of ML applied, like supervised machine learning (SML), unsupervised machine learning (uSML), DL, reinforcement learning (RL), deep reinforcement learning (DRL), and federated learning (FL), as depicted in DNNs for the efficient channel model in Figure 1.3.

In [32], allocation of resource for wireless/cellular edge systems is presented. Edge machine learning (EML) is designed to optimize energy usage of system, edge-to-edge latency, and learning efficacy. The presented strategies are extremely flexible to supervised, half-supervised, and unsupervised educational scenarios. Authors of [33] explored DL for resource allocation in wireless edge networks. A supervised learning statistical model predicts proficiency in learning over multiple activities via the volume of data for training available. Authors in [34] employed transfer learning to handle mixed integer nonlinear programming (MINLP) to optimize allocation of resource in wireless systems. In [35], a paradigm utilizing uSML is offered to oversee subcarrier allocation in power domain non-orthogonal multiple access (NOMA) systems. RL strategies are used for resource management in several application events, notably network heterogeneity, hyper-dense systems, NOMA-driven systems, and IoT [36]. DRL-reliant strategy for NOMA uplink transmissions [38] and a smart time division duplex (TDD) strategy for the upstream and downstream HetNets [39] are presented for allocation of resource. In [40], the authors have present the management of resources and smart network construction in Multi-RAT HetNets and phase selection for small base station (SBS) employs multi-agent DL (MADL) [41].

For lower time required for execution for optimum task off-loading and allocation of resource, [43] offers federated reinforcement learning (FRL).

1.4.2 Task off-loading

6G wireless networks will entail low latency and high computation tasks driven by multiple broadcast access techniques, segments, systems, and emerging automobile scenarios, for instance, driver-less vehicles and vehicle-to-everything connection, along with vehicular multimedia programs [45]. The key obstacle is to create present-time choices about off-loading that prevent delay and energy usage. Moreover, outsourcing choices must be chosen within milliseconds, as channel conditions and other system factors are continually evolving. When the quantity of subscribers and activities grows, defining each conceivable choice becomes difficult. Recent research explores outsourcing tasks in multi-access edge computing networks [46–48]. This section examines current works on ML-assisted task off-loading, organizing them as SML, uSML, DL, RL, DRL, and FL, as well as dealing with numerous task off-loading instances depicted in Figure 1.4.

1.4.3 Energy efficiency

Engineering 6G equipment places a strong emphasis on energy efficiency, especially because of its usage of higher frequency bands. Autonomous vehicles and connected drones, along with other intelligent, self-healing machinery, are dependent on direct D2D communication and base station connections, prompting concerns regarding sustainable wireless infrastructure and increased energy consumption. Moreover, in mobile edge computing (MEC)-enabled IoT networks, fog node networks perform an essential

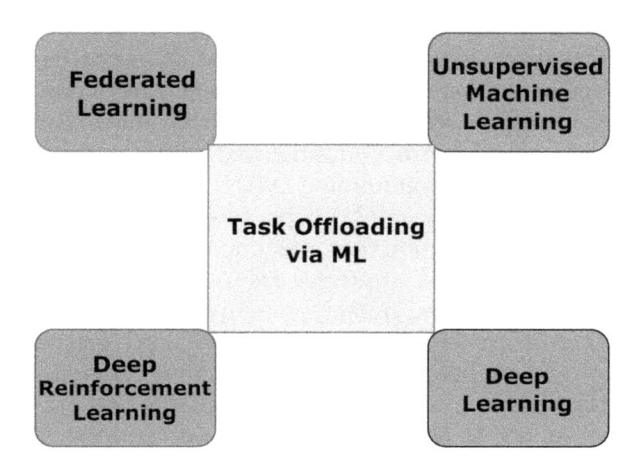

Figure 1.4 Diverse ML techniques for task off-loading.

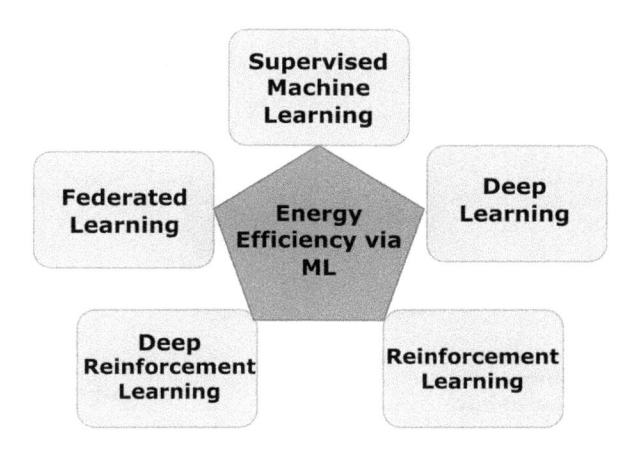

Figure 1.5 Different ML techniques for energy efficiency.

role in off-loading low-latency computations with restricted storage and power resources. To overcome these challenges, conventional algorithms for supervised education and reinforcement education, as well as current FL techniques, are crucial. Such technologies may drastically decrease energy consumption while boosting communication efficacy across 6G systems. Integrating a minimal-powered system for communication together with an energy-efficient evaluating mechanism merely improves the QoS and QoE for subscribers; additionally, it also contributes to the creation of an economically viable and environmentally friendly infrastructure for communication for wireless networks with 6G. This section delves into recent research on ML-reliant strategies for energy-efficient communications, classifying them as SML, uSML, DL, RL, DRL, and FL employed to improve efficiency of energy in wireless systems [4], as illustrated in Figure 1.5.

1.4.4 Latency reduction/minimization

The forthcoming 6G communication networks aspire to attain data speeds of terabit per second while retaining exceptionally low latency during transmission. Recent advancements in massive MTCs and URLLCs have resulted in a revolution in cellular communication. MTC facilitates diverse intelligent IoT connectivity, including applications which include self-driving vehicles, no-touch cognitive systems, and crowd monitoring. Every single of these MTC circumstances has different QoS necessities. Nevertheless, present challenges in such circumstances include the chance that a distant information center or server in the cloud might fail to satisfy the low delay requirements for accessing content in programs like this. To overcome these challenges, powerful ML techniques need to be integrated alongside 6G cellular system topologies and smart optimization strategies. This integration

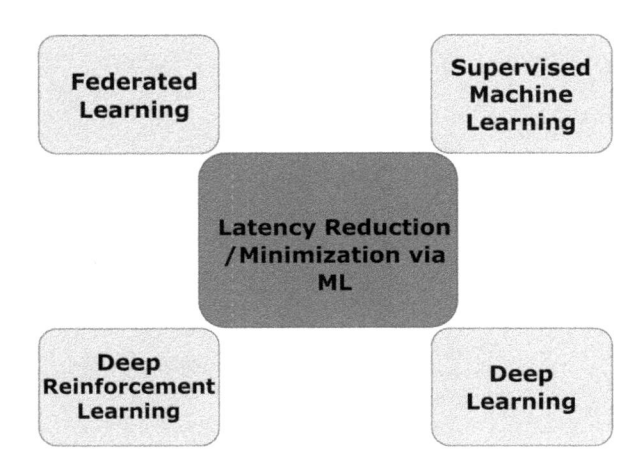

Figure 1.6 Different recent ML techniques for latency reduction/minimization.

is critical for satisfying the diverse QoS needs of various MTC instances. Recent research has focused on ML-driven latency reduction/minimization as a remedy to these obstacles. This section summarizes current research, sorting it into several forms of ML, namely SML, DL, DRL, and FL. Figure 1.6 depicts how different ML techniques are employed to mitigate latency in wireless networks [4].

1.4.5 Handover supervision

Efficient handover supervision is fundamental for preserving QoS in 6G communication networks, which are experiencing a variety of challenges, including dropped throughput and disruptions in service. Moreover, the expanded deployment of mmWave base stations in 6G networks is aimed at establishing a direct link across all portable devices with installed base stations. Consequently, a pivotal field of research focuses on developing ML-driven choice systems that can discern the optimal base station to commence the handover process. This section gives an overview of current research on ML-driven handover supervision, categorizing them according to ML sorts, notably DL, RL, DRL, and FL, as depicted in Figure 1.7, and focusing on their implementation in wireless networks [4].

1.5 CHALLENGES OF AI IN WIRELESS COMMUNICATION

In the subsequent section, the authors highlight the key challenges that AI confronts in the wireless communication networks.

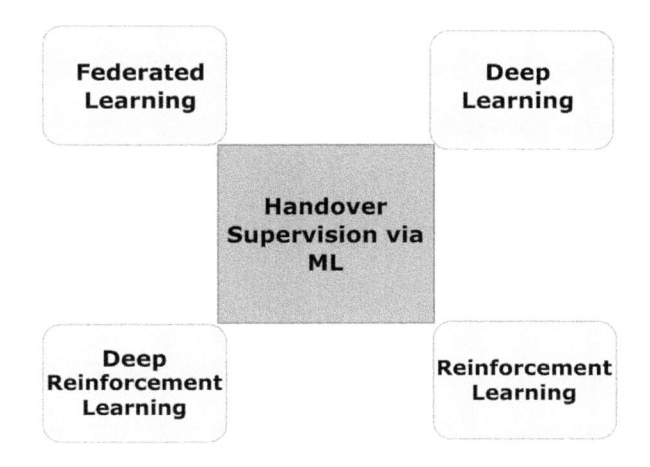

Figure 1.7 Diverse ML techniques for handover supervision.

1.5.1 Increased/greater communication overhead

Leveraging AI for boosting the efficacy of IoT devices, their ability to perform, or the operational efficiency of the underpinning IoT network will lead to increased/greater transmission overhead. This increased/greater transmission overhead triggered via AI might have been traced back toward the core underlying principles of operation of AI networks. As an illustration, ML networks extract vital insights through extensive datasets, petitioning for transmission of data across gadgets executing ML methods. In order to exemplify the higher expenses engaged with transmission, let us examine the learning device depicted in Figure 1.8. The aforementioned device employs transmission bandwidth and spectrum for a wide range of motives including monitoring, communicating interpreter results, and exchanging function space information among all IoT gadgets in this vicinity. Accordingly, the transmission expenditures related to ML algorithms may generically be evaluated based on three factors: (1) the number of steps for the ML technique to reach the convergent, (2) the frequency of channels employed during each transmission cycle, and (3) spectrum consumed per channel during a transmission [49].

In the realm of IoT, the vast and varied data generated by the multitude of interconnected IoT gadgets demand significantly greater processing and storage capacities [50]. As an outcome, the constrained capacity of IoT devices compels the offloading of ML processing and storage onto additional feasible resources, the most notable of which are nodes at the edge, that are positioned near IoT gadgets primarily for fulfilling delay requirements [51]. Nevertheless, as outlined in [51], effectively handling the massive amounts of big data produced via the IoT at the edge is a tremendous undertaking.

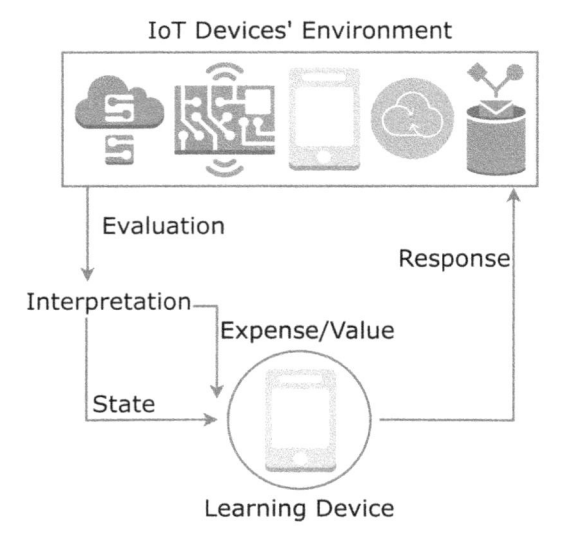

Figure 1.8 A streamlined system for RL.

Consequently, there are likely two predominant strategies for tackling this challenge. Firstly, there is concurrently shared processing across multiple nodes at the edge, followed by consolidated processing within a cluster comprising bigger memory and processing networks, akin to those commonly found in centralized networks for cloud computing. In all instances, the transmission overhead is expected to climb beyond what was previously estimated.

1.5.2 Challenges in achieving latency-critical IoT systems

This evolving characteristics for emerging IoT offerings shall necessitate lightning-fast computations, conveniently situated in close proximity to users, or alternatively, without noticeable delays [52]. Unfortunately, owing to their restricted capacity, AI processing within IoT devices is projected to take substantially more time. In terms of edge processing, the authors of [53] constructed that several communication rounds will be essential, even in settings with low assumptions, to achieve worst-case optimum utilization. In essence, the procedure of accumulating data in its raw form, evaluating it, and educating it, as well as the continual response cycle in ML, can greatly increase delays. This is a challenge not only for classic iterative or feedback systems without ML (non-ML) but also for satisfying the ever-changing needs for dynamic offerings along with extremely cellular consumers because of computation and network delays [54]. More precisely, when it comes to preparing ML models, such as for streamed programs, the introduced delay is a significant barrier in achieving the necessary latency standards [55].

Moreover, it is essential to consider the time span constraints that pertain to data relevance for AI processing as well as AI outcome accuracy. Evaluate intrusion detection systems to gain a greater awareness of these constraints. The timeliness of data analysis is of the utmost significance in intrusion detection, and any delays introduced by the communication channel may potentially render the entire process ineffective [56]. Exploring distributed ML systems [57] illustrates that the time spent on network communication is orders of magnitude more relative to the period spent on computing to train ML models. Hence, in these systems, latency is equally vital just like system legality and correctness. That is reinforced additional via the case of identifying an object employing DL, as stated in [58]. Manually processed pictures for identification of objects demand 7 W of energy, but cloud-processed images consume just 2 W. Still, the latency exceeds 500 ms. Cloud processing yields results within 2–5 seconds. Hence, the conclusion derived in [58] is suitable to actual time DL actions; the web remains an unfeasible alternative due to its elevated latency.

1.5.3 Inadequate standardized case studies and datasets for AI evaluation and development

Perhaps the most significant challenges for enormous-scale investigation and implementation of AI and hindering the provision of diverse from top to bottom services in upcoming mobile networks are the inadequate standardized case studies and datasets. It is crucial for global standard organizations and researchers in academia and industry to prioritize AI standardization, with a special emphasis on developing standards for test cases, datasets, parameter sets, and associated protocols [59]. This endeavor is vital for easing the implementation of AI algorithms into forthcoming wireless networks. Furthermore, active engagement and contributions from network administrators and providers of services were essential for the success of AI standardization initiatives. For instance, at the beginning of 2017, ETSI launched the initial interconnected AI research group, the Experiential Networked Intelligence (ENI) Industrial Specification Group (ISG). Until July 2019, there were nearly thirty teammates, comprised of carriers (like Vodafone, China Telecom, Telefonica, China Unicom, Verizon, and SK Telecom) as well as equipment and chip vendors (like Intel, Huawei, Samsung, and ZTE). The ENI-ISG endeavors at offering an intelligent management of systems paradigm utilizing AI as well as content-sensitive rules enable operations that respond to evolving user demands, environmental conditions, and organizational objectives. By March 2019, a new ISG called zero-touch system and management of services (ZSM) was built. ZSM aims to automate every business procedures along with duties (such as provisioning, installation, arrangement, reliability, and optimization) preferably through full automation. Initially, ISG ZSM will focus on edge-to-edge system and management of services for 5G, including network segmentation. Over time, its scope will expand to encompass network management for future generations [59].

1.5.4 Obstacles in routing and controlling traffic across network

Although AI has proved recommended for routing, conventional AI/ML methods comparable to artificial neural networks (ANNs) encounter significant restrictions in scalability and computational efficacy whenever employed for routing [60]. Evaluating the potential advantages of employing DL-reliant routing compared to the conventional Open Shortest Path First (OSPF) protocol routing technique, the findings presented [61] illustrate that OSPF protocol achieves identical throughput and average lag, while the signaling interval across routers exceeds a specific limit. Nevertheless, when taking into account computational and storage resources, opting for basic OSPF proves to be a much better choice for core and backhaul networks. This is especially true in environments where fluctuations are more unpredictable than in the dynamic nature of access networks. Additionally, the act of tweaking addresses of Internet Protocol or information of packet header, whether for safety reasons or as a preventive precaution against such assaults [62], might pose additional challenges to the learning process. It may result in continual feedback loops in the quest to determine the optimal route.

Numerous research projects are devoted to the integration of AI in dynamic networks, which are somewhat characterized by periodically fluctuating topologies that necessitate persistent information flow between network nodes. Mobile ad hoc network (MANET) is a type of dynamic network rendered up for restricted resources such as mobile gadgets. MANET emerges spontaneously from the random and erratic connections of independent nodes. As a result, the routing protocols they implement usually have more overhead. This is primarily due to the ongoing broadcast of topological information and data exchange, especially while experiencing transient breakdowns in routing protocol convergence [63]. Nonetheless, the dynamic nature of their topologies leads to a constant influx of new information. These systems respond similarly to closed-loop systems, posing challenges for methods of learning that converge underneath the defined delay limitations.

1.5.5 Challenges of system complexity

Integrating AI into communication networks will undoubtedly lead to higher system complexity, especially when the current method continues: employing case-specific ML for specific targets without regard for larger objectives or overall network goals. Consequently, a major challenge is evident—present research on employing AI in wireless networks strives to optimize specific goals while neglecting crucial constraints that involve delay, network quality, storage, and evaluating overhead. An illustration of this is the endorsement in [64] to improve spectral efficiency via RL. The expense related to exchange, preserve, and evaluate information while utilizing the recommended technique in actual or extensive systems is not specified.

1.5.6 Network caching challenges

Network caching schemes store details or data temporarily in close proximity to subscribers in order to prevent needless network traffic [65]. Previously, routers typically cached data that acquired a high volume of responses or were frequently passed through them. Anyway, the rise of massive data from the IoT presents a significant challenge to the classic principles of in-network caching. There have been suggestions for solutions based on AI that permit the network to learn and discern to reveal data or information that should be cached [66]. Nonetheless, the integration of AI into network devices, for example, switching gadgets as well as routers, would drain the resources needed for preserving routing processes, pathways, lists of access controls, and other functions.

In this regard, the authors of [67] suggested employing DL for content caching in software-defined networking (SDN). The examination focused on the OpenFlow SDN standard, uncovering that OpenFlow switches face constraints in retaining unforeseen flows until the controller refreshes the flow tables. Additionally, as highlighted in [68], these switches have a restricted capacity for storing flow rules in certain instances. Moreover, as outlined in [68], scalability challenges are visible in SDN controllers, prompting the proposal of several hierarchical as well as decentralized interface designs. Leveraging AI algorithms on network material will necessitate significant memory and processing capacity to handle current time offerings. Moreover, information caching at extremes has its own constraints and challenges [67].

1.5.7 Challenges related to safety and privacy

A fundamental challenge for applying AI in IoT safety is developing a top-notch set of training data that encompass common threat types along with trends. The preciseness of AI schemes depends heavily on the quality of the training dataset. To guarantee the effective execution of AI strategies regarding IoT safety, a diversified training dataset including knowledge that represents many techniques utilized in actual threats is essential. Nonetheless, the challenge persists in attaining immediate, excellent information transmission and acquisition due to the enormous quantity of gadgets providing enormous amounts of data. Moreover, because of the broad variety of IoT devices, accumulating a trustworthy dataset by means of device collaboration might be challenging. It is worth noting that the majority of articles regarding AI for IoT security emphasize applications that employ high-quality data [49]. For instance, the AI system in [69] employs NSL-KDD datasets to train and validate its intrusion detection technique. Nonetheless, NSL-KDD might not accurately convey the properties of real networks. The presence of a high amount of heterogeneous gadgets in IoT networks might cause noise and interference, thus compromising the integrity of the dataset. Hence, executing AI strategies for IoT safety with high-quality datasets is exceedingly unfeasible. It is vital to note that collecting a set of data for training in the realm of IoT safety poses greater challenges compared to gathering datasets for image or language processing.

1.6 FUTURE DIRECTION OF AI IN WIRELESS COMMUNICATION

Given the aforementioned challenges, we will now delve into potential opportunities and unresolved matters. In the subsequent section, the authors highlight the future direction of AI in wireless communication.

Numerous approaches, when employed in accordance with their settings, network architecture, and available resources, can assist us in creating more efficient utilization of AI processes. Consequently, an in-depth knowledge of the worldwide communication infrastructure along with readily accessible assets would be extremely valuable. Nevertheless, extensive investigation needs to be conducted across the subsequent transmission network-specific domains for maximizing the numerous advantages from AI.

1.6.1 Harnessing AI strategies within restrictions related to bandwidth, spectrum, and latency

The sheer number of end-user devices in the next generation of wireless networks is expected to grow significantly. These devices are likely to exhibit numerous traffic patterns and may be vulnerable to security flaws [70]. Therefore, identifying the most efficient way to optimize obtainable bandwidth and spectrum resources while preserving data rates, service quality, and user experience is an enormous challenge. Harnessing AI strategies to predict traffic increases and abrupt surges, preemptively shifting facilities across peripheral and controlled systems in the cloud, along with continuously allocating capabilities is likely to produce superior outcomes. Yet, further research will be required to accumulate AI platforms which might be instructed efficiently as well as swiftly utilizing fewer bits of information, hence consuming fewer bandwidth or spectrum resources [49].

1.6.2 Streamlined test cases and datasets for effective merely reputable validation of AI

The operational instances beyond 5G (B5G)/6G are projected to be more challenging and diverse than those found in 5G networks. This expectation is reliant on three application types: eMBB, mMTC, and URLLC. These challenges may extend beyond our current understanding. However, prominent AI and wireless communications specialists emphasize the critical importance of standardizing test cases, datasets, parameter settings, interfaces, and protocols. This standardization is deemed essential to facilitate the integration of AI into future wireless networks [71].

1.6.3 Simplify complexity through network abstraction

It has proven feasible to simplify the network for varied services by segregating the network's fundamental construction through the features that it supports. SDN [72] makes this feasible by providing granular, influenced by events regulation of system components by employing extremely high rules, thereby eliminating the need for modest level setups. Similar strategies, for example, functional segregation along with conceptualization, were recently recommended for satellite access systems. However, further research must be performed, particularly into MIMO aspects. Abstraction has been an important domain in IoT research and industrial activities, exemplified by the Pelion IoT platform. Employing AI in a comparable manner and leveraging AI as a service seamlessly across networks may yield equivalent benefits without expanding total system complexity. This avenue warrants further exploration and research [49].

1.6.4 Combating caching needs in the era of big data

Utilizing big data analytic tools requires a variety of resources, namely data storage, computing power, and network bandwidth. It is critical to judge early on if the data qualify as big data, as stated in [56], in order to make educated decisions regarding the suitable technology. Expanding big data capabilities should become less problematic in the future as edge clouds, or resources situated closer to users in communication networks, evolve. Nonetheless, further research must be conducted to allow the application of AI to big data at the data source while functioning within constraints on resources [49].

1.6.5 Developing AI-driven security strategies to resolve AI challenges in security

Historically, substantial research has been conducted on employing AI to improve network security. On the contrary, there is an urgency to investigate the deployment of AI to detect threats among connected entities. AI-driven attacks are extremely difficult to identify as well as counter, which is illustrated in [73] and detailed in [74]. Likewise, AI relies on information, as well as preserving the confidentiality of these data is essential. As an outcome, tackling the challenges of embracing AI-driven strategies aimed at safeguarding assets and information from AI-driven safety hazards along with confidentiality concerns is an intriguing area that warrants further investigation.

1.7 CONCLUSION

The integration of AI into wireless communication systems marks a significant advancement that promises to reshape the landscape of telecommunications. This chapter provides an overview of AI-based channel modeling, decoding, and signal detection, as well as AI-driven resource management and its associated research. Leveraging AI algorithms enables wireless communication systems to dynamically allocate resources, mitigate interference, and forecast network behavior, thereby enhancing user experiences and elevating service quality. Nonetheless, this integration also presents significant hurdles that must be surmounted for the successful implementation of AI-powered 6G networks. The chapter outlines these challenges and future avenues for research. Moreover, it identifies several open research areas that demand further attention. To date, substantial investments and efforts from both academia and industry have been dedicated to integrating AI into the framework of wireless communication systems. However, there remains a considerable journey ahead to address challenges and achieve the objectives of next-generation communication systems.

REFERENCES

1. S. J. Russell and P. Norvig, *Artificial Intelligence a Modern Approach.* Pearson: London, 2010.
2. M. Agiwal, A. Roy, and N. Saxena, "Next generation 5G wireless networks: A comprehensive survey," *IEEE Communications Surveys & Tutorials,* vol. 18, no. 3, pp. 1617–1655, 2016.
3. A. L. Imoize, H. I. Obakhena, F. I. Anyasi, and S. N. Sur, "A review of energy efficiency and power control schemes in ultra-dense cell-free massive MIMO systems for sustainable 6G wireless communication," *Sustainability,* vol. 14, p. 11100, Sept. 2022.

4. H. M. F. Noman, E. Hanafi, K. A. Noordin, K. Dimyati, M. N. Hindia, A. Abdrabou, and F. Qamar, "Machine learning empowered emerging wireless networks in 6G: Recent advancements, challenges and future trends," *IEEE Access*, vol. 11, pp. 83017–83051, 2023.
5. J. Wang, J. Liu, J. Li, and N. Kato, "Artificial intelligence-assisted network slicing: Network assurance and service provisioning in 6G," *IEEE Vehicular Technology Magazine*, vol. 18, pp. 49–58, Mar. 2023.
6. E. Coronado, R. Behravesh, T. Subramanya, A. Fernandez-Fernandez, M. S. Siddiqui, X. Costa-Perez, and R. Riggio, "Zero touch management: A survey of network automation solutions for 5G and 6G networks," *IEEE Communications Surveys & Tutorials*, vol. 24, no. 4, pp. 2535–2578, 2022.
7. V. Kristem, C. U. Bas, R. Wang, and A. F. Molisch, "Outdoor wideband channel measurements and modeling in the 3–18 ghz band," *IEEE Transactions on Wireless Communications*, vol. 17, pp. 4620–4633, July 2018.
8. Z. Jiang, Z. He, S. Chen, A. F. Molisch, S. Zhou, and Z. Niu, "Inferring remote channel state information: Cramer-Rae lower bound and deep learning implementation," in *2018 IEEE Global Communications Conference (GLOBECOM)*, pp. 1–7, IEEE, Abu Dhabi, Dec. 2018.
9. C.-K. Wen, W.-T. Shih, and S. Jin, "Deep learning for massive MIMO CSI feedback," *IEEE Wireless Communications Letters*, vol. 7, pp. 748–751, Oct. 2018.
10. A. Decurninge, L. G. Ordonez, P. Ferrand, H. Gaoning, L. Bojie, Z. Wei, and M. Guillaud, "CSI-based outdoor localization for massive MIMO: Experiments with a learning approach," in *2018 15th International Symposium on Wireless Communication Systems (ISWCS)*, pp. 1–6, IEEE, Lisbon, Aug. 2018.
11. M. Arnold, S. Dorner, S. Cammerer, J. Hoydis, and S. ten Brink, "Towards practical FDD massive MIMO: CSI extrapolation driven by deep learning and actual channel measurements," in *2019 53rd Asilomar Conference on Signals, Systems, and Computers*, pp. 1972–1976, IEEE, Pacific Grove, CA, Nov. 2019.
12. Y. Zhang, J. Wen, G. Yang, Z. He, and X. Luo, "Air-to-air path loss prediction based on machine learning methods in urban environments," *Wireless Communications and Mobile Computing*, vol. 2018, pp. 1–9, June 2018.
13. Y. Huangfu, J. Wang, C. Xu, R. Li, Y. Ge, X. Wang, H. Zhang, and J. Wang, "Realistic channel models pre-training," in *2019 IEEE Globecom Workshops (GC Wkshps)*, pp. 1–6. IEEE, Waikoloa, HI, July 2019.
14. Y. Huangfu, J. Wang, R. Li, C. Xu, X. Wang, H. Zhang, and J. Wang, "Predicting the mumble of wireless channel with sequence-to-sequence models," in *2019 IEEE 30th Annual International Symposium on Personal, Indoor and Mobile Radio Communications (PIMRC)*, pp. 1–7, IEEE, Istanbul, Sept. 2019.
15. L. Bai, Q. Xu, Z. Huang, S. Wu, S. Ventouras, G. Goussetis, and X. Cheng, "An atmospheric data-driven Q-band satellite channel model with feature selection," *IEEE Transactions on Antennas and Propagation*, vol. 70, pp. 4002–4013, June 2022.
16. X. Zhao, F. Du, S. Geng, Z. Fu, Z. Wang, Y. Zhang, Z. Zhou, L. Zhang, and L. Yang, "Playback of 5G and beyond measured mimo channels by an ANN-based modeling and simulation framework," *IEEE Journal on Selected Areas in Communications*, vol. 38, pp. 1945–1954, Sept. 2020.

17. V. V. Ratnam, H. Chen, S. Pawar, B. Zhang, C. J. Zhang, Y.-J. Kim, S. Lee, M. Cho, and S.-R. Yoon, "Fadenet: Deep learning-based mm-wave large-scale channel fading prediction and its applications," *IEEE Access*, vol. 9, pp. 3278–3290, 2021.

18. A. H. Sodhro, G. H. Sodhro, M. Guizani, S. Pirbhulal, and A. Boukerche, "AI-enabled reliable channel modeling architecture for fog computing vehicular networks," *IEEE Wireless Communications*, vol. 27, pp. 14–21, Apr. 2020.

19. N. Simmons, S. B. F. Gomes, M. D. Yacoub, O. Simeone, S. L. Cotton, and D. E. Simmons, "AI-based channel prediction in D2D links: An empirical validation," *IEEE Access*, vol. 10, pp. 65459–65472, 2022.

20. D. Neumann, T. Wiese, and W. Utschick, "Learning the MMSE channel estimator," *IEEE Transactions on Signal Processing*, vol. 66, pp. 2905–2917, June 2018.

21. H. Huang, J. Yang, H. Huang, Y. Song, and G. Gui, "Deep learning for super-resolution channel estimation and DOA estimation based massive MIMO system," *IEEE Transactions on Vehicular Technology*, vol. 67, pp. 8549–8560, Sept. 2018.

22. T. Wen, G. Xie, Y. Cao, and B. Cai, "A DNN-based channel model for network planning in train control systems," *IEEE Transactions on Intelligent Transportation Systems*, vol. 23, pp. 2392–2399, Mar. 2022.

23. C. Huang, R. He, B. Ai, A. F. Molisch, B. K. Lau, K. Haneda, B. Liu, C.-X. Wang, M. Yang, C. Oestges, and Z. Zhong, "Artificial intelligence enabled radio propagation for communications—Part II: Scenario identification and channel modeling," *IEEE Transactions on Antennas and Propagation*, vol. 70, pp. 3955–3969, June 2022.

24. S. Li, L. D. Xu, and S. Zhao, "The Internet of Things: A survey," *Information Systems Frontiers*, vol. 17, pp. 243–259, Apr. 2014.

25. E. Ordentlich, G. Seroussi, S. Verdu, and K. Viswanathan, "Universal algorithms for channel decoding of uncompressed sources," *IEEE Transactions on Information Theory*, vol. 54, pp. 2243–2262, May 2008.

26. H. Ye, G. Y. Li, and B.-H. Juang, "Power of deep learning for channel estimation and signal detection in OFDM systems," *IEEE Wireless Communications Letters*, vol. 7, pp. 114–117, Feb. 2018.

27. M. Soltani, V. Pourahmadi, A. Mirzaei, and H. Sheikhzadeh, "Deep learning-based channel estimation," *IEEE Communications Letters*, vol. 23, pp. 652–655, Apr. 2019.

28. Y. Yang, F. Gao, X. Ma, and S. Zhang, "Deep learning-based channel estimation for doubly selective fading channels," *IEEE Access*, vol. 7, pp. 36579–36589, 2019.

29. Y. Yang, F. Gao, G. Y. Li, and M. Jian, "Deep learning-based downlink channel prediction for FDD massive MIMO system," *IEEE Communications Letters*, vol. 23, pp. 1994–1998, Nov. 2019.

30. T. Wang, L. Zhang, and S. C. Liew, "Deep learning for joint MIMO detection and channel decoding," in *2019 IEEE 30th Annual International Symposium on Personal, Indoor and Mobile Radio Communications (PIMRC)*, pp. 1–7, IEEE, Istanbul, Sept. 2019.

31. Q. Hu, F. Gao, H. Zhang, S. Jin, and G. Y. Li, "Deep learning for channel estimation: Interpretation, performance, and comparison," *IEEE Transactions on Wireless Communications*, vol. 20, pp. 2398–2412, Apr. 2021.

32. M. Merluzzi, P. D. Lorenzo, and S. Barbarossa, "Wireless edge machine learning: Resource allocation and trade-offs," *IEEE Access*, vol. 9, pp. 45377–45398, 2021.
33. L. Zhou, Y. Hong, S. Wang, R. Han, D. Li, R. Wang, and Q. Hao, "Learning centric wireless resource allocation for edge computing: Algorithm and experiment," *IEEE Transactions on Vehicular Technology*, vol. 70, pp. 1035–1040, Jan. 2021.
34. Y. Shen, Y. Shi, J. Zhang, and K. B. Letaief, "LORM: Learning to optimize for resource management in wireless networks with few training samples," *IEEE Transactions on Wireless Communications*, vol. 19, pp. 665–679, Jan. 2020.
35. M. A. Jamshed, F. Heliot, and T. W. C. Brown, "Unsupervised learning based emission-aware uplink resource allocation scheme for non-orthogonal multiple access systems," *IEEE Transactions on Vehicular Technology*, vol. 70, pp. 7681–7691, Aug.2021.
36. R. Alkurd, I. Y. Abualhaol, and H. Yanikomeroglu, "Personalized resource allocation in wireless networks: An AI-enabled and big data-driven multi-objective optimization," *IEEE Access*, vol. 8, pp. 144592–144609, 2020.
37. E. Kim, H.-H. Choi, H. Kim, J. Na, and H. Lee, "Optimal resource allocation considering non-uniform spatial traffic distribution in ultra-dense networks: A multi-agent reinforcement learning approach," *IEEE Access*, vol. 10, pp. 20455–20464, 2022.
38. W. Ahsan, W. Yi, Z. Qin, Y. Liu, and A. Nallanathan, "Resource allocation in uplink NOMA-IoT networks: A reinforcement-learning approach," *IEEE Transactions on Wireless Communications*, vol. 20, pp. 5083–5098, Aug. 2021.
39. F. Tang, Y. Zhou, and N. Kato, "Deep reinforcement learning for dynamic uplink/-downlink resource allocation in high mobility 5G HetNet," *IEEE Journal on Selected Areas in Communications*, vol. 38, pp. 2773–2782, Dec. 2020.
40. M. S. Allahham, A. A. Abdellatif, N. Mhaisen, A. Mohamed, A. Erbad, and M. Guizani, "Multi-agent reinforcement learning for network selection and resource allocation in heterogeneous multi-RAT networks," *IEEE Transactions on Cognitive Communications and Networking*, vol. 8, pp. 1287–1300, June 2022.
41. D. Bega, M. Gramaglia, A. Banchs, V. Sciancalepore, and X. Costa-Perez, "A machine learning approach to 5G infrastructure market optimization," *IEEE Transactions on Mobile Computing*, vol. 19, pp. 498–512, Mar. 2020.
42. Z. Shi, J. Liu, S. Zhang, and N. Kato, "Multi-agent deep reinforcement learning for massive access in 5G and beyond ultra-dense NOMA system," *IEEE Transactions on Wireless Communications*, vol. 21, pp. 3057–3070, May 2022.
43. Q. Zhang, H. Wen, Y. Liu, S. Chang, and Z. Han, "Federated-reinforcement-learning-enabled joint communication, sensing, and computing resources allocation in connected automated vehicles networks," *IEEE Internet of Things Journal*, vol. 9, pp. 23224–23240, Nov. 2022.
44. R. Zhong, X. Liu, Y. Liu, Y. Chen, and Z. Han, "Mobile reconfigurable intelligent surfaces for NOMA networks: Federated learning approaches," *IEEE Transactions on Wireless Communications*, vol. 21, pp. 10020–10034, Nov. 2022.

45. P. Zeng, A. Liu, C. Zhu, T. Wang, and S. Zhang, "Trust-based multi-agent imitation learning for green edge computing in smart cities," *IEEE Transactions on Green Communications and Networking*, vol. 6, pp. 1635–1648, Sept. 2022.

46. H. Wu, Z. Zhang, C. Guan, K. Wolter, and M. Xu, "Collaborate edge and cloud computing with distributed deep learning for smart city Internet of Things," *IEEE Internet of Things Journal*, vol. 7, pp. 8099–8110, Sept. 2020.

47. H. Liao, Y. Mu, Z. Zhou, M. Sun, Z. Wang, and C. Pan, "Blockchain and learning-based secure and intelligent task offloading for vehicular fog computing," *IEEE Transactions on Intelligent Transportation Systems*, vol. 22, pp. 4051–4063, July 2021.

48. I. Khan, X. Tao, G. M. S. Rahman, W. U. Rehman, and T. Salam, "Advanced energy-efficient computation offloading using deep reinforcement learning in MTC edge computing," *IEEE Access*, vol. 8, pp. 82867–82875, 2020.

49. I. Ahmad, S. Shahabuddin, T. Sauter, E. Harjula, T. Kumar, M. Meisel, M. Juntti, and M. Ylianttila, "The challenges of artificial intelligence in wireless networks for the Internet of Things: Exploring opportunities for growth," *IEEE Industrial Electronics Magazine*, vol. 15, pp. 16–29, Mar. 2021.

50. A. L'Heureux, K. Grolinger, H. F. Elyamany, and M. A. M. Capretz, "Machine learning with big data: Challenges and approaches," *IEEE Access*, vol. 5, pp. 7776–7797, 2017.

51. H. Li, K. Ota, and M. Dong, "Learning IOT in edge: Deep learning for the Internet of Things with edge computing," *IEEE Network*, vol. 32, pp. 96–101, Jan. 2018.

52. I. Ahmad, T. Kumar, M. Liyanage, M. Ylianttila, T. Koskela, T. Braysy, A. Anttonen, V. Pentikinen, J.-P. Soininen, and J. Huusko, "Towards gadget-free internet services: A roadmap of the naked world," *Telematics and Informatics*, vol. 35, pp. 82–92, Apr. 2018.

53. Y. Arjevani and O. Shamir, "Communication complexity of distributed convex learning and optimization," in Part of *Advances in Neural Information Processing Systems 28 (NIPS 2015)*, vol. 28, pp. 1756–1764, Montreal, 2015.

54. H. Sun, X. Chen, Q. Shi, M. Hong, X. Fu, and N. D. Sidiropoulos, "Learning to optimize: Training deep neural networks for wireless resource management," in *2017 IEEE 18th International Workshop on Signal Processing Advances in Wireless Communications (SPAWC)*, pp. 1–6, IEEE, Sapporo, July 2017.

55. C. Augenstein, N. Spangenberg, and B. Franczyk, "Applying machine learning to big data streams: An overview of challenges," in *2017 IEEE 4th International Conference on Soft Computing & Machine Intelligence (ISCMI)*, pp. 25–29, IEEE, Mauritius, Nov. 2017.

56. S. Suthaharan, "Big data classification: Problems and challenges in network intrusion prediction with machine learning," *ACM SIGMETRICS Performance Evaluation Review*, vol. 41, pp. 70–73, Apr. 2014.

57. P. Sun, Y. Wen, T. N. Binh Duong, and S. Yan, "Timed dataflow: Reducing communication overhead for distributed machine learning systems," in *2016 IEEE 22nd International Conference on Parallel and Distributed Systems (ICPADS)*, pp. 1110–1117, IEEE, Wuhan, Dec. 2016.

58. J. Tang, D. Sun, S. Liu, and J.-L. Gaudiot, "Enabling deep learning on IOT devices," *Computer*, vol. 50, no. 10, pp. 92–96, 2017.

59. M. Lin and Y. Zhao, "Artificial intelligence-empowered resource management for future wireless communications: A survey," *China Communications*, vol. 17, pp. 58–77, Mar. 2020.

60. N. Kato, Z. M. Fadlullah, B. Mao, F. Tang, O. Akashi, T. Inoue, and K. Mizutani, "The deep learning vision for heterogeneous network traffic control: Proposal, challenges, and future perspective," *IEEE Wireless Communications*, vol. 24, pp. 146–153, June 2017.

61. B. Mao, Z. M. Fadlullah, F. Tang, N. Kato, O. Akashi, T. Inoue, and K. Mizutani, "Routing or computing? The paradigm shift towards intelligent computer network packet transmission based on deep learning," *IEEE Transactions on Computers*, vol. 66, pp. 1946–1960, Nov. 2017.

62. J. H. Jafarian, E. Al-Shaer, and Q. Duan, "An effective address mutation approach for disrupting reconnaissance attacks," *IEEE Transactions on Information Forensics and Security*, vol. 10, pp. 2562–2577, Dec. 2015.

63. M. Suchara, D. Xu, R. Doverspike, D. Johnson, and J. Rexford, "Network architecture for joint failure recovery and traffic engineering," in Suchara, Martin, Dahai Xu, Robert Doverspike, David Johnson, and Jennifer Rexford (eds.), *ACM SIGMETRICS Performance Evaluation* Review, vol. 39, no. 1, pp. 97–108, ACM, June 2011.

64. M. Bennis, S. M. Perlaza, P. Blasco, Z. Han, and H. V. Poor, "Self-organization in small cell networks: A reinforcement learning approach," *IEEE Transactions on Wireless Communications*, vol. 12, pp. 3202–3212, July 2013.

65. X. Wang, M. Chen, T. Taleb, A. Ksentini, and V. Leung, "Cache in the air: Exploiting content caching and delivery techniques for 5G systems," *IEEE Communications Magazine*, vol. 52, pp. 131–139, Feb. 2014.

66. Z. Chang, L. Lei, Z. Zhou, S. Mao, and T. Ristaniemi, "Learn to cache: Machine learning for network edge caching in the big data era," *IEEE Wireless Communications*, vol. 25, pp. 28–35, June 2018.

67. W.-X. Liu, J. Zhang, Z.-W. Liang, L.-X. Peng, and J. Cai, "Content popularity prediction and caching for ICN: A deep learning approach with SDN," *IEEE Access*, vol. 6, pp. 5075–5089, 2018.

68. I. Ahmad, S. Namal, M. Ylianttila, and A. Gurtov, "Security in software defined networks: A survey," *IEEE Communications Surveys & Tutorials*, vol. 17, no. 4, pp. 2317–2346, 2015.

69. H. H. Pajouh, R. Javidan, R. Khayami, A. Dehghantanha, and K.-K. R. Choo, "A two-layer dimension reduction and two-tier classification model for anomaly-based intrusion detection in IOT backbone networks," *IEEE Transactions on Emerging Topics in Computing*, vol. 7, pp. 314–323, Apr. 2019.

70. M. Frustaci, P. Pace, G. Aloi, and G. Fortino, "Evaluating critical security issues of the IOT world: Present and future challenges," *IEEE Internet of Things Journal*, vol. 5, pp. 2483–2495, Aug. 2018.

71. P. Chemouil, P. Hui, W. Kellerer, Y. Li, R. Stadler, R. Stadler, Y. Wen, and Y. Zhang, "Special issue on artificial intelligence and machine learning for networking and communications," *IEEE Journal on Selected Areas in Communications*, vol. 37, pp. 1185–1191, June 2019.

72. B. A. A. Nunes, M. Mendonca, X.-N. Nguyen, K. Obraczka, and T. Turletti, "A survey of software-defined networking: Past, present, and future of programmable networks," *IEEE Communications Surveys & Tutorials*, vol. 16, no. 3, pp. 1617–1634, 2014.

73. M. Abadi and D. G. Andersen, "Learning to protect communications with adversarial neural cryptography," Oct. 2016., arXiv preprint arXiv:1610.06918.
74. L. Huang, A. D. Joseph, B. Nelson, B. I. Rubinstein, and J. D. Tygar, "Adversarial machine learning," in *Proceedings of the 4th ACM Workshop on Security and Artificial Intelligence, CCS'11*, pp. 43–58, ACM, Chicago, IL, Oct. 2011.

Chapter 2

AI-empowered social internet of things and crowdsensing applications

Koushik Karmakar, Sujoy Das, Arka Gain, Jyoti Sekhar Banerjee, and Panagiotis Sarigiannidis

2.1 INTRODUCTION

During the past several years, extensive research has been done on various smart technologies and machine learning methodologies [1–23]. The Internet of Things (IoT) [24–31] is one such emerging technology that is gaining a reputation. It will connect billions of objects with one another. It is an example of a paradigm shift where bridges are created between humans, devices, and other objects. It explains how the physical world acts intelligently in the digital world. In this magnificent world even objects like actuators and sensors have their own separate identities too. Actually, it is a combination of different emerging modern technologies. For example, in the IoT, wireless sensor networks (WSNs), artificial intelligence and machine learning (AI and ML), data analytics and big data, is used. WSNs are based on a combination of different sensors used for different purposes. It is still in its infancy stage and significant research works are going on for improvement over the previous ones.

One advanced development of IoT is known as the Social Internet of things (SIoT) [32–36]. It is capable of making relationships between humans and objects. Through the representative architecture of the SIoT, one navigation may be established where one device initializes the system and the other connected devices will be navigated and thus autonomous relationships will be created among humans and objects. In this modern time, the combination of social networks (SNs) and the IoT has a large number of intelligence services and applications that can solve the problems faced by different organizations and individuals in their daily lives.

This chapter focuses on another important aspect. For running a city properly, efficient monitoring of urban and community dynamics is very much important. For that, sometimes sensor networks are used, which are generally distributed over a large area to monitor the real-time conditions there. But in reality, commercial sensors could never be deployed in large areas due to some practical problems like higher installation charges, insufficient spatial coverage, and many others. Mobile crowd sensing and computing (MCSC) is thus an important one with respect to large scale sensing at a minimum of infrastructure and cost.

MCSC allows collecting local information like noise level, and traffic conditions and sharing them with a remote cloud for further processing and analysis. The mobility of MCSC devices makes it a versatile platform, and thus it is possible to update the existing data with the updated one. Its coverage area is also huge, including environmental monitoring, mobile social recommendation, traffic planning, public safety, and many others.

Now the question is what is MCSC? Actually, it is a new method for data collection that allows everyone to collect data from their mobile phones, sum it up, and extract information out of it. MCSC is used as a distributed problem-solving model. MCSC uses the technique of crowd-powered data collection methods. Several research works have been done on the concept of crowd powered problem-solving. A few publications and books are also written on it.

In this chapter, a detailed discussion is made on these emerging technologies discussed above. The organization of this chapter is as follows: In Section 2.2, the origin and evaluation of the technology are described. Section 2.3 presents an AI-based approach in the relevant field. A literature survey is given in Section 2.4. SIoT building blocks are described in Section 2.5. Applications of SIoT are described in Section 2.6. Challenges in SIoT are described in Section 2.7. MCSC is described in Section 2.8. MCSC frameworks and applications are described, respectively, in Sections 2.9 and 2.10. Data reliability is explained in Section 2.11. Factors of the MCSC are described in Section 12. The conclusion is drawn in Section 2.13.

2.2 ORIGIN AND EVALUATION

The number of human beings living on earth as well as the number of objects in the world is increasing gradually. Hence, the connectivity between humans and objects is very much required to enhance the quality of life of the various services of the objects. Initially, a few years ago, the internet was known to be an e-commerce and information services network. But with the discovery of Facebook and YouTube, the concept has been changed thoroughly. Aston Kevin from MIT coined the term 'IoT' as 'Internet of Things'. A new journey has begun since then. The idea of the creation of SIoT came into picture when a small world phenomenon was introduced by Kleinberg. SIoT imitates the structure of human SN. It tells about different activities like friend selection and finding out better services. Every object that has an interaction with another object must have a few characteristics like interoperability, scalability, and trustworthiness. As well as some pre-set objectives. They work based on mutual understanding of each other. Devices are improving each day as research is going on to make them smarter and more intelligent.

The future of global computing is based on the various types of intelligent applications and services that include different problems that are faced by different organizations as well as individuals regularly through the connection between objects and people anywhere and anytime. With this invention, digital worlds are becoming smaller, and more people have access to them. It is a reality now and is gaining popularity day by day. It delivers high quality services. There are plenty of opportunities for new research and development in this arena. In Table 2.1 a comparative study of the state-of-the-art works is described in brief.

2.3 AI BASED APPROACH

AI is the major backbone of the (SIoT) and Crowd sensing applications. It enhances the functionality of these systems and facilitates a more seamless user experience overall. It works on different areas like decision-making, data processing, in many other fields. AI has a huge impact on the use of applications for the IoT and Crowd sensing. A few of them are mentioned below (Figure 2.1).

2.3.1 Awareness of the context

AI may help crowdsenses and the SIoT system in understanding the requirement properly for data collection. Using AI, user's location, choice, and behavioral patterns can be predicted so that customized data collection becomes possible. As such, the required settings of the IoT devices can be changed internally.

2.3.2 Forecast analysis

AI is also used for future prediction. The previous dataset is collected, and then, using AI and other techniques, future events can be predicted. It can be used to predict traffic and weather conditions, among many other purposes. It has different applications in real-life conditions.

2.3.3 Modular Control

It is possible to develop self-learning systems using AI that can train themselves as per changing conditions. It can be used in the SIoT for an improvement of the transportation system to use the available energy to its optimal level.

2.3.4 Pattern recognition and data analysis

AI systems can recognize patterns as well as trends in the large volume of data generated by IoT devices and crowd sensing applications. It helps in

Table 2.1 Comparison table between SIoT vs. IoT

Solutions Type	Social Internet of Things (SIoT)	Internet of Things (IoT)
Scope and purpose	The integration of social media and information sharing IoT frameworks is often linked to the SIoT, highlighting the collaborative and social aspects of related devices	The IoT encompasses a wide range of applications and devices that communicate with the internet to collect and distribute data for various uses, such as automation, expertise, and enhanced self-direction
Interaction paradigm	SIoT emphasizes interactions between people and machines as well as between people and machines that are mediated by machines inside an IoT context	Conventional IoT centers' on robotization and machine-to-machine communication
Data Sharing and Collaboration	Places a strong emphasis on interpersonal relationships and cooperative information sharing, potentially using interpersonal organizations to enhance IoT applications	Mainly concentrates on information sharing between devices and platforms for effective actions
User-Centric Design	Combines customer-driven plan requirements, considering the social context and customer experience as essential components of the IoT framework design	Consistently organized with a focus on productivity and usefulness
Privacy and security	Despite common IoT worries, SIoT may take social protection and social collaboration, security within the IoT system into consideration	While information assurance and device security are often at the centre of security and protection discussions, they are still the most important ones
Application domains	May find use in situations where social ties play a significant role, such as in cooperative leadership, local area detection, or social welfare observation	Used in a variety of contexts, such as gardening, contemporary machinery, medical services, clever homes, and so on
Communication Protocols	May organize socially connected conventions, like the correspondence and informing customs used in interpersonal organizations	Employs many correspondence protocols that have been enhanced for communication between machines
Community and SN integration	Rotates between connecting devices and systems	Looks into approaches to better coordinate IoT with local area stages and interpersonal organizations to enhance social efforts and communications

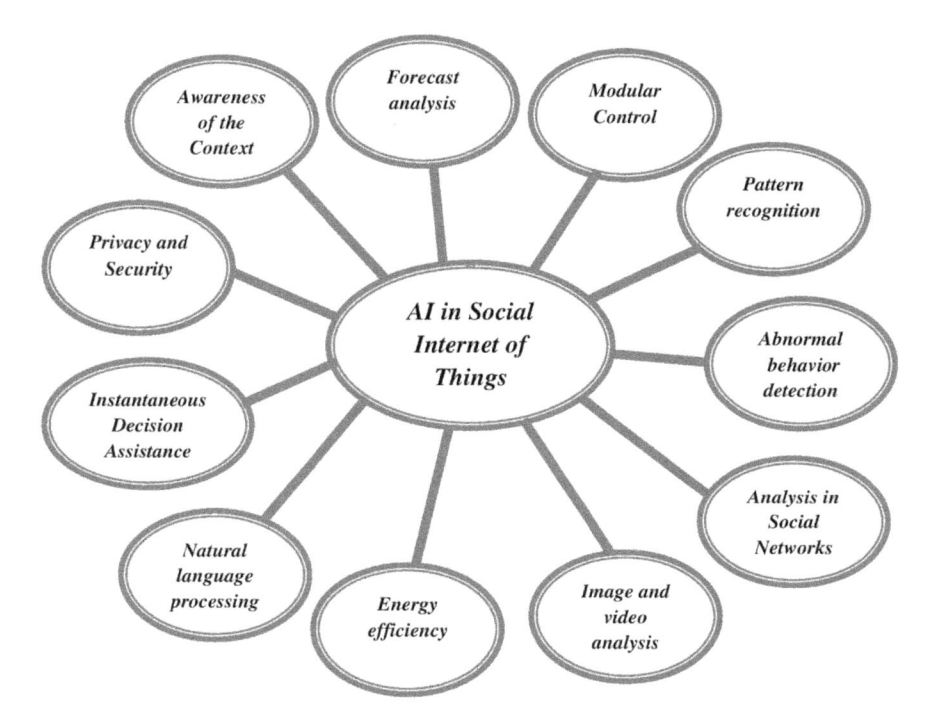

Figure 2.1 Role of AI in SIoT.

the extraction of the required data from large volumes and also facilitates the interpretation.

2.3.5 Abnormal behavior detection

AI techniques like ML and deep learning methods can identify anomalies in the generated data from IoT devices and crowd sensing methods. Such anomalies may have different types, like security breaches, failures in equipment, or other types.

2.3.6 Analysis in social networks

AI can inspect social media data for understand the details of the participants along with their relationships in crowd sensing applications. This helps in understanding current trends, emerging topics, and different other related issues.

2.3.7 Image and video analysis

AI can recognize and analyze different objects, human faces, pictures in still photos and videos captured by IoT devices, and crowdsourced data. As a result, virtual reality and other similar applications may be possible.

2.3.8 Energy efficiency

All such devices used in the IoT must be very much energy efficient. The use of AI can predict usage patterns and, as such, modify the internal settings of the devices for optimal use of energy consumption. This helps in saving energy of the devices.

2.3.9 Natural language processing

SIoT devices will be more user-friendly and approachable, provided they can communicate in their natural languages with the users. It is possible for users to make voice commands and have conversational interactions.

2.3.10 Instantaneous decision assistance:

It is possible to respond in emergencies through providing real-time data and recommendations of SIoT and crowd sensing systems. It also helps in decision support systems.

2.3.11 Privacy and security

AI may help in identifying possible security lapses or unwanted entries. Also, it can detect and reduce privacy and security risks related to crowd sensing and SIoT applications.

AI-based SIoT and crowd sensing applications are constantly increasing with the development of computer vision, natural language processing, ML and deep learning methods. As such, these systems have become much more intelligent and smart. These technologies are spearheading innovations in numerous other fields where IoT and crowd sensing are used, like smart cities, healthcare, and transportation.

2.4 LITERATURE SURVEY

Several applications on the sensor-based smart systems have been developed and widely used worldwide in different purposes, including smart home based systems. Such applications can be divided into several types. Classification is done based on their applications and the technology used. In Table 2.2, a comparative study of the state-of-the-art works is described in brief.

Table 2.2 Literature survey of the state-of-the art works

Author	Summary	Technology	Applications
Diaz et al. [37]	A multi-facet self-mending algorithm called Disjoint Path and Clustering Algorithm (DACA) was created to address security flaws in IoT applications. Nevertheless, low hub thickness and constrained internet access present problems	NA	Centralized architecture based IoT applications
Raja et al. [38]	It covered the necessity of D-D communication and middleware, as well as problems and various challenges associated with evaluating IoT applications. Future IoT architectures should feature self-patch, confidence, personal progress, and self-plan	Software	D-D communication and middleware, issues and, various research challenges of IoT applications
Cao et al. [39]	In order to address portability issues, the study suggests two ways for battery-powered IoT applications: blended whole number direct programming and QoS adaptable web-based approaches. Additionally, it suggests heuristic solutions based on the cross-entropy methodology and MILP computation for longer battery life as well as an infinitely flexible calculation offloading mechanism	NA	Mobility aware network for battery-powered IoT applications
Guan et al. [40]	This process actually expands the area of the location; it was discovered that occasionally quickly sitting down and walking while standing up would trigger the false discovery	NA	Smart homes based on sensors
Sethi et al. [41]	To obtain data, different drivers must activate the different sensors. First and foremost, various sensors will face significant challenges as a result of the device drive's advancement. Second, it is costly to set up many smart living space sensors.	Gadget drive, various sensors	Smart homes based on sensors
Virmani et al. [42]	In order to identify human movements, the creator looked into using the Wi-Fi banner. This could lead to opportunities for developing new client PC interfaces for sophisticated home automation systems	Wi-Fi,-and PC	Smart home automation with Wi-Fi

(Continued)

Table 2.2 (Continued) Literature survey of the state-of-the art works

Author	Summary	Technology	Applications
Yang et al. [43]	In light of IoT and its techniques, the creator proposed a torture monitoring application intended for patients' continuous information monitoring in the crisis centre	Pain Monitoring Machine	Smart Health Pain Monitoring
Guan et al. [44]	Its range of feelings is limited. It is therefore beneficial for acknowledging movement that is solely reliant on the route	sensors	Smart homes based on sensors
Liu et al. [45]	The inventor focused on using Wi-Fi movement to track respiration and detect heartbeat. Because developers could use this information to launch a dangerous attack, this could also give rise to serious security concerns	Wi-Fi	Smart home automation with Wi-Fi
Lijun [46]	In light of the IoT, creators suggested a standard helpful getting marker. This framework connects a PC, a cell phone, and a blood glucose expert	Therapeutic Securing device,-and Smartphone	Monitoring blood glucose in smart wellness
Xin et al. [47]	The inventor focused on the IoT with his clever site-based pulse control terminal	BP Monitoring Machine	BP monitoring of smart wellness
Liu et al. [48]	The author provided an IoT based electrocardiogram noticing structure that included a minimized remote getting spreader and a remote getting centralized server. In order to reliably perceive heart limit, the structure arranges a computerization strategy to recognize surprising data	ECG Machine	ECG surveillance in Smart Health

2.5 SIoT BUILDING BLOCK

The SIoT needs a large number of devices in order to operate. Some essential elements needed for a successful SIoT application are as follows:

2.5.1 ID

To identify an object in a typical system, this is the method assigned to it. MACID, IPv6ID, and universal product are a few examples of identification.

2.5.2 Meta-information

The structure and functions of a device are referred to as meta-information in a system. In the world of IoT devices, devices must be placed appropriately to establish relationships with one another.

2.5.3 Security controls

It is the constraints placed on device connections by the device's owner. Owner controls are another term for it occasionally.

2.5.4 Service discovery

These specific directories have the capacity to hold information about gadgets that offer particular services. When the directories are current, devices may be able to learn about one another.

2.5.5 Relationship management

It speaks about the interaction and administration of devices. Keeping track of the connection between a light sensor and a light controller is one example.

2.5.6 Service composition

It gives the users a better service through enabling the system to interact with an analytics engine, where a vast amount of data is examined to learn about usage patterns for future enhanced output or services.

2.5.6.1 Architecture

There is no standard architecture for SIoT because the structure of the architecture provided by various publications varies. In maximum publications, a three-layer architecture is shown. Among these three layers, the object layer is shown as the base layer, the network layer is portrayed as the middle layer, and the application layer works as the top layer.

2.5.6.1.1 Object layer

This layer consists of items or apparatus used for data collection and environmental monitoring. Numerous publications provide various structures regarding the architecture of the SIoT.

2.5.6.1.2 Network layer

This layer has transport and networking capabilities. Whenever any device wants to send data to another device, both of them require an active Internet connection. Different protocols and gateways are used for such data transfer.

2.5.6.1.3 Application layer:

This layer exists at the top most position of the SIoT structure. It contains many applications like smart homes, smart cities, smart health, and other industrial uses.

It consists of different components, like an intelligent system, an interface, and an actor. Users and objects make up the actor layer. To exchange information, an intelligent system is required, as is an interface for command and service management control. To provide users with object-oriented services, the internet is required.

2.5.6.2 Relationship management (RM)

If every object works on the SN instead of working independently, people will receive more accurate answers to their requests. As such, devices and objects must form a relationship for better performance. Initially, IoT wants to create a SN among devices, which are normally remain isolated. For the exchange of information, objects must fix their type of relationship with other objects. As such, the type of relationship that exists between the objects decides how relationships are managed among different objects. However, it can be divided into five categories (Figures 2.2), which are described as follows:

2.5.6.2.1 Parental OR (POR)

Parental OR relationship refers to the relationship that exists between an object and its manufacturing plant. Every object that cooperates in this relationship is homogeneous.

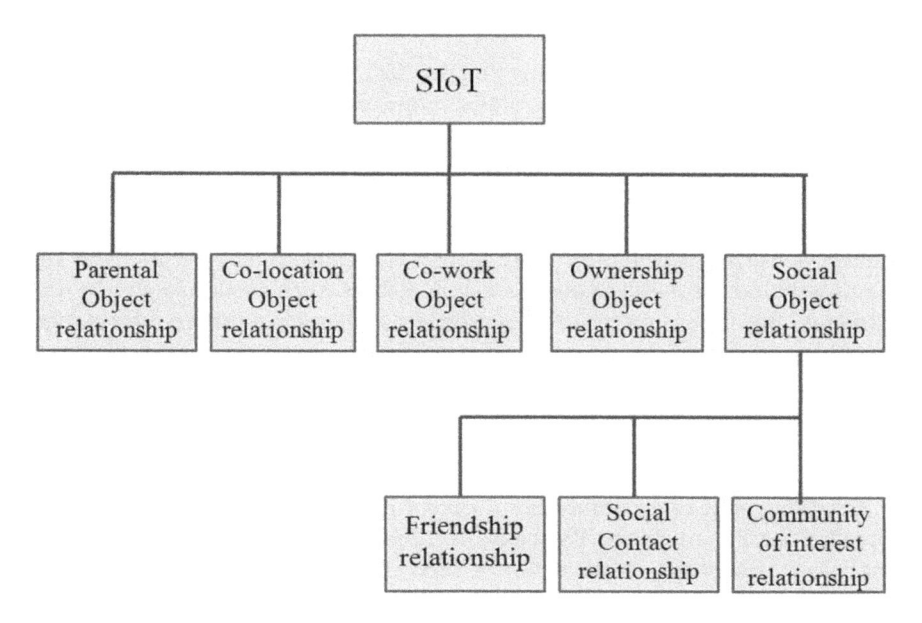

Figure 2.2 Relation management diagram in SIoT.

2.5.6.2.2 Owner OR (OOR)

An owner-object relationship is defined as the relationship between an object and its current owner. This is analogous to the diverse software installed on an individual's computer.

2.5.6.2.3 Co-Work OR (CWOR)

Co-Work OR relationships are those in which every object collaborates with another to achieve a common goal. Even for objects that are located in different locations, this relationship can exist.

2.5.6.2.4 Social OR (SOR)

Social OR relationships refer to the relationships between the owners of the objects. Their devices cooperate as a result of this relationship.

2.5.6.2.5 Co-Locate OR (CLOR):

Co-locate OR relationships are those that exist between objects that are in the same location. Both homogeneous and heterogeneous relations are possible.

2.5.6.3 Trust Management (TM)

There is no standard architecture for the SIoT because the structure of the architecture provided by various publications varies. Trust is a strong belief in something's veracity, truth, or ability. It typically builds a connection between the two objects concerned (trustor and trustee). Both parties will benefit from this relationship. Regarding collaboration, trust is the most vital factor. Since many items on the IoT operate in partnerships, trust is the most important factor. Sensitive information is frequently transferred between items in the modern world through various electronic devices, such as mobile phones and cameras. But it is a matter of great concern that such sensitive information may be misused if it is shared among arbitrarily. Lack of trust in SIoT can lead to issues like exchange, privacy, safety, and access loss. Furthermore, the owners of the items have the ability to launch damaging attacks like on-of attacks, self-promotion, and badmouthing. Thus, for customers and providers to interact as effectively as possible, trust is necessary.

Confidence in every linked SIoT object is necessary to establish a relationship based on trust. Trustworthy objects can generate requests and malicious things within the network through this trustworthy connection, which can also lead to higher trustworthiness in communication. Since every object is vulnerable, security is one of the primary concerns with trust management. Thus, a control system is required to stop unwanted access to data and network resources. In order to protect the network from attacks and implement an access control system, a security policy should be put in place. Another crucial element is that, in order to meet requests and improve the security and safety of SIoT networks, each object should only exchange data and services with other trustworthy objects that foster stable and reliable communication.

2.5.6.4 Web services

Web services development is one of the primary obstacles to the advancement of modern technologies like cloud, fog, and SIoT. A web service is a process that generates messages for use by client and server applications. Its intention is to perform a particular set of tasks. There are three primary sub processes that make up web services. The initial one is called "service discovery," followed by "service selection" and "service composition."

2.5.6.4.1 Service discovery:

It's a method for looking for things that can offer the desired service.

2.5.6.4.2 Service selection

The process involves choosing a service that is comparable to the desired services.

2.5.6.4.3 Service composition

It is the procedure that combines various services to generate a suitable response to the requested services. A service's inability to satisfy every user's need may lead to improvements in its functionality, quality, and viability. Thus, selecting appropriate services and combining them could be one of the biggest issues in the SIoT environment.

2.5.6.5 Information

The three key components of the Big Data concept—volume, velocity, and variety—are what drive the majority of data exchange in the IoT. The three central figures of the Big Data concept possess the ability to practically gather, organize, and oversee information or data. It can also perform the process of examining a vast quantity of data in order to assess and characterize the information and behaviors of the objects. Due to the vast volume of data in this field that was incorporated into the connections between various objects and devices, an exact solution is required to gather and aggregate data, and specific software is required to process this enormous volume of data. An illustration of a four-tiered framework for SIoT Big Data would be as follows.

2.5.6.5.1 A collector server

It is employed to gather sensed data from items in our surroundings that are part of a common community.

2.5.6.5.2 A storage server

Diverse forms of data originating from different devices and objects are stored in it.

2.5.6.5.3 A filtering server

It serves as the main processing unit for separating noisy data from raw data before sending it to the process server for use in IoT communications.

2.5.6.5.4 A process server

It is employed for processing and data analysis.

It consists of different components, like an intelligent system, an interface, and an actor. Users and objects make up the actor layer. To exchange information, an intelligent system is required, as is an interface for command and service management control. To provide users with object-oriented services, the internet is required.

2.5.6.6 SIoT tools

IoT platforms are basically a collection of software tools. It is used to create connections between objects in an IoT system. The objective is to create new applications. A few of these tools are described below:

2.5.6.6.1 CRAWADAD

In order to gather and examine data, it is necessary to store it in this dataset from a variety of sources. The primary purpose is to gather wireless network data resources for the community.

2.5.6.6.2 Reality mining dataset

The location, communication, and device usage patterns of the owner are represented by this dataset. It is a nine-month study involving one hundred human participants.

2.5.6.6.3 Washington State University's CASAS dataset

By storing crucial information from the surrounding environment, the CASAS project is referred to as a real smart environment with intelligent agents like sensors that are used to provide a safer and more comfortable life in every environment. Based on how it is used, this dataset can be categorized into a wide range of areas, such as real-time Smart Home Stats and Assisted Care Apartments Real-Time Activity Update.

2.5.6.6.4 SNAP

The Stanford Large Network Dataset is an enormous collection of diverse datasets covering a wide range of topics. One example of this collection is the SN dataset, which is concerned with user interaction with online SN.

2.6 APPLICATIONS OF SIoT

There are numerous uses for the IoT [49–51]. We have been concentrating on smart cities in this context. Research on IoT applications in the context of smart cities has been done in great detail. Numerous studies were conducted, and publications were released as a result. It can be divided into different categories like governance, living and infrastructure, mobility and transportation, economy, industry and production, energy, environment, and healthcare.

1. **Smart living and infrastructures:** The smart living domain includes all components involved in developing smarter city infrastructures, such as smart homes, buildings, etc., as well as the management and enhancement of public services, such as cultural events and educational programs, which are focused on improving the general quality of life of citizens.

2. **Smart governance:** Smart governance seeks to enhance decision-making processes and expedite bureaucratic and administrative procedures in cities through the smarter collaboration of diverse stakeholders and social factors, including public administrations, city officers, private companies, and citizens. Using social media and ICT-based tools, for instance, citizens can take part in city governance and decision-making processes.

3. **Smart economy:** A smart economy is one that makes innovative use of ICT to link local and global marketers, enabling e-commerce services that improve productivity and deliver. Another choice is peer-to-peer labor services, where users and members contribute their knowledge and abilities for specific tasks.

4. **Smart building:** The IoT makes it possible to implement a wide range of building facilities, such as tools for air conditioning and rainwater drainage, monitoring human activity, and monitoring the structural integrity of buildings. Depending on its use, different IoT techniques are implemented.

5. **Smart energy:** Through the intelligent integration and efficient distribution of decentralized renewable and suitable energy sources, smart energy systems optimize power consumption. Smart grids use ICT and IoT technologies to enhance the management of power generation and distribution. As an illustration, prediction models that are created from gathered consumption data are frequently employed, and the energy network supply typically self-heals.

6. **Smart healthcare:** Within the hospital industry and management, one of the most promising areas is smart health care. It might take a sensor to save a patient's life. One example is the ability to connect a heartbeat sensor to a cloud computer or smartphone, enabling a doctor thousands of miles away to diagnose and treat serious cardiac conditions in their patients.

7. **Smart tourism:** Two new applications of the IoT are cultural heritage management and tourism management. One way to introduce more customizations is to grant tourists greater control via their phones. For example, some well-known locations (like museums) allow guests to book a time slot ahead of time to avoid the crowds and queues and to enjoy a better experience. However, the travel industry kept track of its customers' information, allowing it to generate customized offers based on those customers' previous search and purchase behavior.

8. **Smart transportation:** IoT applications in automotive services will be very beneficial. For example, someone may find it difficult to resolve a problem with their new automobile. A profile of the issue and the vehicle is constructed using the data gathered by the junction box from sensors installed in that vehicle. After that, this profile is shared throughout the SIoT to search for issues that are comparable and have previously been resolved by other vehicles connected to the network, enabling the lookup of any nodes that could be useful in resolving this problem.

2.7 CHALLENGES IN SIoT

Although there are a lot of benefits of SIoT, there are still some limitations that are required to be solved. A list of such challenges is given below:

1. **Heterogeneity:** Millions of objects with various features across platforms, protocols, and standards comprise the IoT and all of the objects and data need to be recovered. Due to these variations, a heterogeneous network of objects has been created. It makes the system more complex. A lot of queries are raised. They require solutions. Examples of these include discovering object identification politics and utilizing POR object relationship.
2. **Mobility and dynamicity:** The location of intelligent objects in dynamic environments is always changing, causing issues like insufficient object search functionality for service selection and provisioning. The way that objects behave dynamically and alter their states is another crucial issue.
3. **Manage dynamic behavior of objects:** A smart economy is one that makes innovative use of ICT objects must be assigned primary rules and protocols by their owners to handle these changes in order to prevent them from altering the network topology. Adaptability, on the other hand, is a problem that results from this dynamically since an object must adjust to these frequent changes.
4. **Tracking objects:** Tracking objects, interactions, and activities is a major problem in large-scale networks and the IoT that was seldom considered before. Here are a few solutions to this issue: using objects' movement patterns and graph models.
5. **Security issues:** Due to the vast array of linked devices in this SIoT, privacy and security are crucial components of sharing information or data. Thus, it continues to be one of the major challenges, despite the fact that many different kinds of research have been done in this field.
6. **Scalability:** The enormous number of SIoT-connected entities needs to be managed properly by the discovery system.

7. **Standardization:** It is necessary to enforce discovery programs using widely recognized conventional communication protocols.
8. **Survivability:** The discovery method's trust protection processes need to be resilient to deliberate attacks. Resources that must be carefully and methodically examined, or even carefully and carefully sifted in order to identify and take into consideration the most relevant ones.

2.8 MOBILE CROWD SENSING AND COMPUTING (MCSC)

In recent years, MCSC has become more and more popular [52–60]. It's turning into a very attractive model for data collection and sensing. Ganti et al. coined the term "mobile crowd sensing" to refer to a model that was broader compared to the traditional mobile phone based sensing method. He claims that MCSC is a novel sensing paradigm that enables regular people to contribute data that they have sensed or generated from their mobile devices. It then gathers, combines, and stores the data in the cloud for the purposes of extracting crowd intelligence and providing people-centric services.

Wearables and smartphones, which are frequently used mobile devices, are the source of sensors and communication interfaces for MCSC systems. The use of different mobile devices is becoming almost compulsory in our modern lives. The most widely used devices are smart watches, glasses, rings, gloves and helmets. Their popularity is reflected in their projected explosive growth in revenue. In the coming years, it is also anticipated that the market for crowd analytics will grow significantly.

MCSC systems need a lot of users to participate and contribute for them to work effectively. Even though a single individual does not want to take part out of selfishness or due to any other reason, entire communities may stand to gain from such a contribution. In recent years, the research community has worked hard to develop appropriate mechanisms for incentives and to investigate privacy concerns in an attempt to lessen this burden. The proliferation of wearables and smartphones, along with their extensive array of integrated sensors, is unquestionably the primary factor contributing to the success of the MCSC paradigm. Urban societies can greatly benefit from MCSC's ability to give citizens new perspectives and greatly enhance their daily lives. Future smart city construction will require the use of MCSC as a key component.

The IoT provides a possible way out of the widespread setup of sensors required to support smart cities. Furthermore, citizens' active participation can enhance the spatial coverage of already-deployed sensing systems without requiring additional funding.

Unlike traditional sensor networks, MCSC uses human intelligence, which has a deeper understanding of context. Cities suffer from a serious lack of infrastructure services, but human intervention can help with better upkeep and monitoring. We present specific use cases to demonstrate this idea. Bridge vibrations can be detected using accelerometer data collected from smartphones traveling over moving vehicles.

Smart traffic management and free parking spot detection are two more potential city services where MCSC could play a significant role. In particular, Park Gauge provides current crowdsensing information regarding the availability of parking slots and uses low-power-driven sensors like accelerometers, which are used to detect driving states, while Park Sense uses Wi-Fi scans of smartphones to identify open parking spaces. A plethora of literature exists regarding MCSC systems. Numerous studies have suggested innovative sensing architectures and, as was already mentioned, looked into particular issues like privacy concerns, incentive mechanisms, and the validity and trustworthiness of the data collection procedure. Because there is so much literature on the subject, there are already surveys covering particular topics like privacy or user recruitment incentives. Others give numerous MCSC facets.

Furthermore, many of the central paradigms in the enormous body of literature remain unclear due to a lack of categorization. For instance, there is disagreement over the definition of "opportunistic sensing." In reality, it is applicable when there is more autonomy in the system with minimum user interaction. But according to some researchers, opportunistic sensing doesn't even need user input because the device itself decides whether or not to perform the sensing. As per the opinion of a few scientists, it is actually a method that exists through the cooperation of more than one smart device like smartphones.

Our manuscript aims to provide comprehensive taxonomies that will facilitate comprehension of existing definitions, techniques, and solutions in the field of molecular and structural anthropology. We intend to classify MCSC works according to the below figure's four-layered architecture (Figure 2.3). The underlying rationale—which is further explained below—is to cover the complete process chain from the point at which the sensor generates a reading until information comes to the top most layer, i.e., the application layer. This is the structure of the four-layered architecture.

The organization of the four layered architecture is given below. The application layer is the top most layer of the system. It is in-charge of higher-level operations like user recruitment, and task distribution. The second layer is also known as the data layer. This layer is responsible for the gathering and processing of the data. The next layer is called the communication layer. This layer is made up of different components for giving delivery of the information sensed through different types of sensors. The sensing layer is the lowest layer of the system. It is located at the bottom side, near to that of the physical layer.

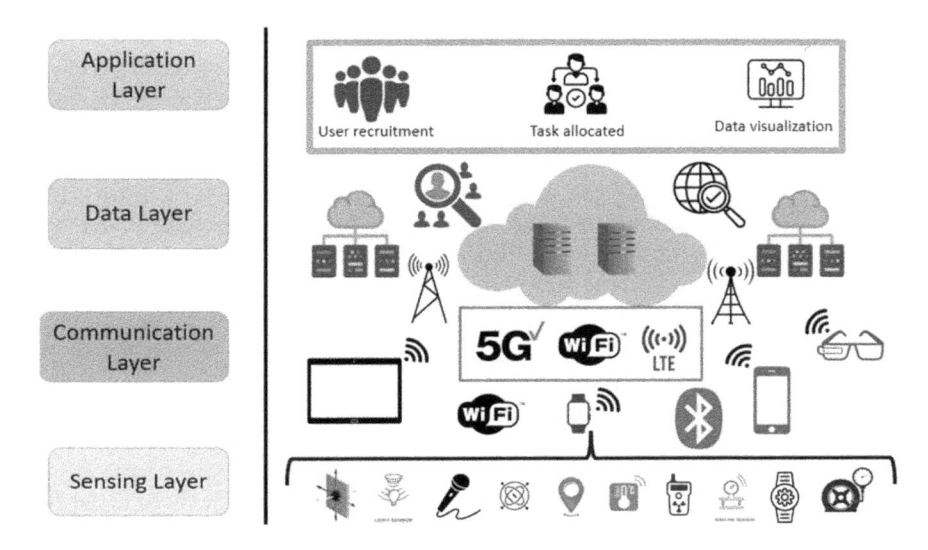

Figure 2.3 The layered architecture of the MCSC system.

2.9 MCSC FRAMEWORK

Regarding the MCSC framework, one reference architecture has been proposed which highlights the main functional components and clarifies the main MCSC techniques. It is meant to serve as the foundation for future research in this emerging field. The proposed architecture is depicted in the following figure (Figure 2.4). There are five different layers in this architecture: applications, crowd data processing, data collection, data transmission, and crowd sensing.

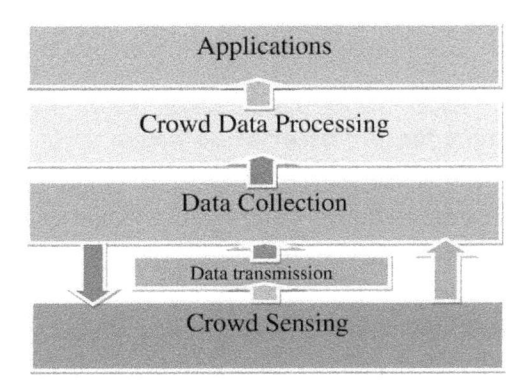

Figure 2.4 A model for MCSC architecture.

2.9.1 Applications

It is the topmost part, which consists of various applications and services that MCSC may be able to enable. Data visualization, user interface, and other related essential functions are included in this layer. The results of crowd computing are presented to users in a readable format through visualization techniques like graphing, mapping, and animation. The purpose of user interface design is to facilitate communication between humans and machines. They both help users like citizens, and service providers, make decisions and share knowledge.

2.9.2 Crowd data processing

The objective is to derive advanced cognitive abilities from unprocessed sensory data through the application of an extensive range of ML and logic-based inference methodologies. Stated differently, the goal of crowd data processing is to identify recurring patterns in data to get three different aspects of crowd intelligence in a comprehensive way.

2.9.2.1 Data processing architecture

One hybrid solution has been proposed instead of a purely centralized or self-supported approach. Certain data processing tasks in HDP can be completed locally. The local results are then sent to servers for additional processing. It improves the flexibility of the entire network. Also, the cost of communication between clients and servers can be significantly reduced by using such a hybrid data processing approach.

2.9.2.2 Data quality maintenance

The data is frequently redundant and varies in quality and credibility among contributors. Therefore, in order to preprocess the data and remove the low-quality data, quality measurement and data selection metrics are required.

2.9.2.3 Cross-space feature association/fusion

Both offline and online locations can provide mobile crowd data. This component investigates complementary feature fusion and cross-space data association techniques.

2.9.2.4 Crowd intelligence extraction

Through the use of numerous data processing techniques, it seeks to extract the three categories of crowd intelligence: ambient awareness, social awareness, and user awareness.

2.9.3 Data collection infrastructure

It collects information using a few specific sources and subsequently gives data to those tools that maintain data privacy. There are the following elements associated with this system.

2.9.3.1 Task allocation

This will evaluate the task of sensing provided by the application system. Then it allocates this to a selected set of nodes based on predetermined parameters, like location, time, and capacity of the device, budget and many such topics.

2.9.3.2 Sensor gateway

Sensor gateways serve as a single interface for all the upper-level parts, including applications cum data processing and offer a standard method like Web services. It facilitates data collection from various crowdsensing sources.

2.9.3.3 Data anonymization

Privacy is a major concern when sharing personal information. This component supports privacy protection tasks. It also performs many other activities by offering anonymization mechanisms prior to data published.

2.9.3.4 Incentive mechanism

It is this component that gives data contributors' strategies for rewards and reputation.

2.9.3.5 Big data storage

There are two important features of the system. They are called large-scale and multi-modality. First of all, the data volume is very large, and thus it is very difficult to store and manage it. Thus, it becomes increasingly difficult to process using the existing tools and systems. Second, the features of different kinds of sensors typically vary greatly. It leads to significant differences in the accuracy of crowd sensing. Consequently, in order to enhance subsequent processing, the raw data collected from various sensors must first be processed, changed and represented in a specific way.

2.9.4 Data transmission

Mobile smart device-sensing data in MCSC will be sent to the backend server for additional processing. Data uploading should be tolerant of poor networking conditions and unavoidable network outages in MCSC applications. Thus, for data transmission, data forwarding and routing protocols become important. The data should be sent to the destination using only opportunistic networking if the network infrastructure is unavailable. Additionally crucial to this layer is the collaboration of heterogeneous networking nodes in order to improve data transmission performance.

2.9.5 Crowd sensing:

This layer includes user-contributed data from mobile Internet applications and different types of data sources for mobile crowd sensing, such as collected data from mobile devices. Participants must have the right to choose whom to share the data with and what to publish. It is especially important in light of the importance of security and privacy concerns in MCSC. Access control thus becomes a crucial component of the participatory sensing clients.

2.10 MCSC APPLICATIONS

We briefly review some of the current mobile crowdsensing applications in this section, which serve as a foundation for illustrating different research challenges in subsequent sections of this article. Depending on the kind of phenomenon being mapped or measured, we divide MCSC applications into three groups. Social, infrastructure, and environmental are a few of these.

The phenomena that are used in environmental MCSC applications come from the natural world. Monitoring wildlife habitats, water levels in creeks, and pollution levels in cities are a few examples. By incorporating the average person, these applications make it possible to map a variety of large-scale environmental phenomena. Common Sense is one example of a prototype deployment for pollution monitoring. Common Sense measures various air pollutants (e.g., CO_2 and NOx) using specialized handheld air quality sensing devices that pair with mobile phones via Bluetooth communication. When spread throughout a sizable population, these devices measure the air quality of a community or a sizable area collectively. In a similar vein, mobile phone microphones can be used to track noise levels in public spaces. The IBM Almaden Research Center's CreekWatch is another example. By compiling user reports—such as images shot at different points along the creek or text messages regarding the quantity of trash present— it keeps an eye on the water levels and quality in creeks. Water control boards can monitor the levels of pollution in water resources by using this information.

Large-scale phenomena connected to public infrastructure are measured in infrastructure applications. Examples include tracking real-time transit and measuring traffic congestion, road conditions, parking availability, and public works outages (such as broken traffic lights or malfunctioning fire hydrants). MIT's CarTel and Microsoft Research's Nericell are two examples of early MCSC deployments that measured traffic congestion levels in cities. CarTel measures a car's location and speed using specialized devices installed in the vehicle and sends the data to a central server via public Wi-Fi hotspots. Then, using cellphones and this central server, information such as least-delay routes and traffic hotspots can be obtained.

In conclusion, social applications represent the third category, whereby people exchange sensed information with one another. To illustrate, people can compare their exercise levels with those of the community by sharing their exercise data, such as how long they spend exercising each day. This analogy can assist them in making improvements to their regular workout schedules. Take BikeNet and DietSense as examples of deployments. People use BikeNet to measure their location and the quality of their bike routes (such as the amount of CO_2 they pass through and how bumpy the ride is), then compile the data to find the routes that are "most" bikeable.

With DietSense, users can compare their eating habits by sharing photos of their meals with other members of the community. A common application for this would be in a diabetic community, where members monitor the foods consumed by other members and manage their own diets or offer advice to others.

2.11 DATA RELIABILITY

With smartphones, we can quickly collect sensing data from various groups of people at various locations thanks to mobile crowd sensing. Given that the sensing tasks have monetary rewards associated with them, participants might try to trick the mobile crowd sensing system in order to obtain money. As a result, mechanisms are needed to properly validate the collected data. In the following, we use several examples of malicious behavior to highlight the necessity of such a mechanism. Maintaining data consistency and reliability is necessary for the following factors: awareness of traffic congestion, data from the general public and environmental concerns.

2.11.1 Traffic jam alerts

Assume that a mobile crowd sensing technology is used by the Department of Transportation to gather alerts from vehicles on busy highways and then disseminate them to other cars. Traffic information is updated in real time, which is advantageous for vehicles on other roads. But, since malevolent users can attempt to proactively divert traffic on upcoming highways in order to clear these routes for themselves, the system must guarantee the authenticity of the notice.

2.11.2 Citizen-journalism

People can provide real-time data from public events or disaster zones in the form of text, images, and video. As soon as an event occurs, the public can access real-time information from anywhere in the world in this fashion. Unfortunately, dishonest people may attempt to make quick money by pretending to be somewhere else when an event is taking place.

2.11.3 Environment

Environmental protection organizations can map the country's pollution zones with great accuracy by using the pollution sensors built into phones. By reporting detected pollution statistics linked to fictitious locations, participants might accuse businesses of creating "fake" pollution to harm their rivals.

In the end, mobile crowd sensing systems need to validate their sensed data in order to provide their clients—who utilize the sensed data—confidence. However, because sensing measurements heavily rely on context, it is difficult to confirm every single sensed data point reported by every participant. To attain a particular level of dependability of the felt data, one way to address this problem is to verify the location linked to the sensed data point. However, there is still a significant obstacle that we must overcome: how can we reliably and economically authenticate the position of data points without the assistance of the wireless carrier? Please take note that wireless carriers might not assist with location validation due to privacy laws.

There exist several conventional methods to attain dependability in the location data of participants, including the utilization of Trusted Platform Modules (TPM) on smartphones or splitting up the tasks among several actors. But, unfortunately, this direct application of these solutions has not always been feasible. Because task replication is not always possible due to a lack of neither additional users, nor it is cost-effective to have TPM on every smartphone. When the participant tries to submit the location of the sensor data, another way to ensure that the location is accurate is to use secure location verification methods in real time. If this solution is used for every sensed data point, it unfortunately necessitates infrastructure support or adds a substantial overhead on user phones.

2.12 FACTORS OF MCSC

MCSC has emerged as a significant data collection and processing technique due to a number of factors (Figure 2.5). They are explained below in detail:

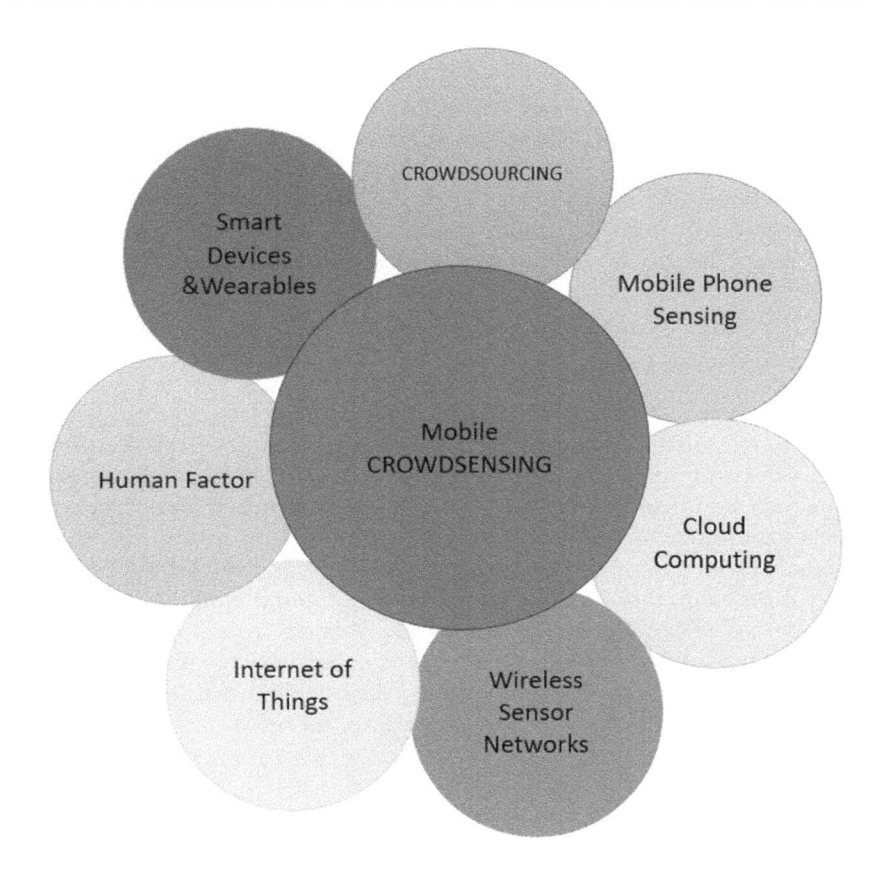

Figure 2.5 Factors of MCSC.

2.12.1 Mobile phone sensing

The origin of MCSC is mobile phone sensing. Individual-level objective phone sensing applications are different from MCSC. For instance, mobile applications for fitness track sessions use the GPS and additional sensors (like a cardio meter).Users of Ubifit are urged to track their regular physical activities, such as cycling, walking, and running. Individual monitoring is also essential in the health care sector for things like diet and recognizing and responding to falls in the elderly. Additional instances include identifying different forms of transportation, identifying speech patterns and driving styles, and using indoor navigation.

2.12.2 Mobile smart devices/wearable

The shift from cell phones to smart mobile devices has expedited MCSC's ascent. Unlike mobile phones, which were limited to text messaging and phone calls, smartphones have sensing, computing, and communication features. Other smart devices with similar characteristics to smartphones are tablets and wearable. The user experience is fully integrated into the technology process with wearables. Body sensor networks and wearables are essential for health-care monitoring, which includes nutrition, medical treatments, and sports activities. For affordable motion tracking that has a significant impact on human-robot interaction, we should concentrate on inertial measurement units (IMUs) and wearable motion tracking systems.

2.12.3 Crowdsourcing

The shift from phone sensing toward crowdsensing has been made possible by crowdsourcing, which leverages widespread user participation for data collection and enforces community-oriented application purposes. Numerous definitions of crowdsourcing have been put forth, according to the discipline. By definition, crowdsourcing is when an organization or business issues an open call to an amorphous network of individuals to complete tasks that were previously completed by employees. When a task is completed collaboratively, this enforces peer-production; however, crowdsourcing is not always a collaborative process. A large number of unmarried people answering the open call still fit the description. Since the early days of crowdsourcing, incentive mechanisms have been proposed.

2.12.4 Human factor

As humans become more integrated into the processes of sensing, computing, and communicating, their complementary roles with machine intelligence become more apparent. For a number of reasons, the human element has a crucial role in MCSC. It provides better results when compared to traditional sensor networks. Furthermore, users clean and occasionally recharge their mobile devices. Human-in-the-loop architecture is difficult to design effectively. By directing actions, for instance, following a period of learning. Probabilistic methods are usually needed for learning and prediction.

2.12.5 Cloud computing

Because smart device's local data storage has limited capacity, processing this kind of data, typically happens in the cloud. It is a popular application that takes place in cloud-based IoT systems. A large amount of data is stored in the cloud storage for processing later.

2.12.6 Internet of Things (IoT)

A significant degree of heterogeneity in end systems, devices, and link layer technologies defines the IoT paradigm. However, MCSC concentrates on urban applications, which reduces the range of IoT's potential applications. Utilizing the most cutting-edge ICT systems, this modern technology must enhance the quality of human life for residents as well as improve community value in order to support the smart city vision. By integrating humans, MCSC enhances the capabilities of current sensing infrastructures in this particular context.

2.12.7 Wireless sensor networks (WSN)

It is a system used for keeping an eye on events. Human mobile devices serve as the sensing nodes in MCSCs. With virtualization, several applications can share a single WSN infrastructure, as WSN nodes have grown in power over time. WSNs have a number of scalability issues. Software-Defined Wireless Sensor Network (SDWSN) is a subset of Software-Defined Networking (SDN) that can be used to improve sustainability and competence by addressing these issues. The widespread use of tiny sensors has produced vast volumes of data in smart communities, which WSNs' inadequate communication capabilities make it impossible to efficiently collect and process. Combining the ideas of cloud computing and WSNs is a viable approach that has been looked into. The authors of this work present the idea of sensor clouds and categorize the current state of WSNs by putting forth a taxonomy based on various factors, including data types and communication technologies.

2.13 CONCLUSION

This chapter has covered the specifics of the SIoT with a thorough discussion of its architectures, linkages, and applications. Several SIoT-related issues have been discussed and solved. In its upcoming work, a framework for addressing privacy and security concerns will be proposed and discussed. The IoT system works in tandem with other technologies, such as fog computing, AI, and ML, to enhance the outcomes of intricate issues like data management and reliability. Additionally, a detailed discussion of MCSC which is basically a cross-space heterogeneous crowd-generated sensing paradigm for large-scale sensing and computing, has been made. The involvement of humans in this system is one of the key components of the MCSC. Also, it will support and expand a wide range of application domains, including mobile social recommendation, intelligent transportation, urban sensing, and environmental monitoring. A framework for creating the MCSC system has also been discussed. These encompass avenues for future investigation.

REFERENCES

[1] Shi, Q., Zhang, Z., He, T., Sun, Z., Wang, B., Feng, Y., and Lee, C. (2020). Deep learning enabled smart mats as a scalable floor monitoring system. *Nature Communications*, 11(1), 4609.

[2] Shirazi, E. and Jadid, S. (2019). Autonomous self-healing in smart distribution grids using multi agent systems. *IEEE Transactions of Industrial Informactics*, 15, 6291–6301.

[3] Acampora, G., Cook, D.J., Rashidi, P., and Vasilakos, A.V. (2013). A survey on ambient intelligence in healthcare. *Proceedings of the IEEE*, 101(12), 2470–2494.

[4] Chakraborty, A., Singh, B., Sau, A., Sanyal, D., Sarkar, B., Basu, S., and Banerjee, J.S. (2022). Intelligent vehicle accident detection and smart rescue system. In: J. K. Mandal, S. Misra, J. Sekhar Banerjee, and S. Nayak (eds.), *Applications of Machine Intelligence in Engineering* (pp. 565–576), Jaipur. CRC Press.

[5] Das, K. and Banerjee, J.S. (2022). Cognitive radio-enabled Internet of Things (CR-IoT): An integrated approach towards smarter world. In: J. K. Mandal, S. Misra, J. Sekhar Banerjee, and S. Nayak (eds.), *Applications of Machine Intelligence in Engineering* (pp. 541–555). Boca Raton, FL: CRC Press.

[6] Banerjee, J.S., Chakraborty, A., Mahmud, M., Kar, U., Lahby, M., Saha, G. (2023). Explainable artificial intelligence (XAI) based analysis of stress among tech workers amidst COVID-19 pandemic. In: M. Lahby, V. Pilloni, J. S. Banerjee, and M. Mahmud (eds.), *Advanced AI and Internet of Health Things for Combating Pandemics* (pp. 151–174). Cham: Springer.

[7] Lahby, M., Pilloni, V., Banerjee, J.S., and Mahmud, M. (eds.) (2023). *Advanced AI and Internet of Health Things for Combating Pandemics*. Springer.

[8] Bhattacharyya, S., Banerjee, J.S., Gorbachev, S., Muhammad, K., and Koeppen, M. (eds.) (2023). *Computer Intelligence Against Pandemics*. Berlin: De Gruyter.

[9] Bhattacharyya, S., Banerjee, J.S., and De, D. (eds.) (2023). *Confluence of Artificial Intelligence and Robotic Process Automation*. Singapore: Springer.

[10] Mandal, J.K., Misra, S., Banerjee, J.S., and Nayak, S. (eds.) (2022). Applications of machine intelligence in engineering. In: *Proceedings of 2nd Global Conference on Artificial Intelligence and Applications (GCAIA, 2021)*, 8–10 September 2021 (pp. 1–597), Jaipur. CRC Press

[11] Chakraborty, A., Banerjee, J.S., Bhadra, R., Dutta, A., Ganguly, S., Das, D., Kundu, S., Mahmud, M., and Saha, G. (2023). A framework of intelligent mental health monitoring in smart cities and societies. *IETE Journal of Research*, 1–14. https://doi.org/10.1080/03772063.2023.2171918

[12] Banerjee, J.S., Mahmud, M., and Brown, D. (2023). Heart rate variability-based mental stress detection: An explainable machine learning approach. *SN Computer Science*, 4(2), 176.

[13] Bhattacharyya, S., Banerjee, J.S., and Köppen, M. (eds.) (2022). *Human-Centric Smart Computing: Proceedings of ICHCSC 2022*, vol. 316. Singapore: Springer Nature.

[14] Bhattacharyya, S., Banerjee, J.S., and Köppen, M. (eds.) (2024). *Human-Centric Smart Computing: Proceedings of ICHCSC 2023*, Singapore: Springer Nature (Press).

[15] Bhattacharyya, S., Banerjee, J.S., De, D., and Mahmud, M. (eds.) (2023). *Intelligent Human Centered Computing. Human 2023. Springer Tracts in Human-Centered Computing.* Springer: Singapore.

[16] Chattopadhyay, J., Kundu, S., Chakraborty, A., and Banerjee, J.S. (2020). Facial expression recognition for human computer interaction. In: *New Trends in Computational Vision and Bio-inspired Computing: Selected works presented at the ICCVBIC 2018* (pp. 1181–1192), Coimbatore, India. Cham: Springer.

[17] Guhathakurata, S., Saha, S., Kundu, S., Chakraborty, A., and Banerjee, J.S. (2021). South Asian countries are less fatal concerning COVID-19: A fact-finding procedure integrating machine learning & multiple criteria decision-making (MCDM) technique. *Journal of the Institution of Engineers (India): Series B*, 102(6), 1249–1263.

[18] Guhathakurata, S., Kundu, S., Chakraborty, A., and Banerjee, J.S. (2021). A novel approach to predict COVID-19 using support vector machine. In: U. Kose, D. Gupta, V. Hugo, C. de Albuquerque, and A. Khanna (eds.), *Data Science for COVID-19* (pp. 351–364). Academic Press.

[19] Ahemad, M.T., Hameed, M.A., and Vankdothu, R. (2022). COVID-19 detection and classification for machine learning methods using human genomic data. *Measurement: Sensors.* 24, 100537.

[20] Pan, C., Banerjee, J.S., De, D., Sarigiannidis, P., Chakraborty, A., and Bhattacharyya, S. (2023). ChatGPT: A OpenAI platform for society 5.0. In *Doctoral Symposium on Human Centered Computing 2023 Feb 23* (pp. 384–397). Singapore: Springer Nature Singapore.

[21] Saif, S., Karmakar, K., Biswas, S., and Neogy, S. (2022). MLIDS: Machine learning enabled intrusion detection system for health monitoring framework using BA-WSN. *International Journal of Wireless Information Networks*, 29(4), 491–502.

[22] Karmakar, K., Saif, S., Biswas, S., and Neogy, S. (2021). A WBAN-based framework for health condition monitoring and faulty sensor node detection applying ANN. *International Journal of Biomedical and Clinical Engineering (IJBCE)*, 10(2), 44–65.

[23] Karmakar, K., Saif, S., Biswas, S., and Neogy, S. (2022). An intelligent vehicular communication-based framework to provide seamless connectivity in WBAN. In: S. Dhar, D-T. Do, S. N. Sur, and H. C-M. Liu (eds.), *Advances in Communication, Devices and Networking: Proceedings of ICCDN 2021* (pp. 583–591). Springer Nature: Singapore.

[24] Shahrour, I. and Xie, X. (2021). Role of Internet of Things (IoT) and crowd-sourcing in smart city projects. *Smart Cities*, 4(4), 1276–1292.

[25] Belli, L., Cilfone, A., Davoli, L., Ferrari, G., Adorni, P., Di Nocera, F., Dall'Olio, A., Pellegrini, C., Mordacci, M., and Bertolotti, E. (2020). IoT-enabled smart sustainable cities: Challenges and approaches. *Smart Cities*, 3(3), 1039–1071.

[26] Ray, P.P. (2016). A survey on Internet of Things architectures. *Journal of King Saud University: Computer and Information Sciences*, 30(3), 291–319.

[27] Al-Fuqaha, A., Guizani, M., Mohammadi, M., Aledhari, M., and Ayyash, M. (2015). Internet of Things: A survey on enabling technologies, protocols, and applications. *IEEE Communications Surveys and Tutorials*, 17(4), 2347–2376.

[28] Biswas, S., Sharma, L.K., Ranjan, R., Saha, S., Chakraborty, A., and Banerjee, J.S. (2021). Smart farming and water saving-based intelligent irrigation system implementation using the Internet of Things. In: S. Bhattacharyya, D. Samanta, I. Pan, P. Dutta, and A. Mukherjee (eds.), *Recent Trends in Computational Intelligence Enabled Research* (pp. 339–354). Academic Press.

[29] Das, K. and Banerjee, J.S. (2022). Green IoT for intelligent cyber-physical systems in industry 4.0: A review of enabling technologies, and solutions. In: J. K. Mandal, S. Misra, J. Sekhar Banerjee, and S. Nayak (eds.), *Applications of Machine Intelligence in Engineering* (pp. 463–478). Boca Raton, FL: CRC Press.

[30] Majumder, R., Dasgupta, M., Biswas, A., and Banerjee, J.S. (2022). IoT-based smart city for the post COVID-19 world: A child-centric implementation emphasis on social distancing. In: J. K. Mandal, S. Misra, J. Sekhar Banerjee, and S. Nayak (eds.), *Applications of Machine intelligence in Engineering* (pp. 599–612). Boca Raton, FL: CRC Press.

[31] Saif, S., Bhattacharjee, P., Karmakar, K., Saha, R., and Biswas, S. (2022). IoT-Based secure health care: Challenges, requirements and case study. In: S. Biswas, C. Chowdhury, B. Acharya, and C-M. Liu (eds.), *Internet of Things Based Smart Healthcare: Intelligent and Secure Solutions Applying Machine Learning Techniques* (pp. 327–350). Springer Nature: Singapore.

[32] Ballesteros, J., Carbunar, B., Rahman, M., Rishe, N., and Iyengar, S.S. (2014). Towards safe cities: A mobile and social networking approach. *IEEE Transactions on Parallel and Distributed Systems*, 25(9), 2451–2462.

[33] Shahab, S., Agarwal, P., Mufti, T., and Obaid, A.J. (2022). SIoT (social Internet of Things): A review. In: S. Fong, N. Dey, and A. Joshi (eds.), *ICT Analysis and Applications* (pp. 289–297). Singapore: Springer Nature.

[34] Atzori, L., Iera, A., Morabito, G., and Nitti, M. (2012). The social Internet of Things (siot)-when social networks meet the Internet of Things: Concept, architecture and network characterization. *Computer Networks*, 56(16), 3594–3608.

[35] Malekshahi Rad, M., Rahmani, A.M., Sahafi, A., and Nasih Qader, N. (2020). Social Internet of Things: Vision, challenges, and trends. *Human-Centric Computing and Information Sciences*, 10(1), 1–40.

[36] Roopa, M.S., Pattar, S., Buyya, R., Venugopal, K.R., Iyengar, S.S., and Patnaik, L.M. (2019). Social Internet of Things (SIoT): Foundations, thrust areas, systematic review and future directions. *Computer Communications*, 139, 32–57.

[37] Diaz, S., Mendez, D., and Kraemer, R. (2019). A multi-layer self-healing algorithm for WSNs. *Journal of Circuits, Systems and Computers*, 29(05), 2050070.

[38] Raja, S.P., Rajkumar, T.D., and Raj, V.P. (2018). Internet of Things: Challenges, issues and applications. *Journal of Circuits, Systems and Computers*, 27(12), 1830007.

[39] Cao, K., Xu, G., Zhou, J., Wei, T., Chen, M., and Hu, S. (2018). QoS-adaptive approximate real-time computation for mobility-aware IoT lifetime optimization. *IEEE Transactions on Computer-Aided Design of Integrated Circuits and Systems*, 38(10), 1799–1810.

[40] Guan, Q., Yin, X., Guo, X., and Wang, G. (2016). A novel infrared motion sensing system for compressive classification of physical activity. *IEEE Sensors Journal*, 16(8), 2251–2259.

[41] Sethi, P. and Sarangi, S.R. (2017). Internet of Things: Architectures, protocols, and applications, *Journal of Electrical & Computer Enginnering*. https://doi.org/10.1155/2017/9324035

[42] Virmani, A. and Shahzad, M. (2017). Position and orientation agnostic gesture recognition using WiFi. In: *Proceedings of the 15th Annual International Conference on Mobile Systems, Applications, and Services (MobiSys)* (pp. 252–264). New York: Association for Computing Machinery.

[43] Yang, G., Jiang, M., Ouyang, W., Ji, G., Xie, H., Rahmani, A.M., Liljeberg, P., and Tenhunen, H. (2018). IoT-based remote pain monitoring system: From device to cloud platform, *IEEE Journal of Biomedical and Health Informatics*, 22(6), pp. 1711–1719, doi: 10.1109/JBHI.2017.2776351.

[44] Guan, Q., Li, C., Guo, X., and Shen, B. (2017). Infrared signal based elderly fall detection for in-home monitoring. In: *2017 9th International Conference on Intelligent Human-Machine Systems and Cybernetics (IHMSC)*. Hangzhou: IEEE.

[45] Liu, J., Wang, Y., Chen, Y., Yang, J., Chen, X., and Cheng, J. (2015). Tracking vital signs during sleep leveraging off-the-shelf WiFi. In: *Procedings of 16th ACM International Symposium on Mobile Ad Hoc Networking and Computing (MobiHoc)* (pp. 267–276). New York: Association for Computing Machinery.

[46] Lijun, Z. (2013). Multi-parameter medical acquisition detector based on Internet of Things. *Chinese Patent*, 202(960), 774.

[47] Xin, T.J., Min, B., and Jie, J. (2013). Carry-on blood pressure/pulse rate/ blood oxygen monitoring location intelligent terminal based on Internet of Things. *Chinese Patent*, 202(875), 31.

[48] Liu, M.L., Tao, L., and Yan, Z. (2012). Internet of Things-based electrocardiogram monitoring system, *Chinese Patent*, 102(764), 118.

[49] Vinod Kumar, T.M. (2016). *Smart Economy in Smart Cities. International Collaborative Research: Ottawa. St Louis, Stuttgart, Bologna, Cape Town, Nairobi, Dakar, Lagos, New Delhi, Varanasi, Vijayawada, Kozhikode, Hong Kong*. Springer: Singapore.

[50] Papernot, N., McDaniel, P., Wu, X., Jha, S., and Swami, A. (2016). Distillation as a defense to adversarial perturbations against deep neural networks. In: *Proceedings of IEEE Symposium on Security and Privacy (SP)* (pp. 582–597). San Jose, CA: IEEE.

[51] Xu, X., Ansari, R., Khokhar, A., and Vasilakos, A.V. (2015). Hierarchical data aggregation using compressive sensing (HDACS) in WSNs. *ACM Transactions on Sensor Networks (TOSN)*, 11(3), 1–25.

[52] Chon, Y., Lane, N.D., Kim, Y., Zhao, F., and Cha, H. (2013). Understanding the coverage and scalability of place-centric crowdsensing. In: *Proceedings of the ACM International Joint Conference on Pervasive and Ubiquitous Computing (UbiComp 13)* (pp. 3–12). ACM, New York, NY, USA.

[53] Chon, Y., Lane, N.D., Li, F., Cha, H., and Zhao, F. (2012). Automatically characterizing places with opportunistic crowd sensing using smartphones. In: *Proceedings of the ACM Conference on Ubiquitous Computing (UbiComp 12)* (pp. 481–490). ACM, New York, NY, USA.

[54] Christin, D., Reinhardt, A., Kanhere, S.S., and Hollick. M. (2011). A survey on privacy in mobile participatory sensing applications, *Journal of System Software*, 84(11), 1928–1946.

[55] Costa, C., Laoudias, C., Zeinalipour-Yazti, D., and Gunopulos, D. (2011). Smart Trace: Finding similar trajectories in smartphone networks without disclosing the traces. In *Proceedings of the 29th IEEE International Conference on Data Engineering* (pp. 1288–1291). Hannover: IEEE.

[56] Kong, X., Liu, X., Jedari, B., Li, M., Wan, L., Xia, F. (2019). Mobile crowdsourcing in smart cities: Technologies, applications, and future challenges. *IEEE Internet Things Journal*, 6(5), 8095–8113.

[57] Bao, X. and Choudhury, R. (2010). MoVi: Mobile phone based video highlights via collaborative sensing. In: *Proceedings of the 8th International Conference on Mobile Systems, Applications, and Services (MobiSys 10)* (pp. 357–370). New York: ACM.

[58] Bengtsson, L., Lu, X., Thorson, A., Garfield, R., and von Schreeb, J. (2011). Improved response to disasters and outbreaks by tracking population movements with mobile phone network data: A post-earthquake geospatial study in Haiti. *PLoS Med*, 8(8), e1001083.

[59] Brabham, D.C. (2008). Crowdsourcing as a model for problem solving: An introduction and cases. *Convergence: The International Journal of Research into New Media Technologies*, 14(1), 75–90.

[60] Calabrese, F., Colonna, M., Lovisolo, P., Parata, D., and Ratti, C. (2011). Real-time urban monitoring using cell phones: A case study in Rome. *IEEE Transactions on Intelligent Transportation Systems*, 12(1), 141–151.

Chapter 3

AI for spectrum intelligence and adaptive resource management

Christopher Akinyemi Alabi, Monday Abutu Idakwo, Agbotiname Lucky Imoize, Talatu Adamu, and Samarendra Nath Sur

3.1 INTRODUCTION AND BACKGROUND CONCEPTS

This chapter delves into the intersection of artificial intelligence (AI) and spectrum management, a field known as spectrum intelligence and adaptive resource management (SIARM). SIARM embodies a transformative approach to spectrum management, leveraging the capabilities of AI and machine learning (ML) to enable dynamic, real-time, and intelligent allocation of spectrum resources. Unlike traditional static spectrum allocation methods, SIARM harnesses the power of AI to analyze complex and dynamic usage patterns, optimize spectrum allocation, and adapt to changing demands instantaneously.

This chapter will lay the groundwork for understanding how AI can be harnessed to revolutionize spectrum management, ushering in an era of unparalleled wireless connectivity and resource utilization efficiency.

3.1.1 Spectrum scarcity and demand

The radiofrequency spectrum, also known as the "spectrum," is a limited and priceless resource that supports contemporary wireless communications. The electromagnetic spectrum is divided into different frequency bands, each of which has unique characteristics suited for various communication systems. However, the exponential growth in wireless devices, applications, and services has resulted in a rising demand for spectrum resources [1]. Increasing demand for wireless connectivity, driven by smartphones, Internet of Things (IoT) devices, autonomous vehicles, and other technologies, is rapidly outpacing the availability of suitable frequency bands.

This scarcity poses a significant challenge for regulators, spectrum managers, and the telecommunications industry. The need to balance diverse and often competing demands, from commercial cellular networks to public safety systems and scientific research, becomes more complex as the spectrum becomes more crowded. The potential for interference between different users and technologies is also a genuine concern [1,2], as it can degrade service quality and disrupt critical operations.

Spectrum management, allocation, and utilization must be approached creatively to address the spectrum shortage [1]. This is where innovations like adaptive resource management (ARM) and spectrum intelligence (SI) come into play. These methods offer dynamic and more effective utilization of the spectrum by leveraging AI, ML, and data analytics [3]. They enable stakeholders to make well-informed decisions on frequency allocation, interference control, and real-time demand adaptation.

3.1.2 Traditional approach to spectrum management

To offer effective monitoring of spectrum use, ensure equitable allocation, and provide support for new communication technologies, the RF spectrum is regulated by the government. At the international level, the spectrum is being apportioned by the International Telecommunication Union into blocks for specific uses. The licenses are assigned for these blocks to specific users at the national level by each country's regulatory agencies such as the National Communication Commission in Nigeria, the Office of Communications (Ofcom) in the United Kingdom (UK), and the Federal Communications Commission in the United States (USA) [4]. Typically, these regulatory authorities adopt the fixed spectrum access (FSA) policy and allocate different parts of the radio spectrum with certain bandwidths to different services. In Nigeria, for instance, the fixed satellite services C-band 3400–6425 MHz is allocated to Radio and Television broadcasting and cannot be used by other services. This static spectrum allocation policy allows only authorized users or licensed users the exclusive right to utilize their assigned spectrum and forbids other users from accessing the spectrum even when the spectrum is free. However, the mission of the FSA is to ensure interference-free operations amidst diverse applications and wireless systems services, which indeed protect against harmful interference in allocated bands.

Nevertheless, this FSA policy is no longer efficient and effective in radio spectrum management as global statistics of spectrum allocation have unveiled that the policy has created spectrum scarcity and spectrum underutilization [5].

3.1.3 Intelligent resource management

SI and ARM are closely interconnected concepts. While SI supplies the intelligence and insights necessary for making adaptive decisions in resource management, ARM translates those insights into real-time actions, resulting in a symbiotic relationship that leads to more efficient, agile, and responsive spectrum management strategies.

A. Spectrum Intelligence

SI refers to the application of advanced data analysis techniques, often utilizing ML and AI to analyze and make informed decisions regarding the allocation, management, and utilization of the radiofrequency spectrum [2]. The radio spectrum is a limited but valuable resource and is used for wireless communication, broadcasting, and various technological applications. SI aims to enhance the efficiency, flexibility, and adaptability of spectrum utilization by dynamically assessing usage patterns, identifying available frequencies, and optimizing the allocated resources in real time.

SI involves the collection of data from different sources like spectrum sensors, communication devices, and network infrastructure [6]. Through AI-driven algorithms, the obtained data are processed and analyzed to extract patterns, trends, and correlations that are not easily discernible through traditional techniques.

By harnessing SI, stakeholders can optimize the utilization of available spectrum bands, accommodate the coexistence of diverse wireless technologies [7], and address the challenges posed by the exponential growth in wireless devices and applications.

B. Adaptive Resource Management

ARM refers to a dynamic and flexible approach to allocating and utilizing resources using real-time conditions, demands, and environmental factors. In the context of wireless communications and spectrum allocation, ARM involves the intelligent and responsive distribution of available frequencies or spectrum resources to various communication systems and technologies.

Unlike traditional static resource management methods that allocate spectrum frequencies based on predetermined assignments, ARM leverages technologies like ML, AI, and data analytics to continuously monitor and analyze the spectrum usage landscape. This approach enables the allocation of spectrum resources in a manner that maximizes efficiency, minimizes interference, and accommodates the varying and sometimes unpredictable demands of different wireless applications.

ARM relies on real-time data collection and analysis to identify underutilized spectrum bands, detect areas of congestion, and recognize emerging communication needs. By doing so, it enables swift adjustments to spectrum allocation, allowing communication networks to dynamically adapt to changing conditions and user requirements. This adaptability is especially critical with rapidly evolving technologies like the 5G and the IoT, which introduce diverse communication needs and usage patterns.

3.2 LITERATURE REVIEW

3.2.1 Cognitive radio (CR) enabling dynamic spectrum management

To aid spectrum resource underutilization and allocation, dynamic spectrum sharing using CR has become the de facto solution. The CR is an intelligent radio system that autonomously and dynamically adapts its transmission strategies, antenna beam or modulation scheme, carrier frequency, transmit power, and bandwidth, among others as they interacts with its surrounding environment and becomes aware of its internal states (e.g., user needs, spectrum use policy, and software and hardware architectures) and achieved optimal performance [8]. The CR is a major enabler of dynamic spectrum access. CRs constantly monitor the radio spectrum to locally detect regions that are free or unused in a primary or licensed band and transmit in those bands. The licensed users or network primary users have sole access to their allocated spectrum bands and are willing to share their spectrum with the unlicensed secondary users. The CR is software-defined, making it easy to reprogram and reconfigure based on SI acquired from the network environment [1]. Dynamic spectrum management in a cognitive radio network (CRN) involves some key network operations [9–11] shown in Figure 3.1.

In every wireless communication environment, the channels are normally subjected to space–time–frequency variations owing to multipath propagation, location-dependent shadowing, and mobility. Hence, dynamic resource allocation (DRA) intelligence is built into CRs to ensure optimal deployment of their transmit strategies for the attainment of maximum secondary network throughput. Therefore, bandwidth, bit rate, antenna beam, and transmit power are allocated dynamically using the available primary and secondary networks' channel state information (CSI) to the CRs [12]. A fundamental key in CRNs is that access to a secondary user is limited as the primary users have the precedence to use the spectrum at all times. Hence, in a CRN, quality of service (QoS) preservation and interference management among all users, especially for the PUs, is an

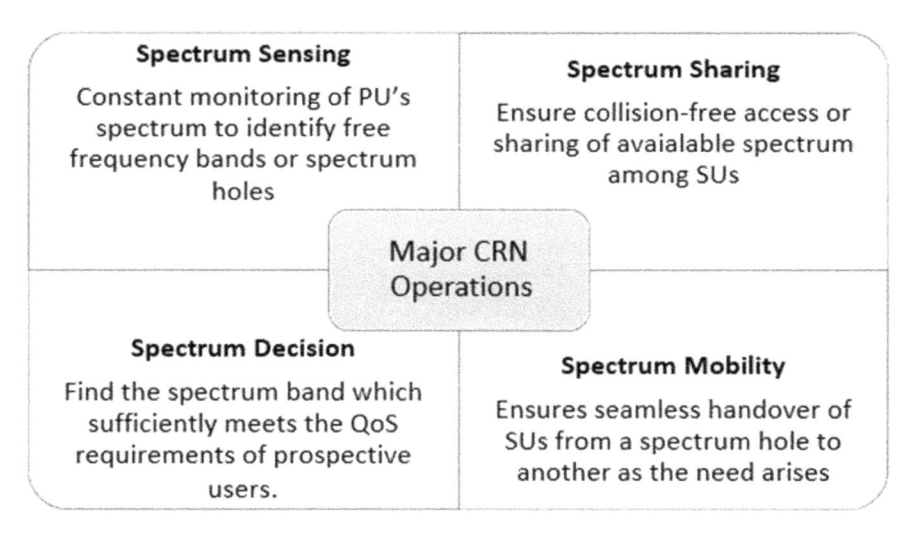

Figure 3.1 Fundamental CRN operations.

essential function. Common QoS optimization techniques and interference mitigation approaches used in a CRN are highlighted in the preceding subsections.

3.2.2 QoS optimization in CRN

QoS is any technology that manages data traffic to reduce latency, packet loss, and jitter on a network. In CRNs, the CR transceivers evaluate whether the network is free or not before any data communication by sending scan signals. Several methods have been proposed to optimize the QoS parameters such as throughput, packet delivery ratio, end-to-end communication delay, and energy consumption. The CR determines the available best spectrum bands that satisfy QoS. The wireless spectrum is opportunistically accessed by secondary users only when it is unused. While a secondary user is using a free channel, the instance that the primary user will come back makes the secondary users stop their communications and free the current channel [13]. Therefore, secondary users' QoS is difficult to ensure. In every wireless communication, several desirable objectives are achievable to obtain quality services. The desirable objectives [14] are as highlighted:

i. **Maximization of throughput:** This objective is responsible for the improvement of the system data throughput. It is an essential objective in areas like computer applications and multimedia. Therefore, throughput is the final rate at which information is generated. More information is produced while maximizing the throughput. Therefore, the fitness function to optimize the maximization of throughput is represented by Equation 3.1.

$$F_{\text{Max\_throughput}} = \frac{\log_2(M)}{\log_{10}(M_{\max})} \tag{3.1}$$

where M refers to the single index's modulation index and $M_{\max}$ is the maximum modulation index.

ii. **Bit error rate (BER) minimization:** It is an objective that minimizes the amount of errors compared with the number of bits being sent. It is used to evaluate the quality of every link in terms of the number of bit errors per unit time. The complexity of the system is reduced while minimizing the errors. The minimization of BER is given by Equation 3.2:

$$F_{\text{Min\_ber}} = \frac{\log_{10} 0.5}{\log_{10} P_{\text{ber}}} \tag{3.2}$$

where P_{ber} refers to the BER modulation type.

iii. **Minimization of power consumption:** This objective is adopted for minimal power consumption. The power fitness function to minimize power consumption is given by Equation 3.3.

$$F_{\text{Min\_Power}} = \frac{P}{\left(P_{\text{max}}\right)} \tag{3.3}$$

where P is the transmitted power average and P_{max} is the maximum available transmit power.

iv. **Interferences minimization:** This objective helps to reduce interference to the minimum level possible. Interferences happen majorly due to the simultaneous usage of the system by the primary and secondary users. The fitness function to optimize the interference is expressed by Equation 3.4

$$F_{\text{Min\_Interference}} = \frac{\left(\left(P + B + T\right) - P_{\text{min}} + B_{\text{min}} + 1\right)}{\left(P_{\text{max}} + B_{\text{max}} + \text{SR}_{\text{max}}\right)} \tag{3.4}$$

where B refers to the needed single carrier bandwidth, B_{max} and B_{min} are the maximum and minimum bandwidth, T is the time division duplex, and SR_{max} is the maximum symbol rate.

v. **Spectral efficiency maximization:** The entire quantity of information transmitted over a given bandwidth is referred to as spectral efficiency. To reduce the spectral space of the transmitted signal, it is important to maximize spectral efficiency. The maximization of spectral efficiency can be expressed using the fitness function of Equation 3.5:

$$F_{\text{Max\_SpectralE}} = \frac{1 - \left(M - B_{\text{min}} + \text{SR}\right)}{\left(B + M_{\text{max}} + \text{SR}_{\text{max}}\right)} \tag{3.5}$$

The weighted sum method in CR is adopted to derive the optimization for the objective function. Since all the highlighted parameters run simultaneously, the weights vector applied to the objective function to reduce the system complexity is given by Equation 3.6.

$$F_{\text{Sys}} = W_1 \times F_{\text{Max\_throughput}} + W_2 \times F_{\text{Min\_ber}} + W_3 \times F_{\text{Min\_Power}}$$
$$+ W_4 \times F_{\text{Min\_Interference}} + W_5 \times F_{\text{Max\_SpectralE}} \tag{3.6}$$

Hence, optimization techniques such as particle swarm optimization algorithm, genetic algorithm, mutated ant colony optimization, and adaptive multi-objective optimization scheme are applied with several iterations on these objectives to ensure QoSs.

3.2.3 Interference management in CRN

Interferences are any disturbances from another signal that starts operating at the same frequency as an active signal operating either by misapprehension or deliberately. The link from the intended receiver to the transmitter is known as a communication link, whereas interference link is the link between unintended receiver and transmitter. An interference within the receiver is an unwanted entity as it indicates a reverse relation with the user's capacity [15]. This can easily be understood through Shannon's channel capacity as given by Equation 3.7.

$$C = B\log_2\left(1 + \frac{RS}{NP + IP}\right) \tag{3.7}$$

where B referred to bandwidth available, RS is the received signal power, NP is the noise power, IP is the interference power, and C is the channel capacity.

From Equation 3.6, an increment in interference power will reduce the channel capacity. Hence, the various scenarios of interferences in CRNs are worth analyzing [15]. The interference scenarios are as highlighted:

i. The usage of licensed spectrum by the secondary user creates an intensive interference avoidance challenge at the primary user network.
ii. The numerous secondary users in CRNs trying to use the same channel group at the same time create data within the cognitive user. This results in multiple access interference.
iii. The interference occurs at the primary user networks or CRN due to secondary user transmission or primary user.

To mitigate these interferences in CRNs, various interference management schemes have been deployed using spectrum prediction, spectrum sensing, transmitting below primary user interference tolerance level, spectrum monitoring, and advanced encoding methods.

A. Spectrum sensing for interference management

For any CR-inspired wireless network to effectively avoid interference, spectrum sensing is pivotal. A CRN detects the free spots' band [15] by sufficiently observing and distributing the spectrum to the demanding users. In a CRN, the secondary user switches to another idle sense channel once the primary user is available to avert interference with the primary user. Several classes of spectrum sensing approaches [16] are shown in Figure 3.2. The classification is based on factors such as spectrum bandwidth, number of users, mode of energy detection, and need for prior channel knowledge.

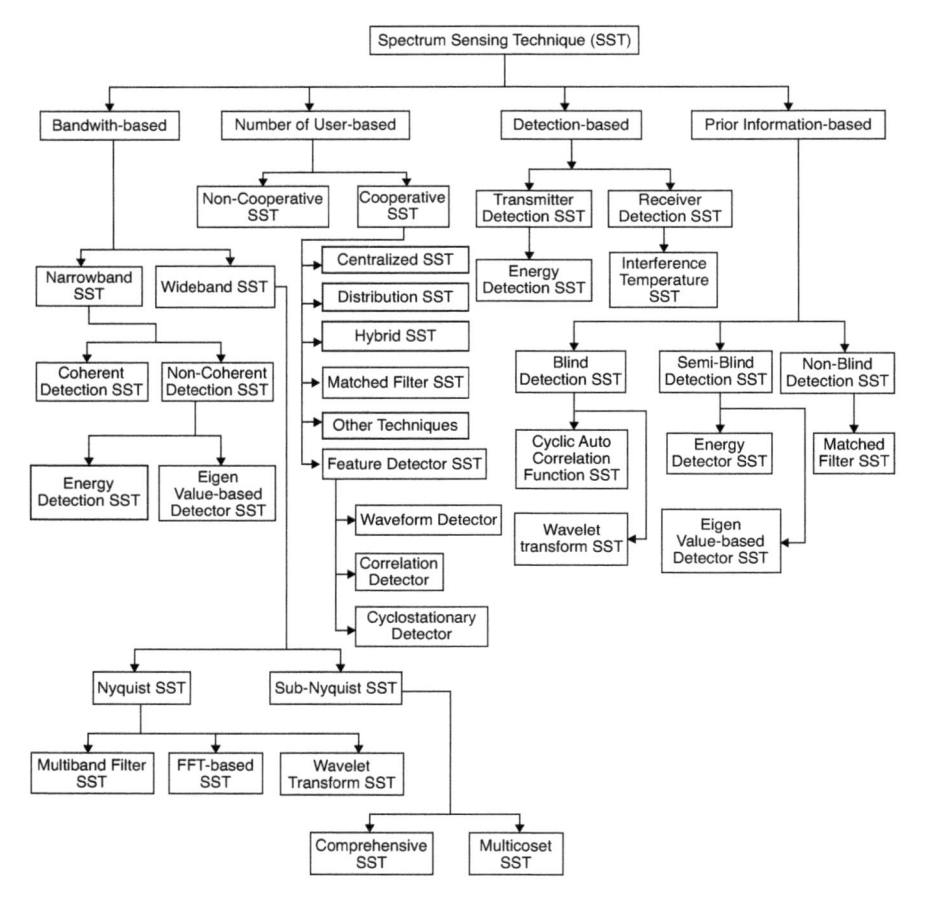

Figure 3.2 Spectrum sensing classification methods.

B. Spectrum prediction

Spectrum prediction entails the usage of the primary user channels' historical information for the forecasting of the channel's future states [12]. It is adopted for interference management to assist spectrum mobility and sensing approach. It predicts the arrival time of the primary user back to its communication channel. Primarily, the secondary user forecasts communication channel states from a highly dense and selects the most idle forecasted channel for spectrum sensing. Data transmission only begins when the sensed channel is idle; otherwise, the next highest idle channel is sensed. This process continues until a channel can be used [17]. In the next phase, the secondary user forecasts the arrival time of the primary user, while the secondary user's transmission is ongoing and enables the secondary user to switch instantly before the arrival of the primary user. This results in proactive spectrum mobility. Hence, spectrum forecasting assists the secondary user in avoiding any interference with the primary user at both the arrival and when the primary user is detected. Different spectrum prediction approaches such as moving-average-based prediction, multilayer perceptron neural-network-based prediction, static neighbor-graph-based prediction [18], hidden Bayesian-inference-based prediction, deep-learning-based prediction, and autoregressive-model-based prediction, and Markov model-based prediction [19] are presented. The fundamental problems with predicting the spectrum are highlighted:

  i. Scare historical information
 ii. Power consumption along with the storage unit for the predictive analysis as large data are needed for effective prediction
iii. The complexity of the prediction method is also a crucial factor considered during any spectrum prediction design due to its effect on power consumption.

C. Transmission below primary users' interference threshold

This is another useful means of preventing any interference from the primary user once the details of the primary user's interference average limit are known. In this method, the secondary user manages its power in a way that the received power at the primary user receiver is lower than the average interference threshold. Nevertheless, the primary problem with this method is the handiness of the interference limit that can be condoned by the primary user. Furthermore, the small data rate for each unit bandwidth limits the application of this method solely in scenarios where the resulting data rate is lower than that of the attained data rate and unsuited for high-data-produced scenarios [20].

D. Using advanced encoding techniques

This approach enables the secondary user to relay data with full power amid the primary user on the medium. The secondary user is expected to have an advanced encoding method like Gelfand–Pinsker binding [21] and dirty paper coding [22]. Nevertheless, the system becomes very complex and requires extra power for processing due to the advanced encoding technique. Hence, this technique is unsuitable for lower power due to high power consumption and complexity. Furthermore, implementing this approach practically is a breakthrough.

E. Spectrum monitoring

Spectrum monitoring is a recent and important analyzed method that is used to detect the arrival of primary users while the secondary users are in communication. The secondary user explores the features of the collected signal and uses the digression from the usually collected signal to detect the arrival of a primary user. On discernment of the arrival of the primary user, the secondary user halts its communication instantly and switches to another idle channel. Several researchers have explored diverse spectrum approaches like energy-based spectrum monitoring, error vector magnitude-based spectrum monitoring, energy-ratio-based spectrum monitoring, received error count-based spectrum monitoring, received error count, and cyclic redundancy check-based spectrum monitoring [17]. The critical issue in adopting the spectrum monitoring method is analyzing the processing power needed by the monitoring method as an increase in the processing power influences the consumed total power.

3.2.4 Spectrum sensing

The key enabling processes for SI or dynamic spectrum management are spectrum awareness and spectrum sensing. Spectrum sensing involves the periodic scanning of a specific frequency spectrum to discover the presence or absence of the primary users [10]. Spectrum sensing, as a fundamental component of SI, enables the detection of available or unused frequency bands. In addendum to detecting the underutilized spectrum, spectrum sensing equally entails determining the spectral resolution of every spectrum hole and approximating the direction as well as the signal function of incoming interfering signals [10]. The various techniques and approaches to spectrum sensing are highlighted in this section.

A. Traditional spectrum sensing techniques

Various spectrum sensing methods have been presented by various researchers. These methods can be categorized based on elements like coherence or incoherence, suitability for wideband or narrowband spectrum, non-cooperative or cooperative mode of operation, and others [10,23]. The common traditional spectrum sensing approaches are presented in Table 3.1 [10,23,24]. A more elaborate classification of the traditional spectrum sensing approaches as presented by [16] is shown in Figure 3.2.

Researchers focus on improving the accuracy of sensing methods, minimizing false alarms, and enhancing the ability to detect weak signals in noisy environments. Given the challenges associated with the traditional spectrum sensing techniques as highlighted in Table 3.1, the introduction of AI-based techniques, which are much more efficient in spectrum management and utilization, has become necessary.

B. ML-based spectrum sensing

The adoption of the ML approach in solving varying problems in different areas is growing by the day. ML-based algorithms are broadly classified into supervised learning, reinforcement learning,

Table 3.1 Common traditional spectrum sensing techniques

Spectrum sensing technique	Application	Description	Challenges
Matched filter detection	Coherent Narrowband Non-cooperative	Requires prior knowledge of PU signal Suitable for narrowband spectrum	Practically not feasible as it requires prior knowledge of PU
Cyclostationary detection	Coherent Narrowband Non-cooperative	Requires prior knowledge of PU signal Suitable for narrowband spectrum	Practically not feasible as it requires prior knowledge of PU [25]. It is complex and slow [11]
Energy detection	Noncoherent Narrow-/wideband Non-cooperative	A priori knowledge of PU signal is not required Suitable for both narrowband and wideband spectrum	Execute poorly at low signal-to-noise ratios (SNRs) and high false alarm rates [26]
Wavelet detection	Noncoherent, wide band Non-cooperative	Previous knowledge of PU signal is not needed Suitable for wideband spectrum	High computational complexity [27]
Compressed detection	Noncoherent, wideband	Previous knowledge of PU signal is not needed Suitable for wideband spectrum	Exhibits sporadic sparsity level [27]

and unsupervised learning. The grail of supervised learning is to learn from training data that have corresponding labels or outputs to forecast the right output for subsequent training samples. On the other hand, unsupervised learning uses unlabeled training data to accept hidden features from the input dataset. Reinforcement learning is a technique that is used when some environment knowledge is available and a decision-making agent learns and adapts to the environment with the primary goal of taking appropriate actions and maximizing reward through interaction with the environment. ML algorithms used in cooperative spectrum sensing (CSS) are itemized in Figure 3.3 [11].

Conventional ML algorithms like support vector machines (SVMs), Naïve Bayes, and K-nearest neighbor have been applied to achieve time reduction in spectrum sensing [9,26,28]. These algorithms reduce spectrum sensing to a binary classification problem and so predict the RF channel status using energy or probability vectors [26]. These ML algorithms are faced with the issues of feature extraction which is manual and time-consuming, thereby negatively impacting the rate of spectrum detection [28].

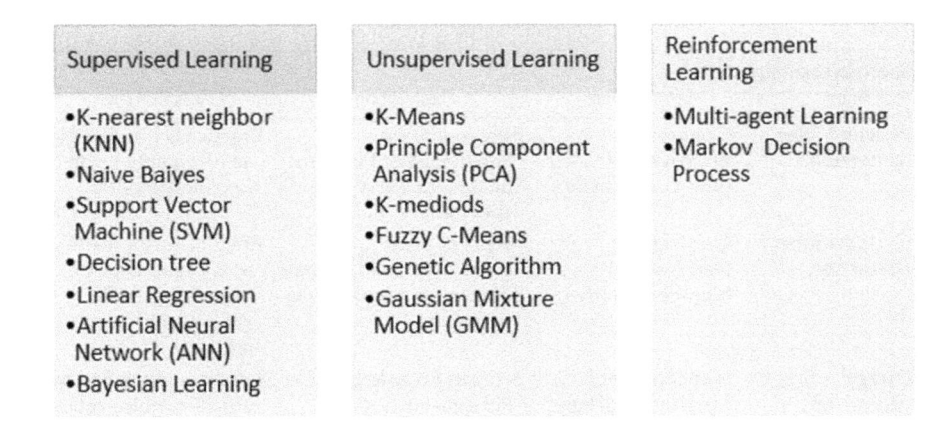

Supervised Learning	Unsupervised Learning	Reinforcement Learning
•K-nearest neighbor (KNN) •Naive Baiyes •Support Vector Machine (SVM) •Decision tree •Linear Regression •Artificial Neural Network (ANN) •Bayesian Learning	•K-Means •Principle Component Analysis (PCA) •K-mediods •Fuzzy C-Means •Genetic Algorithm •Gaussian Mixture Model (GMM)	•Multi-agent Learning •Markov Decision Process

Figure 3.3 ML algorithms suitable for CSS.

Deep learning is an unconventional data-driven ML approach capable of autonomously drawing inferences from complex data patterns and features [29]. Deep learning precepts are robust enough to support or adapt to the quick changes that characterize a radio environment. Based on the model architecture, [9] categorized deep learning algorithms suitable for spectrum sensing into
  i.   Convolution neural networks (CNNs)
 ii.   Long short-term memory (LSTM) networks
iii.   Multilayer perceptrons (MLPs)
 iv.   CNN-LSTM hybrid network

3.2.5 Spectrum awareness

Spectrum awareness goes beyond spectrum sensing by considering a broader context of spectrum utilization. Literature explores methods to characterize the spectral environment, including identifying different types of signals, their modulation schemes, and transmission behaviors. This context awareness is essential for making informed decisions about spectrum access and allocation. Techniques like ML and data fusion are employed to process data from multiple sources, such as spectrum sensors, databases, and historical usage patterns, to build a comprehensive view of the spectrum landscape.

The critical part of CR technology is the ability of a secondary user to reliably detect unused frequencies over multiple dimensions (like frequency, time, angle, code, and space), exploit them efficiently, and dynamically update the transmission parameters while mitigating interference from the primary users of the spectrum. Hence, it becomes imperative for secondary users to have an efficient and robust spectrum sensing capability to detect

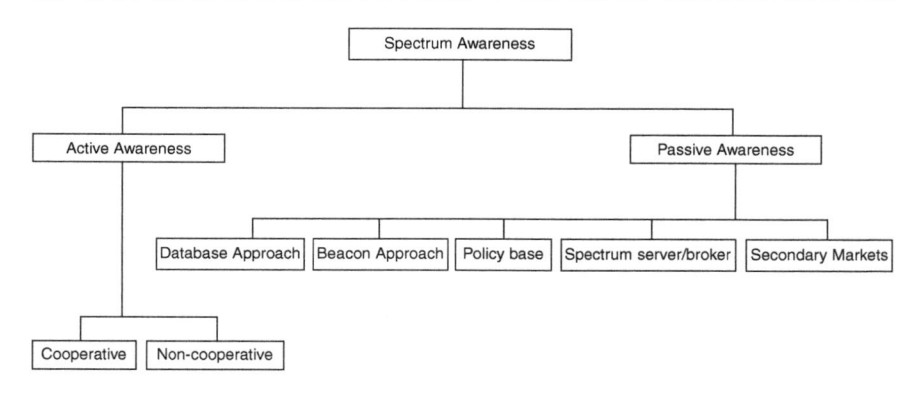

Figure 3.4 Spectrum awareness classification.

free frequency bands. Furthermore, the secondary user must be free of the spectrum immediately after the licensed user is available. Hence, the secondary user must be spectrum-aware to utilize the free spectrum efficiently [30]. Spectrum awareness techniques are grouped into active and passive awareness. In the passive awareness approach, the spectrum use pattern can be obtained by negotiating with the primary user or obtained from a server or database, whereas in the active awareness approach, the radio environment is constantly monitored and the secondary users adapt their transmission to suit the obtained value. [31]. The classical spectrum awareness is further categorized as shown in Figure 3.4.

Traditional CR spectrum awareness techniques are inefficient in detecting the occupancy status of the sensed spectrum bands. Furthermore, the spectrum access approaches cannot determine network changes or observe unlicensed users' requirements. This has led to excessive latency and poorer QoS [32]. As user-specific demands are pivotal in meeting wireless communication network demands, it becomes essential for conventional CR spectrum access to be more effective and agile. Interestingly, AI holds great potential for improving spectrum awareness and optimal utilization. To accurately and optimally improve detections while minimizing false alarms and missed detection, many AI methods have been deployed. The AI techniques are deployed at the fusion center. Conventionally, at the fusion center, Majority, OR, and AND rules are used for the final decision [30]. To improve the decision process, [33] exploited an adaptive neuro-fuzzy interference system that outperformed the conventional approach. For context-aware reasoning, fuzzy logic has been proven to be the best [34]. The fuzzy operators NOT, AND, and OR enable the combination of different spectrum awareness conditions such as signal-to-noise ratio, priori information, probability, and available time duration for effective decision. Evolutionary algorithms' random nature makes them suitable for spectrum conditions to distinguish secondary or primary spectrum users efficiently [35]. Thus, they are equally employed for CR spectrum sensing to provide the utmost detection of spectrum holes. In [36], the performance of a backpropagation artificial neural network for spectrum forecasting was improved using a genetic algorithm in the training phase, as conventional back-propagation artificial neural is easily trapped in local optimum [37]. The authors of [32] predicted succeeding traffic loads of the distinct radio access technologies utilizing individual bands of the spectrum employing artificial neural networks. A virtual wideband sensing method was proposed to predict relative traffic loads in wideband to enable narrowband sensing. Furthermore, Q-learning was exploited to minimize the sensing latency while satisfying other user requirements. Therefore, AI methods entail self-monitoring, self-protection, optimization, self-repair, self-healing, and self-adaptation of CRNs. Other spectra future occupancy states' prediction methods such as the hidden Markov model [38], autoregressive model [39], multilayer perceptron with backpropagation [40], LSTM [41], principal component analysis, k-means clustering, Bayesian learning, and AI-inclined spectrum-aware and energy cluster-based routing protocol for CR sensor networks [42] have been widely exploited. The varying AI methods employ in CR engines to ameliorate cognition capabilities in CRNs unluck great opportunities for future wireless networks. Nevertheless, the high computation time of AI algorithms must be minimized for effective power management and QoSs.

3.2.6 AI techniques for ARM

AI integration into CR resource management has evolved to solve the hard-code rules implemented on policy-based radios to control their behavior in certain scenarios. AI enables secondary users to fix issues by mimicking human biological procedures like reasoning, learning, self-organization, decision-making, self-stability, and self-adaptation. Hence, AI is deployed to discover spectrum holes, examine the parameters, and allocate the best available spectrum band among available bands to satisfy the user transmission requirements with improved throughput. Researchers have proposed different AI approaches like genetic algorithms, fuzzy logic, game theory, neural networks, support vector machines, reinforcement learning, Markov model, decision tree, multi-agent systems, and Bayesian, artificial bee colony, and entropy algorithms to resolve interference and spectrum scarcity in CR [43]. The various implemented AI-based approaches are shown in Figure 3.5.

AI enhances radio learning abilities in a shared environment as it exploits and utilizes free frequencies in the licensed band without any interference.

A. Reinforcement learning for resource allocation

The reinforcement learning method is a branch of ML that allows inexperienced agents to learn continuously through trial and error

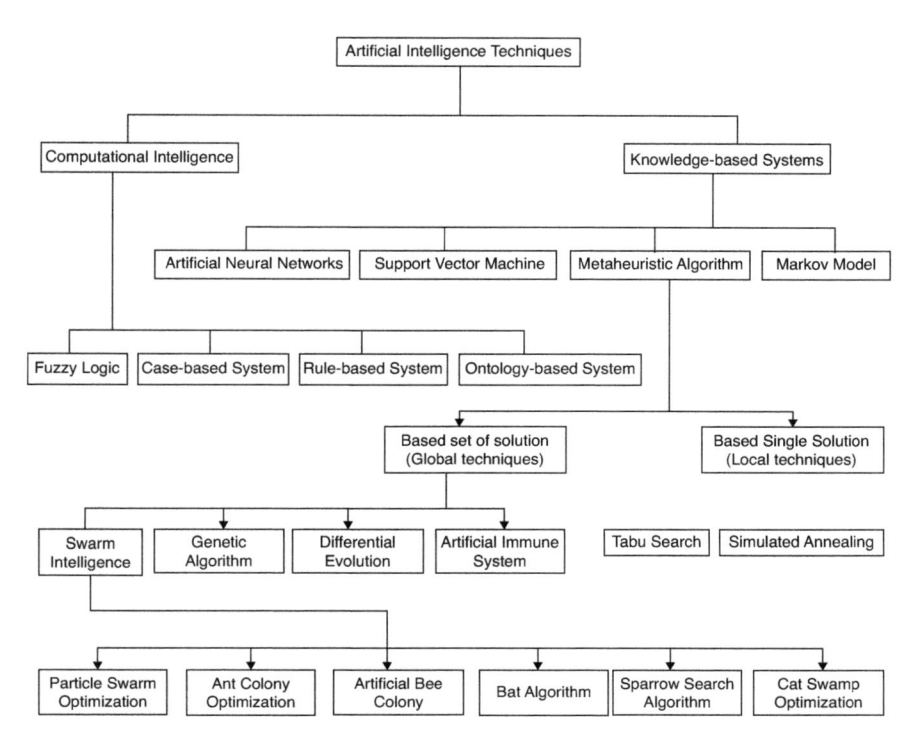

Figure 3.5 AI approaches for CRNs.

and obtain optimal strategies or maximize the reward function. Delay reward and trial and error are the two main features of reinforcement learning. The agent explores the environment without any prior knowledge. In the delayed reward approach, the agents obtained feedback from the actions it has taken in the environment. The Q-learning algorithm is the most adopted reinforcement learning technique. The algorithm updates the Q-table values constantly to determine the actions to be selected using the current state of the agent. The Q-table values are updated using the current, past, and others' experience making Q-learning an off-policy algorithm. The Q-learning searching strategies allow secondary users to not always choose the channel with the highest values but instead select another channel with a given probability to explore the environment for optimal long-term rewards [44]. This allows the reinforcement learning algorithm to achieve channel and power allocation autonomy while improving the channel efficiency and throughput [21]. For effective spectrum sensing, [45] modeled a secondary user with a Q-learning algorithm that learns every activity of multiple secondary users and subsequently selects individual users while maintaining a higher sensing accuracy.

Nevertheless, the sensing ability of secondary users was not considered while selecting neighboring nodes for cooperation, and this resulted in lower sensing accuracy. A Multiarmed bandit (MAB) [45] which modifies the performance gain of the existing greedy [46] approach as a reward for the MAB was applied to compressive sensing by Kawaguchi and Togami [47]. The result obtained showed an improvement over the existing greedy approach. Likewise, [48] proposed deep Q-learning (DQL) to select sensing channels and eventually access the maximum sensing rate. In [49], also deep Q-network (DQN) was adopted for spectrum sharing and power control. The authors in [50] presented a DQN experiment on a reservoir computing framework adopting a recurrent neural network for temporal prediction of the dynamic patterns of the spectrum sensing. In [44], the authors proposed a CNN based on CSS and assessed energy-controlled data protection in CRs. Furthermore, [51] hybridized the DQN and replicator dynamic (DQN-RD) algorithm, the RD utilized the evolutionary game theory principles in processing the reward function of the reinforcement learning to analyze the dynamic spectrum access strategy. Generally, spectrum sensing using deep reinforcement learning is achieved through the learning and training of agents which is in contrast with the conventional spectrum sensing schemes that achieve spectrum sensing by designing alternative or iterative algorithms.

3.3 AI-ENABLED SPECTRUM MANAGEMENT SYSTEMS

3.3.1 Spectrum databases and access control

Spectrum sensing alone does not provide enough assurances for free interference in a coexisting frequency band operation of wireless systems. The use of spectrum databases has been proposed to enable a more reliable dynamic spectrum management [52]. To have smarter and more effective spectrum use, policymakers as well as cognitive systems must be equipped with adequate spectrum utilization insights across frequency, time, and geography [53]. Spectrum databases serve as repositories of comprehensive information regarding spectrum availability, occupancy, and licensing regulations. These databases consist of geospatial, temporal, and spectral data, creating a robust framework for managing the intricacies of spectrum allocation and usage [54].

With the assistance of AI, spectrum databases can be constantly updated to maintain accurate records of spectrum usage. Utilizing ML algorithms, these databases can predict spectrum availability patterns based on historical data, usage trends, and environmental factors. This predictive capability enhances proactive spectrum management [55], allowing regulators and service providers to anticipate and prepare for fluctuations in demand, ultimately optimizing spectrum allocation and utilization.

While spectrum databases provide insight for effective spectrum use, access control strategies offer a means of ensuring that only authorized users are granted access. AI-driven access control mechanisms will help the regulators in managing and enforcing spectrum usage policies. These mechanisms ensure fair and efficient spectrum sharing among diverse stakeholders while preventing harmful interference and maintaining service quality [56].

3.3.2 AI-assisted spectrum auctions

Historically, spectrum auctions have been complex, multifaceted processes, involving numerous stakeholders, bidding strategies, and regulatory compliance. The traditional approach relied heavily on human decision-making, often leading to inefficiencies, suboptimal allocation, and potential revenue loss for governments.

The integration of AI technologies has evolved the way spectrum auctions are conducted. AI algorithms, powered by ML and data analytics, offer unparalleled capabilities in processing vast amounts of data, predicting bidder behavior, and optimizing auction designs. These advancements enable governments and regulatory bodies to orchestrate auctions that are more transparent, competitive, and economically efficient.

The adoption of AI for spectrum auctions signifies a monumental leap in the efficient allocation of wireless resources. By leveraging the capabilities of AI, governments and regulatory bodies can orchestrate auctions that optimize spectrum utilization, foster innovation, and meet the ever-growing demands for connectivity in the digital age. As AI technologies evolve, they will undoubtedly continue to revolutionize spectrum allocation, shaping the future of global communication networks.

3.3.3 Implications, benefits, and challenges

The application of AI in spectrum management promises to redefine the efficiency, adaptability, and fairness of spectrum allocation and usage, ushering in a new era of efficiency in wireless communication. While the integration of AI into spectrum management systems introduces several benefits, it also presents notable challenges

Striking a balance between harnessing AI's capabilities and addressing its associated challenges will be instrumental in realizing the full potential of AI-enabled spectrum management systems in fostering better wireless connectivity and an innovative future.

Some of the key challenges of AI in spectrum management are highlighted below [56]:

    i. **Data privacy:** Large volumes of data, containing sensitive information, are needed for effective AI model development. In the AI-enabled spectrum management regime, network operators and regulators must share data on spectrum usage. One of the biggest challenges is finding a balance between protecting user privacy and gaining insights from data. It is anticipated that the integration of AI into the spectrum management space will give rise to fresh privacy and security issues [1], necessitating more investigation and learning. The authors of [7] demonstrated the necessity of a novel physical-layer privacy strategy based on AI for the upcoming 6G network and its various uses [57].

ii. **Complexity:** Although AI holds enormous promise for solving complex issues, there are inherent limitations that vary depending on the application's complexity. Due to the dynamic RF environment in wireless communication, an allocation that is fair one moment may swiftly become unfair the next [58]. AI models and algorithms become more complicated due to the requirement to adjust to a rapidly changing spectrum environment. Furthermore, received signals in wireless networks, along with their higher-order statistics, are usually complicated numbers that are difficult for neural networks to directly analyze.

iii. **Algorithm bias:** The issue of bias in AI algorithms is becoming more prominent as AI continues to find broad applications in various fields [59,60]. AI algorithms may unintentionally perpetuate current spectrum allocation biases, which could result in unequal access or favoritism of particular apps or users. If not given due attention, algorithm bias in AI systems has the potential to reinforce past injustices and societal prejudice in spectrum allocation and management [60]. According to [61], to surmount this difficulty, AI models need to be built on more transparent and diverse datasets.

iv. **Regulatory frameworks:** Regulatory authorities continue to face the difficulty of modifying existing spectrum regulatory frameworks to allow AI-driven spectrum management while maintaining equity and transparency. System complexity and objective clarity determine the efficiency of AI-enabled regulatory operations: The more complex a field, the more difficult it is to operationalize objectives and carry out regulatory and control functions [58]. In an AI regime, the detection of the underutilized or wrongly used spectrums is expected to be more effective. Hence, regulatory provisions for appropriate sanctions must also be in place to enforce compliance.

3.4 CASE STUDIES AND APPLICATIONS

3.4.1 AI-based spectrum allocation in 5G and beyond networks

The 5G technology, coupled with IoT applications, is the fourth industrial revolution's major drivers [62]. The key characteristics of 5G technology include the idea of hardware agnosticism, virtualization at various network levels, and orchestration at various levels, utilizing various cutting-edge technologies such as network function virtualization and software-defined networking on a cloud-based platform [62,63].

5G networks require very low latency, great reliability, and high transmission rates. The integration of AI will be very instrumental in meeting these lofty requirements. Also, the 5G networks, especially in their IoT applications, are anticipated to generate unprecedented traffic and large amounts of data [64], paving the way for the development of intelligent ML models that can predict user demands. To accommodate various enterprise applications and business services, 5G and beyond networks adopt network slicing. Network slicing is the technique of dividing the actual network infrastructure into virtual networks. On such shared infrastructure, several services with various architectural and QoS requirements complicate the network environment [65]. ML models have been used to support similar complex problems; hence, they offer a solution to network management and resource allocation issues in a complex network environment [66]. Several researchers have reported the use of AI techniques [65] in various aspects of 5G network design, including radio resource allocation [64].

3.4.2 SI for enhanced IoT deployment

The rapid advancement in sensing technologies and communication has rapidly evolved the IoT. In [67], IoT was defined as a global network of interconnected objects uniquely addressable, based on standard communication protocols that allow people and things to be connected at any time, any place, and with anything and anyone, ideally using any path/network and any service. With the rapid deployment of IoT objects, spectrum band allocation becomes difficult. Therefore, CR with its spectrum-sharing advantage is a boost for object interaction. The IoT objects are equipped with cognition to learn, think, and make decisions through understanding both social and physical worlds. Additional requirements include massive data analytics, intelligent decision-making, perception-action cycles, semantic derivation, on-demand service provisioning, and knowledge discovery. In [68], a CR-based IoT with several primary user base stations and secondary user devices as the IoT smart objects was developed. The CRs perform combined optimal allocation and spectrum sensing to the requesting secondary user IoT devices in the network through an intelligent fusion center. Equally, SVM was integrated into the network for adaptive learning of the network dynamics and identification of the primary spectrum user.

3.4.3 CR application in unmanned aerial vehicle (UAV) networks

Unmanned aerial vehicles (UAVs) require technologies for autonomous flight while communicating with other UAVs, computers, base stations, flight controllers, or other devices. The UAVs operate in unlicensed spectrum bands.

Hence, they compete with the rapidly growing WSNs and mobile devices. Interestingly, the integration of the CR into UAVs creates the potential for speedy executions of missions when compared to conventional UAV performances [69]. CR offers smart wireless communication that uses software transmission instead of the inherent transmission frequency. In [70], a UAV network equipped with a CR in which the UAV communicates with ground terminals in an underlay mode using a licensed spectrum band was investigated. The total network was optimized using interference and throughput as constraints. The results obtained show the potential of integrating UAV and CR.

3.4.4 Improved energy efficiency in wireless sensor networks (WSNs)

Wireless sensor network (WSN) communications are event-driven as they use wireless sensor nodes to monitor and observe sensitive and critical activities [71]. Existing WSNs using the industrial, scientific, and medical bands share bands with other successful communication technologies [72]. The wide range covered by IEEE 802.11 devices, large transmit power, wide deployments, and other proprietary devices degrades WSN performances once operated within an overlapped frequency band. WSN devices can be interferer most times and not always a victim [73]. Frequency, time, and space are intelligently used to avoid coexistence interference in conventional WSNs. A wireless sensor network equipped with a CR is a candidate area that allows cognitive approaches for opportunistic spectrum access. In [74], a game theory approach that clusters nodes in CR WSNs that outperformed the benchmarked low-energy adaptive uneven clustering hierarchy, and spectrum-aware clustering algorithm protocol was presented. Also, in [75], energy-aware Q-learning ad hoc on-demand vector routing for CR sensor networks has been successfully deployed on low-powered wireless sensor networks. The Q-learning allows optimal path selection amidst residual energy, licensed channel, common channel, communication range number of hops, and trust factor while using the ad hoc on-demand vector routing protocol. The comparative analysis of the method shows an improved performance in terms of network lifetime, energy consumption, and delay.

3.5 FUTURE DIRECTIONS AND CONCLUSIONS

3.5.1 Future direction

The fusion of AI with spectrum management not only addresses the issues posed by the ever-growing request for wireless connectivity but also unlocks the potential for unprecedented efficiency gains and the coexistence of diverse wireless technologies. While AI presents diverse opportunities that must be explored, it also poses several challenges some of which are inherent and others associated with the domain of application, spectrum management. Some of these opportunities and challenges for future research are highlighted:

i. **Energy efficiency:** The issue of energy efficiency in a cognitive network environment remains an open one. As performance-enhancing algorithms and schemes are developed, the computational and, by extension, power requirements for implementation also increase, thereby constituting a challenge to energy-limited edge devices [10].

ii. **System integration:** In developing intelligent systems for spectrum use and management, several ML and AI algorithms have been proposed and experimented with. Implementing an efficient and cost-effective integrated intelligent system for spectrum detection, allocation, and regulation constitutes a significant challenge worthy of further research consideration [76].

iii. **Security and privacy:** The implementation of novel AI-enabled spectrum access technologies and their realization presents security and privacy challenges that are hitherto unexplored. Policymakers and regulators are faced with the issue of determining what spectrum use information may be gathered in accessing spectrum utilization without violating the privacy of users [1]. Future networks must prioritize security and privacy. With the advent of big data and AI, traditional encryption methods and other privacy management approaches have been rendered less secure [77]. Therefore, innovations in the privacy of physical layers must be developed for AI-based future communication networks [7].

3.5.2 Conclusion

This chapter explores the foundational concepts of SIARM, delving into the underlying AI techniques, the benefits they offer, and the potential challenges that must be overcome. It provides a comprehensive review of various approaches toward achieving intelligent spectrum detection, allocation, and management. Highlight of SI and ARM use cases, covering scenarios in 5G communication and beyond, IoT, and UAV applications were also presented. The chapter logically provides theoretical concepts as well as a review of relevant works of literature in SI and ARM, thereby offering researchers and students full vital information and facts for future work.

REFERENCES

1. A. Kaur and K. Kumar, "A comprehensive survey on machine learning approaches for dynamic spectrum access in cognitive radio networks," *J. Exp. Theor. Artif. Intell.*, vol. 34, no. 1, pp. 1–40, 2022. doi: 10.1080/0952813X.2020.1818291.

2. "AI in spectrum management," India, ISO 9001:2008, 2021 [Online]. Available: https://www.tec.gov.in/pdf/Studypaper/AI_in_Spectrum_management.pdf

3. E. Alozie et al., "Intelligent process of spectrum handoff in cognitive radio network," *SLU J. Sci. Technol.*, vol. 4, no. 1, p. 205, 2022.

4. O. H. Toma, M. Lopez-Benitez, D. K. Patel, and K. Umebayashi, "Estimation of primary channel activity statistics in cognitive radio based on imperfect spectrum sensing," *IEEE Trans. Commun.*, vol. 68, no. 4, pp. 2016–2031, 2020.

5. F. Qamar, M. U. A. Siddiqui, M. N. Hindia, R. Hassan, and Q. N. Nguyen, "Issues, challenges, and research trends in spectrum management: A comprehensive overview and new vision for designing 6G networks," *Electronics*, vol. 9, no. 9, p. 1416, 2020.

6. X. Liu, Q. Sun, W. Lu, C. Wu, and H. Ding, "Big-data-based intelligent spectrum sensing for heterogeneous spectrum communications in 5G," *IEEE Wirel. Commun.*, vol. 27, no. 5, pp. 67–73, 2020.

7. H. F. Alhashimi et al., "A survey on resource management for 6G heterogeneous networks: Current research, future trends, and challenges," *Electron. MDPI, Basel, Switz.*, vol. 12, no. 3, pp. 1–42, 2023.

8. A. Patil, S. Iyer, and R. J. Pandya, "A survey of machine learning algorithms for 6g wireless networks," 2002. doi: 10.48550/arXiv.2203.08429.

9. S. N. Syed and V. Holmes, "Deep learning approaches for spectrum sensing in cognitive radio networks," in *2022 25th International Symposium on Wireless Personal Multimedia Communications*, pp. 480–485, 2022. doi: 10.1109/WPMC55625.2022.10014805.

10. G. Eappen and T. Shankar, "A survey on soft computing techniques for spectrum sensing in a cognitive radio network," *SN Comput. Sci.*, vol. 1, no. 6, p. 352, 2020. doi: 10.1007/s42979-020-00372-z.

11. D. Janu, K. Singh, and S. Kumur, "Machine learning for cooperative spectrum sensing and sharing," *Trans. Emerg. Telecommun. Technol.*, vol. 33, no. 1, p. e4352, 2022. doi: 10.1002/ett.4352.

12. Y. Li, T. Wang, Y. Wu, and W. Jia, "Optimal dynamic spectrum allocation-assisted latency minimization for multiuser mobile edge computing," *Digit. Commun. Networks*, vol. 8, no. 3, pp. 247–256, 2022.

13. S. S. Chopade and S. S. Dalu, "Improving security with optimized QoS in cognitive radio networks using AI backed blockchains," in *ICCCE 2021: Proceedings of the 4th International Conference on Communications and Cyber Physical Engineering*, pp. 629–638, Singapore: Springer, 2021.

14. P. A. Sandeep, "Comparative analysis of optimization techniques in cognitive radio (QoS)," *Int. J. Eng. Adv. Technol.(IJEAT)*, vol. 6, pp. 2249–8958, 2017.

15. S. Pandit and G. Singh, *Spectrum Sharing in Cognitive Radio Networks*. Berlin, Germany: Springer, 2017.

16. N. Chaudhary and R. Mahajan, "Comprehensive review on spectrum sensing techniques in cognitive radio. *Engineering Review: Međunarodni časopis namijenjen publiciranju originalnih istraživanja s aspekta analize konstrukcija, materijala i novih tehnologija u području strojarstva, brodogr,"* vol. 42, no. 1. pp. 88–102, 2022.

17. P. Thakur, A. Kumar, S. Pandit, G. Singh, and S. N. Satashia, "Performance analysis of cognitive radio networks using channel-prediction-probabilities and improved frame structure," *Digit. Commun. Netw.*, vol. 4, no. 4, pp. 287–295, 2018.

18. J. Isabona et al., "Development of a multilayer perceptron neural network for optimal predictive modeling in urban microcellular radio environments," *Appl. Sci.*, vol. 12, no. 11, p. 5713, 2022, doi: 10.3390/app12115713.

19. S. Jain, A. Goel, and P. Arora, "Spectrum prediction using time delay neural network in cognitive radio network," in *Smart Innovations in Communication and Computational Sciences: Proceedings of ICSICCS-2018*, pp. 257–269, Singapore: Springer, 2019.

20. K. Sato and T. Fujii, "Kriging-based interference power constraint: Integrated design of the radio environment map and transmission power," *IEEE Trans. Cogn. Commun. Netw.*, vol. 3, no. 1, pp. 13–25, 2017.

21. Z. Goldfeld and H. H. Permuter, "A useful analogy between wiretap and gelfand-pinsker channels," in *2018 IEEE International Symposium on Information Theory (ISIT)*, pp. 121–125, Vail, CO: IEEE, 2018.

22. Y. Mao and B. Clerckx, "Beyond dirty paper coding for multi-antenna broadcast channel with partial CSIT: A rate-splitting approach," *IEEE Trans. Commun.*, vol. 68, no. 11, pp. 6775–6791, 2020.

23. R. Garg, "Review of cooperative sensing and non cooperative sensing in cognitive radios," *Int. J. Eng. Technol. Sci. Res.*, vol. 4, no. 5, pp. 229–234, 2017.

24. Y. Arjoune and N. Kaabouch, "A comprehensive survey on spectrum sensing in cognitive radio networks: Recent advances, new challenges, and future research directions," *Sensors (Switzerland)*, vol. 19, no. 1, p. 126, 2019. doi: 10.3390/s19010126.

25. V. Sharma and S. Joshi, "A literature review on spectrum sensing in cognitive radio applications," in *2018 Second International Conference on Intelligent Computing and Control Systems (ICICCS)*, pp. 883–893, Madurai: IEEE, 2018.

26. Y. Arjoune and N. Kaabouch, "A comprehensive survey on spectrum sensing in cognitive radio networks: Recent advances, new challenges, and future research directions," *Sensors*, vol. 19, no. 01, p. 126, 2019.

27. A. Haldorai, J. Sivaraj, M. Nagabushanam, and M. Kingston Roberts, "Cognitive wireless networks based spectrum sensing strategies: A comparative analysis," *Appl. Comput. Intell. Soft Comput.*, vol. 2022, p. 6988847, 2022. doi: 10.1155/2022/6988847.

28. D. Janu, K. Singh, and S. Kumar, "Machine learning for cooperative spectrum sensing and sharing: A survey," *Trans. Emerg. Telecommun. Technol.*, vol. 33, no. 01, p. e4352, 2022.

29. H. Xing, H. Qin, S. Luo, P. Dai, L. Xu, and X. Cheng, "Spectrum sensing in cognitive radio: A deep learning based model," *Trans. Emerg. Telecommun. Technol.*, vol. 33, no. 1, p. e4388, 2022.

30. N. H. A. Siddique, *Computational Intelligence: Synergies of Fuzzy Logic, Neural Networks, and Evolutionary Computing*. West Sussex, United Kingdom: John Wiley & Sons, 2013.

31. M. M. Mabrook, H. A. Khalil, and A. I. Hussein, "Artificial intelligence based cooperative spectrum sensing algorithm for cognitive radio networks," *Procedia Comput. Sci.*, vol. 163, pp. 19–29, 2019.

32. M. Ozturk, M. Akram, S. Hussain, and M. A. Imran, "Novel QoS-aware proactive spectrum access techniques for cognitive radio using machine learning," *IEEE Access*, vol. 7, pp. 70811–70827, 2019.

33. M. M. Mabrook, H. A. Taha, and A. I. Hussein, "Cooperative spectrum sensing optimization based adaptive neuro-fuzzy inference system (ANFIS) in cognitive radio networks," *J. Ambient Intell. Humaniz. Comput.*, vol. 13, pp. 3643–3654.

34. R. Ganesh Babu and V. Amudha, "A survey on artificial intelligence techniques in cognitive radio networks," in *Emerging Technologies in Data Mining and Information Security: Proceedings of IEMIS 2018*, vol. 1, pp. 99–110, Singapore: Springer, 2019.

35. E. Fadel et al., "Spectrum-aware bio-inspired routing in cognitive radio sensor networks for smart grid applications," *Comput. Commun.*, vol. 101, pp. 106–120, 2017.

36. J. Yang, H. Zhao, and X. Chen, "Genetic algorithm optimized training for neural network spectrum prediction," in *2016 2nd IEEE International Conference on Computer and Communications (ICCC)*, pp. 2949–2954, Chengdu: IEEE, 2016.

37. X. Sun, G. Gui, Y. Li, R. P. Liu, and Y. An, "ResInNet: A novel deep neural network with feature reuse for Internet of Things," *IEEE Internet Things J.*, vol. 6, no. 1, pp. 679–691, 2018.

38. H. Eltom, S. Kandeepan, Y. C. Liang, and R. J. Evans, "Cooperative soft fusion for HMM-based spectrum occupancy prediction," *IEEE Commun. Lett.*, vol. 22, no. 10, pp. 2144–2147, 2018.

39. A. Eltholth, "Forward backward autoregressive spectrum prediction scheme in cognitive radio systems," in *2015 9th International Conference on Signal Processing and Communication Systems (ICSPCS)*, pp. 1–5, Cairns: IEEE, 2015.

40. V. K. Tumuluru, P. Wang, and D. Niyato, "A neural network based spectrum prediction scheme for cognitive radio," in *2010 IEEE International Conference on Communications*, pp. 1–5, Cape Town: IEEE, 2010.

41. P. Zuo, X. Wang, W. Linghu, R. Sun, T. Peng, and W. Wang, "Prediction-based spectrum access optimization in cognitive radio networks," in *2018 IEEE 29th Annual International Symposium on Personal, Indoor and Mobile Radio Communications (PIMRC)*, pp. 1–7, Bologna: IEEE, 2018.

42. T. Stephan, F. Al-Turjman, K. S. Joseph, B. Balusamy, and S. Srivastava, "Artificial intelligence inspired energy and spectrum aware cluster based routing protocol for cognitive radio sensor networks," *J. Parallel Distrib. Comput.*, vol. 142, pp. 90–105, 2020.

43. B. Benmammar and A. Amraoui, "Artificial intelligence application to cognitive radio networks," in *Intelligent Network Management and Control: Intelligent Security, Multi-criteria Optimization, Cloud Computing, Internet of Vehicles, Intelligent Radio*, pp. 217–243, 2021. https://doi.org/10.1002/9781119817840.ch9

44. F. Obite, A. D. Usman, and E. Okafor, "An overview of deep reinforcement learning for spectrum sensing in cognitive radio networks," *Digit. Signal Process.*, vol. 113, p. 103014, 2021.

45. B. F. Lo and I. F. Akyildiz, "Reinforcement learning-based cooperative sensing in cognitive radio ad hoc networks," in *21st Annual IEEE International Symposium on Personal, Indoor and Mobile Radio Communications*, pp. 2244–2249, Istanbul: IEEE, 2013.

46. S. Vakili, K. Liu, and Q. Zhao, "Deterministic sequencing of exploration and exploitation for multi-armed bandit problems," *IEEE J. Sel. Top. Signal Process.*, vol. 7, no. 5, pp. 759–767, 2013.

47. Y. Kawaguchi and M. Togami, "Adaptive boolean compressive sensing by using multi-armed bandit," in *2016 IEEE International Conference on Acoustics, Speech and Signal Processing (ICASSP)*, pp. 3261–3265, Shanghai: IEEE, 2016.

48. X. Li, J. Fang, W. Cheng, H. Duan, Z. Chen, and H. Li, "Intelligent power control for spectrum sharing in cognitive radios: A deep reinforcement learning approach," *IEEE Access*, vol. 6, pp. 25463–25473, 2018.

49. A. Zhang, M. Sun, J. Wang, Z. Li, Y. Cheng, and C. Wang, "Real-time data transmission scheduling algorithm for Wireless Sensor Networks based on deep Q-learning," *Electronics*, vol. 11, no. 12, p. 1877, 2022.

50. H. Song, L. Liu, J. Ashdown, and Y. Yi, "A deep reinforcement learning framework for spectrum management in dynamic spectrum access," *IEEE Internet Things J.*, vol. 8, no. 14, pp. 11208–11218, 2021.

51. T. Mai, H. Yao, N. Zhang, L. Xu, M. Guizani, and S. Guo, "Cloud mining pool aided blockchain-enabled Internet of Things: An evolutionary game approach," *IEEE Trans. Cloud Comput.*, vol. 11, no. 1, pp. 692–703, 2021.

52. M. Höyhtyä, J. Ylitalo, X. Chen, and A. Mämmelä, "Chapter 14- Use of databases for dynamic spectrum management in cognitive satellite systems," in S. Chatzinotas, B. Ottersten, and R. De Gaudenzi, Eds. *Cooperative and Cognitive Satellite Systems*, Academic Press, 2015, pp. 453–480. doi: 10.1016/C2013-0-12952-2.

53. L. A. Stefani, *Artificial Intelligence and the Future of Spectrum Management: FCC Adopts Notice of Inquiry on Spectrum Use Data*. Venable LLP, 2023. https://www.venable.com/insights/publications/2023/08/ai-and-the-future-of-spectrum-management

54. D.A., Guimarães, E.J. Pereira, A.M. Alberti, and J.V. Moreira, "Design guidelines for database-driven Internet of Things-enabled dynamic spectrum access". *Sensors*, vol. 21, no. 9, p. 3194, 2021.

55. J. Zhang, Y. Chen, Y. Liu, and H. Wu, "Spectrum knowledge and real-time observing enabled smart spectrum management," *IEEE Access*, vol. 8, pp. 44153–44162, 2020. doi: 10.1109/ACCESS.2020.2978005.

56. C. A. Alabi, M. A. Giwa, S. T. Tersoo, A. L. Imoize, N. Faruk, and A. E. Ehime, "Artificial Intelligence in Spectrum Management: Policy and Regulatory Considerations," in *The 2nd International Conference on Multidisciplinary Engineering and Applied Sciences (ICMEAS-2023)*, pp. 2–7, Abuja: IEEE, 2023.

57. A. L. Imoize, O. Adedeji, N. Tandiya, and S. Shetty, "6G enabled smart infrastructure for sustainable society: Opportunities, challenges, and research roadmap," *Sensors (Switzerland)*, vol. 21, no. 5, p. 1709, 2021, doi: 10.3390/s21051709.

58. L. B. Weissinger, "AI, complexity, and regulation," pp. 1–22, 2021 [Online]. Available: https://papers.ssrn.com/sol3/papers.cfm?abstract_id=3943968

59. I. Bousquette, "Rise of AI puts spotlight on bias in algorithms," *The Wall Street Journal*, 2023. https://www.wsj.com/articles/rise-of-ai-puts-spotlight-on-bias-in-algorithms-26ee6cc9

60. B. Mitra and R. Anand, "Understanding algorithmic bias and how to build trust in AI," *Price Water Coopers (PwC)*, 2023.

61. C. M. P. Jacoba et al., "Bias and non-diversity of big data in artificial intelligence: Focus on retinal diseases," *Semin. Ophthalmol.*, vol. 38, no. 5, pp. 433–441, 2023. doi: 10.1080/08820538.2023.2168486.

62. E. Esenogho, K. Djouani, and A. M. Kurien, "Integrating artificial intelligence Internet of Things and 5G for next-generation smartgrid: A survey of trends challenges and prospect," *IEEE Access*, vol. 10, pp. 4794–4831, 2022. doi: 10.1109/ACCESS.2022.3140595.

63. J. Holtom, A. Herschfelt, I. Lenz, O. Ma, H. Yu, and D. W. Bliss, "WISCANet: A rapid development platform for beyond 5G and 6G radio system prototyping," *Signals*, vol. 3, no. 4, pp. 682–707, 2022. doi: 10.3390/signals3040041.

64. M. Al-Khafaji and L. Elwiya, "ML/AI empowered 5G and beyond networks," in *2022 International Congress on Human-Computer Interaction, Optimization and Robotic Applications (HORA)*, pp. 1–6, Ankara: IEEE, 2022. doi: 10.1109/HORA55278.2022.9799813.

65. C. Ssengonzi, O. P. Kogeda, and T. O. Olwal, "A survey of deep reinforcement learning application in 5G and beyond network slicing and virtualization," *Array*, vol. 14, no. January, p. 100142, 2022. doi: 10.1016/j.array.2022.100142.

66. M. A. Adelabu, A. L. Imoize, and M. B. Ugwu, "Radio resource management of WLAN hotspot access points in next generation wireless networks," *SN Comput. Sci.*, vol. 4, no. 3, p. 313, 2023.

67. A. A. Khan, M. H. Rehmani, and A. Rachedi, "Cognitive-radio-based Internet of Things: Applications, architectures, spectrum related functionalities, and future research directions," *IEEE Wirel. Commun.*, vol. 24, no. 3, pp. 17–25, 2017.

68. R. Ahmed, Y. Chen, B. Hassan, and L. Du, "CR-IoTNet: Machine learning based joint spectrum sensing and allocation for cognitive radio enabled IoT cellular networks," *Ad Hoc Networks*, vol. 112, p. 102390, 2021.

69. D. Santana, G. Marcel, R. S. Cristo, and K. R. L. J. C. Branco, "Integrating cognitive radio with Unmanned Aerial Vehicles: An overview," *Sensors*, vol. 21, no. 3, p. 830, 2021.

70. S. K. Nobar, M. H. Ahmed, Y. Morgan, and S. A. Mahmoud, "Resource allocation in cognitive radio-enabled UAV communication," *IEEE Trans. Cogn. Commun. Netw.*, vol. 8, no. 1, pp. 296–310, 2021.

71. I.M. Abutu, U.J. Imeh, T.M. Abdoulie, A.E. Adewale, and M.M. Bashir, "Real time universal scalable Wireless Sensor Network for environmental monitoring application." *Int. J. Comput. Netw. Inform. Secur.*, vol. 10, no. 6, p. 68, 2018.

72. D. Kruger, R. Heynicke, and G. Scholl, "Wireless sensor/actuator-network with improved coexistence performance for 2.45 GHz ISM-band operation," in *International Multi-Conference on Systems, Signals & Devices,* Chemnitz, Germany, IEEE, 2012, pp. 1–5, doi: 10.1109/SSD.2012.6198099.

73. D. Yang, Y. Xu, and M. Gidlund, "Wireless coexistence between IEEE 802.11 and IEEE 802.15.4-based networks: A survey," *Int. J. Distr. Sens. Netw.*, vol. 7, no. 1, pp. 1–17, 2011.

74. P. Rai, M. K. Ghose, and H. K. D. Sarma, "Game theory based node clustering for cognitive radio Wireless Sensor Networks," *Egypt. Informatics J.*, vol. 23, no. 2, pp. 315–327, 2022.

75. R. Joon and P. Tomar, "Energy aware Q-learning AODV (EAQ-AODV) routing for cognitive radio sensor networks," *J. King Saud Univ. Inf. Sci.*, vol. 34, no. 9, pp. 6989–7000, 2022.

76. M. Banafaa et al., "6G mobile communication technology: Requirements, targets, applications, challenges, advantages, and opportunities," *Alexandria Eng. J.*, vol. 64, pp. 245–274, 2023. doi: 10.1016/j.aej.2022.08.017.

77. M. G. Aruna, M. K. Hasan, S. Islam, K. G. Mohan, P. Sharan, and R. Hassan, "Cloud to cloud data migration using self sovereign identity for 5G and beyond," *Cluster Comput.*, vol. 25, no. 4, pp. 2317–2331, 2022. doi: 10.1007/s10586-021-03461-7.

The future of mobile management

Predictive maintenance and fault detection with AI

Showkat Ahmad Dar and P. Sakthivel

4.1 INTRODUCTION

The rise of mobile devices has transformed the operations of organizations, allowing them to enhance efficiency and productivity. Managing these devices has become more intricate due to many obstacles, like software updates, security risks, and hardware faults. Many organizations are utilizing artificial intelligence (AI) for predictive maintenance and defect detection to tackle these challenges (Ai et al. 2015). The study explores the future of mobile management, focusing on the revolutionary power of AI in predictive maintenance and defect detection. The study emphasizes the significant impact of AI technology on mobile ecosystems, leading to increased productivity, less downtime, and higher device reliability (Ayvaz & Alpay 2021). We intend to transform the management and maintenance of mobile devices by utilizing AI algorithms and models, with the goal of creating a smoother and more secure mobile future.

The study investigates how AI can impact mobile device management (MDM) and change how firms handle their mobile devices. Utilizing AI for MDM offers the key benefit of predictive maintenance (Chen et al. 2021). This entails utilizing AI algorithms to forecast the potential failure of a device and preemptively intervening to avoid it. AI systems can detect trends in data from sensors and other sources to anticipate issues and activate automated measures to avert them. Implementing this strategy can greatly decrease downtime, enhance device performance, and prolong the lifespan of devices, leading to cost savings for the business. Lee et al. (2019) did a study investigating the application of AI in predictive maintenance of mobile devices within the healthcare sector. The research revealed that implementing AI for predictive maintenance led to a 40% decrease in maintenance expenses and a 15% decrease in device downtime. Zonta et al. (2020) discovered that implementing AI for predictive maintenance in a manufacturing setting led to a 25% decrease in unexpected downtime, a 30% decrease in maintenance expenses, and a 15% enhancement in total equipment efficiency. AI in MDM is crucial for fault detection (Divya et al. 2023). This entails utilizing AI algorithms to detect problems with

    DOI: 10.1201/9781003517689-4

equipment and autonomously address them. AI algorithms can detect patterns suggesting a failure by analyzing data from sensors, logs, and other sources and then initiate automated steps to address the issue. This method can greatly decrease the necessity for manual involvement and enhance the overall dependability of gadgets.

Zheng et al. (2020) investigated the application of AI for identifying faults in mobile devices, with a specific emphasis on battery performance. The research revealed that employing AI for fault identification led to a 20% enhancement in battery life and a 25% decrease in battery-related problems. Wellsandt et al. (2022) investigated the application of AI for identifying faults in the transportation sector, specifically concentrating on engine functionality. The research revealed that employing AI for issue identification led to a 35% decrease in maintenance expenses and a 20% enhancement in total equipment efficiency.

The advantages of implementing AI in MDM are substantial, and numerous organizations are already starting to integrate this technology. Lee et al. (2020) conducted a case study where a healthcare organization implemented AI for predictive maintenance of mobile devices. This led to a 30% decrease in maintenance expenditures and a 25% increase in device uptime. Chuang et al. (2019) conducted a case study where a manufacturing company implemented AI for problem detection, leading to a 20% decrease in maintenance expenses and a 15% enhancement in equipment uptime. Nevertheless, there are obstacles linked to implementing AI in MDM. The primary challenge is the requirement for data of good quality.

AI systems depend on extensive data to precisely forecast errors and identify flaws. Organizations must secure access to top-notch data from sensors and other sources to optimize the efficiency of AI in MDM. Another obstacle is the requirement for proficient workers. Proficiency in data science, machine learning, and software development is essential for implementing and overseeing AI systems. Organizations must invest in training and development programs to ensure they possess the essential skills to effectively implement and manage.

4.2 SIGNIFICANT CONTRIBUTION OF THE STUDY

The study makes significant contributions to the fields of mobile management and predictive maintenance with AI. It introduces advanced predictive maintenance techniques enhanced by AI, optimizes mobile management processes, emphasizes data-driven decision-making, highlights innovative fault detection methodologies (Su et al. 2018), and explores the integration of the IoT and AI. These contributions enhance mobile device reliability, improve operational efficiency, reduce downtime, and enable more targeted troubleshooting. The study equips businesses with the tools to thrive in a rapidly evolving mobile technology landscape, ultimately improving customer satisfaction and market competitiveness.

The importance of this study lies in its potential to revolutionize MDM. Enhanced device reliability, operational efficiency, and cost savings are vital for organizations across various industries. By adopting proposed methodologies, businesses can reduce maintenance costs, minimize unexpected downtime, and gain a competitive edge in their respective markets. Its emphasis on sustainability aligns with the global push for eco-friendly practices by extending the lifespan of mobile devices and reducing electronic waste. It has the power to transform MDM practices across industries; ensuring organizations remain agile and responsive to emerging technological trends. The study serves as a valuable resource for future-proofing organizations against the evolving challenges and opportunities in the mobile technology landscape.

4.3 ORGANIZATION OF THE CHAPTER

In an era dominated by mobile devices, effective mobile management has become paramount for organizations across industries. The evolution of technology brings with it new challenges, such as ensuring device reliability, minimizing downtime, and optimizing resource allocation. To address these challenges, the study explores the "The Future of Mobile Management: Predictive Maintenance and Fault Detection with AI." At its core, the study explores the potential of AI-powered predictive maintenance and fault detection techniques to revolutionize mobile management practices.

The study addresses critical aspects of MDM. Firstly, it seeks to understand the current landscape of mobile management practices and the challenges organizations encounter in maintaining their mobile devices. Secondly, the study explores the potential of AI-powered predictive maintenance and fault detection techniques in mobile management systems, scrutinizing various AI algorithms and their ethical implications. Ultimately, the study intends to provide practical recommendations to organizations seeking to implement AI-powered solutions, drawing from industry best practices and real-world case studies. These objectives collectively contribute to enhancing device reliability, reducing operational costs, and ensuring ethical responsibility in the integration of AI within mobile management systems.

The rapid proliferation of mobile devices has made them indispensable tools for businesses, but it has also introduced complexities in managing and maintaining these devices. Organizations face daunting tasks, such as tracking device health, preventing unexpected failures, and ensuring data security. As a result, there is a pressing need to explore innovative approaches to mobile management that can alleviate these challenges. AI emerges as a promising solution, offering the potential to transform mobile management from reactive to proactive.The role of AI models in mobile management is pivotal. AI models encompass various algorithms, like machine learning, which predict device failures, reducing downtime. They also optimize resource allocation, saving costs. Ethically, they require transparency and fairness to prevent bias and protect user privacy

The study focuses on the current landscape and prospects of AI-powered mobile management, future research endeavors in this domain could delve deeper into specific AI models and their practical applications. Exploring the scalability and adaptability of AI-driven solutions across diverse industries and contexts is a promising avenue for future investigation. The ethical considerations surrounding the use of AI in mobile management warrant continued exploration and discussion. Within the scope of the study, a diverse range of AI models and techniques has been examined and evaluated for their suitability in predictive maintenance and fault detection. These models are not only in terms of their technical prowess but also with a keen eye on ethical implications. The significant contributions of study lie in its potential to offer a roadmap for organizations in embracing AI-powered solutions, enhancing device reliability, and reducing operational costs, all while being mindful of ethical responsibilities.

4.4 RESEARCH OBJECTIVES

The study aims to examine the current state of mobile management practices and the challenges faced by organizations in managing and maintaining mobile devices. It seeks to explore the potential benefits of AI-powered predictive maintenance and fault detection techniques in mobile management systems, analyzing the different AI algorithms and techniques and their ethical implications. The paper intends to provide recommendations to organizations seeking to implement AI-powered predictive maintenance and fault detection in their mobile management systems based on best practices and real-world case studies.

4.5 METHODOLOGY

The study employed a combination of primary and secondary sources of data, as well as documentary and analytical methods. It was qualitative in nature, and the secondary sources were selected based on their relevance to the study. The researchers carefully analyzed these sources to identify key themes and patterns, which enabled them to gain a deeper understanding of the research and identify important insights and trends. The rigorous and systematic approach to data collection and analysis involved utilizing a range of primary and secondary sources, as well as documentary and analytical methods. Overall, the study is able to generate valuable insights and contribute to a broader understanding of the research by utilizing thematic patterns and other analytical techniques.

4.6 RESULT AND DISCUSSION

4.6.1 Mobile device management practices and the problems organizations confront

Mobile devices are becoming essential in modern organizations, allowing employees to access important business applications and data from anywhere at any time. Organizations may find monitoring and maintaining these devices to be a tough Endeavor. Sarker et al. (2021) note that current mobile management methods encompass strategies such as MDM, mobile application management (MAM), and mobile content management (MCM). MDM is a centralized method for managing mobile devices by defining settings, enforcing policies, and protecting devices via a MDM platform. MAM specializes in overseeing the apps on mobile devices, including app delivery, upgrades, and security. MCM is the practice of overseeing the content accessed or produced on mobile devices, including tasks like document sharing and collaboration. Nevertheless, organizations have many obstacles in efficiently managing mobile devices while implementing these mobile management techniques. The multiplicity of mobile devices and operating systems poses a significant issue. Tripathy et al. (2015) highlight the challenge organizations face in creating a cohesive mobile management strategy due to the variety of mobile devices with diverse operating systems. An organization could utilize distinct MDM solutions for iOS and Android devices, resulting in discrepancies in policy implementation. Data security concern Li et al. (2017) state that mobile devices are susceptible to security breaches that may jeopardize important company data.

Organizations need to employ strong security measures like device encryption, remote wipe, and mobile threat defense to tackle this challenge. Organizations may face issues due to the swift technology advancements in mobile business. Chen et al. (2014) contend that the rise of new technologies like mobile cloud computing and the IoT has heightened the intricacy of MDM. IoT devices may necessitate distinct management approaches compared to smartphones or tablets, which might complicate mobile management procedures. Organizations face a serious difficulty managing and sustaining mobile devices. Mobile management methods currently encompass several strategies, such as MDM, MAM, and MCM. Organizations encounter issues in efficiently managing mobile devices due to the variety of devices and operating systems, data security concerns, and the fast technical advancements in the mobile business. Organizations need to implement strong mobile management practices and keep current with the newest advancements in mobile technology to overcome these problems.

4.6.2 AI's predictive maintenance and problem detection for cost-effective and efficient mobility management systems

Within the realm of mobile management systems, the application of AI strategies, such as defect detection and predictive maintenance, has the potential to confer a multitude of advantages upon organizations. A significant benefit of AI in mobile management is its capacity to boost productivity while simultaneously lowering expenses. The term "predictive maintenance" refers to a method that employs machine learning algorithms to analyze data collected from sensors installed on mobile devices. This method enables organizations to anticipate when maintenance repairs will be necessary before a failure takes place. By taking preventative measures to solve maintenance concerns, organizations can cut down on downtime, extend the lifespan of devices, and minimize the expenses associated with maintenance. For instance, Chen et al. (2016) conducted a study on a fleet of smartphones and demonstrated that predictive maintenance is helpful in reducing fleet maintenance costs.

According to their findings, predictive maintenance was able to cut maintenance expenses by as much as 35%. An additional application of AI that can assist businesses in identifying and resolving problems with mobile devices before they become significant is called fault detection. Learning algorithms can be trained on past data to recognize patterns of behavior that are indicative of problems or failures. This can be accomplished through the use of machine learning. When organizations are able to detect errors at an early stage, they are able to limit the amount of time that their systems are offline and the fees that are incurred for repairs. According to Yang et al.'s (2020) findings, the use of machine learning algorithms for problem identification resulted in a reduction of the mean time to repair by as much as 30%. Additionally, AI can assist organizations in optimizing their mobile management processes by offering insights on the patterns of device usage, application performance, and user behavior associated with mobile devices. Using data collected from mobile devices, businesses are able to pinpoint areas that require enhancement and put into action solutions that are specifically designed to boost productivity while simultaneously lowering costs. It is possible for organizations to reap enormous benefits from the application of AI techniques in mobile management systems. These techniques include predictive maintenance and defect detection. Organizations are able to better manage their mobile devices and give better service to their employees if they improve their efficiency and reduce their costs.

Table 4.1 reveals the effectiveness of AI within the mobile management ecosystem. AI successfully detects a substantial majority of faults,

Table 4.1 Percentages of faults detected by AI

Category	Percentage (%)
Faults detected by AI	80
Faults not detected by AI	20
Partially detected by AI	5
False alarms by AI	3
Human-verified faults	12

accounting for 80% of identified issues. This high detection rate signifies AI's crucial role in improving efficiency and reducing the workload on human operators. However, it's important to note that AI is not infallible, as evidenced by the 20% of faults that it does not detect. Additionally, in 5% of cases, AI only partially detects issues, necessitating further human intervention for resolution. Furthermore, AI generates false alarms in 3% of instances, highlighting the need for ongoing refinement and optimization of AI algorithms to minimize unnecessary disruptions. In 12% of cases, human verification remains an essential component of the mobile management process, underscoring the significance of human expertise and oversight. AI significantly enhances mobile management by detecting the majority of faults; there is room for improvement in reducing false alarms and increasing the comprehensiveness of detection. A harmonious collaboration between AI and human operators is essential to maintaining the reliability and efficiency of mobile management systems (Figure 4.1).

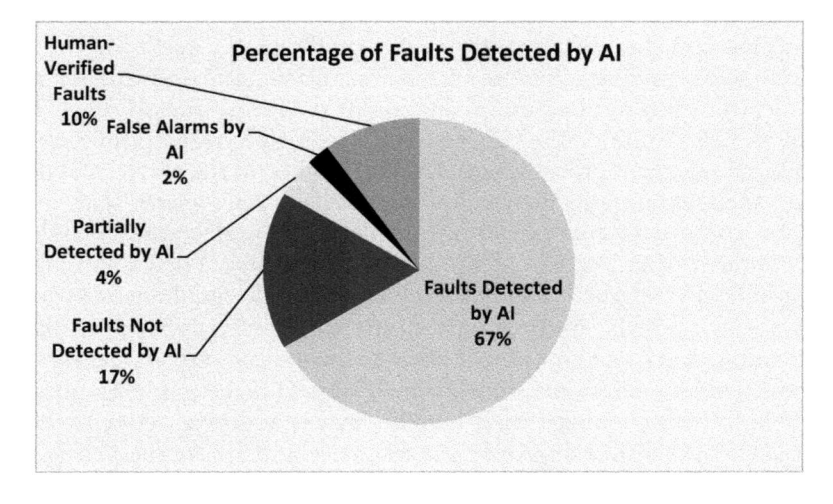

Figure 4.1 Percentage of faults detected by AI.

4.6.3 Privacy, security, and bias issues in mobile management systems with AI

The growing application of AI in mobile management systems has resulted in a number of important benefits; yet, it has also given rise to ethical concerns around privacy, data security, and algorithmic prejudice. Given the sources that are listed below, the ethical implications are as follows. When it comes to the application of AI in mobile management systems, privacy is an extremely important concern. A serious privacy risk is raised by Dabla et al. (2021), who point out that AI has the ability to acquire and analyze personal data without the user's awareness. In addition, the implementation of AI in mobile management systems can result in the development of comprehensive user profiles, which can then be offered for sale to third-party advertising agencies. Users could be put at risk of targeted advertising and other privacy violations as a result of this behavior.

One other ethical concern that arises from the application of AI in mobile management systems is the protection of data (Del Carmen Rodríguez-Hernández & Ilarri 2021). According to McAuliffe et al. (2021), AI systems are susceptible to cyber attacks, which could lead to data breaches and the loss of critical information. Thus, it is of the utmost importance to guarantee that the AI systems utilized in mobile management are safe and comply with the legislation governing data protection.

The application of AI in mobile management systems raises a number of serious ethical concerns, one of which is algorithmic prejudice. Park & Park (2020) express concern that the algorithms that are utilized in AI systems can have a bias that discriminates against particular groups of individuals, such as those based on their gender, race, or religion. It is possible that this problem could result in the unfair treatment of individuals or groups, which will be detrimental to the principles of equality and justice. It has been suggested by Prabowo et al. (2018) that one method for addressing the issue of algorithmic bias is to make use of varied datasets and train algorithms using data that is more inclusive. This strategy has the potential to assist in mitigating the impacts of bias in AI systems. However, it is of the utmost importance to make certain that the data that is utilized in the training of AI models is correct, pertinent, and objective.

AI in mobile management systems has a wide range of ethical considerations, including the impact technology has on employment. According to Munir et al. (2019), the implementation of AI in the workplace may lead to the loss of jobs and a widening of the skills gap. As a result, it is essential to put into effect laws that will guarantee the fair distribution of the advantages brought about by AI while decreasing the adverse effects it will have on employment.

A number of ethical considerations are raised by the implementation of AI in mobile management systems. These considerations include privacy, data security, algorithmic bias, and the influence on employment. It is necessary

Table 4.2 Components of mobile management ecosystem and AI integration

Component	Description
Mobile devices	Smartphones, tablets, IoT devices, and other mobile hardware
Mobile applications	Software applications running on mobile devices
MDM	Software for configuring, securing, and managing mobile devices
Cloud infrastructure	Remote servers and data storage for mobile data and services
AI algorithms and models	Machine learning models and algorithms for AI capabilities
Predictive maintenance	AI-driven system for predicting device maintenance needs
Fault detection	AI algorithms to identify and flag device faults or issues
Data analytics	Analyzing mobile data for insights and performance metrics
User Interface (UI)	Interfaces for users and administrators to interact with the system
Security measures	Measures for securing mobile data and AI systems

to give careful consideration to these ethical issues and to take a proactive strategy in order to reduce the potential adverse effects that they may have. It is imperative that users, policymakers, and technology developers work together to guarantee that AI is utilized in mobile management systems in a responsible and ethical manner at all times (Table 4.2).

Integrating AI into the mobile management ecosystem improves different aspects of MDM and operation. The ecosystem consists of various essential components, each serving a crucial role in supporting the effective and secure operation of mobile devices and applications. Mobile devices and applications play a crucial role as the hardware and software foundations for AI-driven solutions, leading to significant effects (Luo et al. 2022; Mao et al. 2022). MDM software facilitates the setup, protection, and supervision of mobile devices, utilizing AI to automate operations and bolster security protocols.

Cloud infrastructure supports data storage and processing, enabling AI to utilize cloud resources for sophisticated analytics and insights. AI algorithms and models are the foundation of predictive maintenance, defect detection, data analytics, and other essential services in the ecosystem. AI-powered predictive maintenance optimizes device maintenance plans, minimizing downtime and extending device lifespan. Fault detection, an AI application, allows for immediate diagnosis and marking of device problems, making troubleshooting and maintenance faster.

AI-driven data analytics offers useful insights on user behavior, performance indicators, and emerging trends to facilitate informed decision-making. AI advancements in user interfaces provide personalized recommendations and enhanced usability for users and administrators.

Strong security measures are crucial in the mobile management ecosystem, and AI is involved in detecting and addressing security risks. AI integration revolutionizes mobile management through automation, predictive features, and improved efficiency to ensure smooth operation of mobile devices and applications while also enhancing security and optimizing performance.

4.6.4 Challenges for implementing AI in mobile management systems

Implementing AI in mobile management systems presents several challenges.

Limited processing power and memory: Mobile devices have limited processing power and memory compared to desktop computers or servers. Implementing AI algorithms on mobile devices requires careful optimization and balancing of computation and memory resources to ensure that the AI models can run smoothly and without consuming too much battery life.

Data security and privacy: AI algorithms often require access to sensitive data, such as location information, contacts, and user behavior patterns. Ensuring the security and privacy of this data is essential to preventing data breaches and maintaining user trust.

Network connectivity: AI models require data to be fed into them to learn and make predictions. This data often needs to be collected from the cloud or other remote sources, which can be challenging in areas with poor network connectivity.

Complexity: Implementing AI algorithms on mobile devices requires specialized knowledge and skills, which may be difficult to find or expensive to acquire. Additionally, integrating AI models with existing mobile management systems can be complex and time-consuming.

User acceptance: Some users may be skeptical of AI-powered mobile management systems or may be concerned about privacy and data security. Educating users about the benefits and limitations of AI can help to alleviate these concerns and increase user acceptance

Battery life: AI algorithms can be computationally intensive and consume significant amounts of battery life. This can be a significant challenge in mobile management systems where battery life is already a scarce resource.

Real-time processing: Some mobile management systems require real-time processing, such as monitoring and responding to security threats. Implementing AI algorithms in real-time environments can be challenging, as they must process data quickly and accurately to provide timely responses.

Data quality and availability: AI algorithms rely on high-quality data to learn and make accurate predictions. However, data quality and availability can be a challenge in mobile environments, where data may be sparse, noisy, or incomplete.

User experience: Mobile management systems must be user-friendly and provide a positive user experience. Implementing AI algorithms that are transparent, understandable, and useful can be a challenge, especially for non-technical users.

Integration with legacy systems: Mobile management systems often rely on legacy systems that may not be compatible with AI algorithms. Integrating AI with legacy systems can be a significant challenge and may require significant changes to existing systems and processes.

Implementing AI in mobile management systems requires careful planning, consideration of the unique challenges of mobile environments, and a willingness to adapt and evolve existing systems and processes to take advantage of the benefits of AI.

4.6.5 Role of AI models in mobile management

AI models are instrumental in achieving the objective of exploring the potential benefits of AI-powered predictive maintenance and fault detection techniques in mobile management systems (Cheng et al. 2020). These AI models encompass a wide range of algorithms and techniques designed to analyze and interpret data from mobile devices. For instance, machine learning models, such as decision trees, random forests, or neural networks, can be employed to predict when mobile devices are likely to experience failures or faults. By training these models on historical data, they can learn patterns and anomalies that indicate potential issues. This predictive capability is essential for proactive maintenance, which can substantially reduce downtime and operational disruptions for organizations. AI models can facilitate efficient resource allocation (Chien et al. 2019). They can analyze usage patterns and device performance data to determine when and where resources should be allocated for maintenance or replacement. This optimization leads to cost savings and ensures that resources are utilized effectively. A significant challenge in developing predictive maintenance systems is the scarcity of failure data, as machines are often repaired before they break down. Digital Twins offer a real-time digital representation of physical machines, generating valuable data on aspects like asset degradation. This data can be leveraged by predictive maintenance algorithms to improve their accuracy and effectiveness (Van Dinter et al. 2022).

Ethical considerations surrounding AI models are also integral to your research. It's important to assess the ethical implications of using AI in mobile management. For example, ensuring fairness and transparency in decision-making processes, preventing bias in AI predictions, and safeguarding user data privacy are critical aspects. Your study should explore these ethical dimensions, providing insights into how AI models can be deployed responsibly in mobile management systems. AI models serve as the technological backbone of your research, allowing you to analyze data, make predictions, and optimize mobile management processes. They are essential tools in achieving the goal of enhancing device reliability, reducing operational costs, and addressing ethical concerns in the context of MDM.

4.6.6 Recommendations AI-based predictive maintenance and fault detection for mobile management systems based on real-world implementations

The rapid proliferation of mobile devices and the ever-expanding landscape of mobile applications have ushered in an era of unprecedented connectivity and productivity. In this dynamic ecosystem, mobile management has emerged as a linchpin, ensuring the seamless operation of devices and applications critical to both individual users and organizations. As mobile technology continues to evolve, so too does the complexity of managing these systems, necessitating innovative solutions to preemptively identify and mitigate potential issues. Enter AI with its transformative potential. AI-based predictive maintenance and fault detection have emerged as promising pillars in the future of mobile management, offering the tantalizing prospect of proactive issue resolution, improved user experiences, and optimized device performance. As we navigate this terrain, it becomes imperative to glean insights from real-world implementations to chart a course forward. This paper aims to bridge the theoretical and practical realms, offering recommendations rooted in actual experiences with AI-driven solutions for mobile management systems. By delving into successful deployments and lessons learned, we endeavor to provide actionable guidance to industry practitioners, researchers, and stakeholders keen on harnessing AI's prowess to unlock the full potential of mobile management in the digital age.

1. **Leverage advanced AI algorithms:** Utilize advanced AI algorithms and machine learning models that have demonstrated success in real-world scenarios. Consider models that can adapt and self-improve over time as they encounter more data.
2. **Continuous model training:** Implement a continuous model training process to keep AI algorithms up-to-date. Mobile environments are dynamic, and models should evolve to capture changing patterns and trends.
3. **Integration with MDM:** Integrate AI-based predictive maintenance and fault detection seamlessly with MDM systems. This integration allows for automated responses to detected issues, reducing the need for manual intervention.
4. **Prioritize critical faults:** Develop a system that can differentiate between minor issues and critical faults. Prioritize the identification and resolution of critical faults to prevent service disruptions or device failures.

5. **Human-AI collaboration:** Promote a collaborative approach between AI systems and human operators. AI can identify issues, but human expertise is often required for complex problem-solving and decision-making.

6. **User-friendly alerts:** Design user-friendly interfaces and alert systems that provide clear and actionable information to both end-users and administrators when issues are detected. Effective communication is crucial for timely responses.

7. **Data security and privacy:** Implement robust data security and privacy measures to protect the sensitive data collected by AI systems. Compliance with data protection regulations is essential.

8. **Performance monitoring:** Continuously monitor the performance of AI-based predictive maintenance and fault detection systems. Regularly evaluate their accuracy, efficiency, and impact on mobile management.

9. **Benchmarking and case studies:** Include benchmarking metrics and real-world case studies in your recommendations. Showcase successful implementations of AI in mobile management and provide benchmarks for measuring success.

10. **Scalability:** Ensure that the AI-based system is scalable to accommodate growing mobile ecosystems. As the number of devices and users increases, the system should be able to handle the added workload.

11. **Feedback mechanisms:** Establish feedback mechanisms for users and administrators to report false positives or negatives in fault detection. Use this feedback to fine-tune AI algorithms.

12. **Cost-benefit analysis:** Conduct a thorough cost-benefit analysis to assess the economic advantages of implementing AI-based predictive maintenance and fault detection. This analysis should consider the potential (Payne, Peltier, & Barger, 2018) reduction in downtime, maintenance costs, and improvements in user satisfaction.

13. **Establish clear goals and objectives:** It is crucial to identify the key performance indicators (KPIs) that the organization seeks to improve with the implementation of AI-powered predictive maintenance and fault detection. This will help in selecting the appropriate tools and technologies and assessing the effectiveness of the implementation. (Munir, Abedin, and Hong, 2019)

14. **Use appropriate AI models:** Organizations should select appropriate AI models based on the type of data and the nature of the problem they seek to solve. For example, machine learning models can be used to analyze historical data and detect patterns, while deep learning models can be used to identify anomalies and predict future events.

15. **Involve domain experts:** Domain experts, such as maintenance engineers and technicians, should be involved in the design and implementation of the AI-powered predictive maintenance and fault detection

system. Their knowledge and expertise can help in identifying relevant data sources, defining KPIs, and validating the accuracy of the AI models.

16. **Test and validate the system:** Before implementing the AI-powered predictive maintenance and fault detection system, organizations should conduct thorough testing and validation to ensure that the system is accurate and effective. This can involve testing the system on a small scale before scaling it up to the entire organization. (Singh et al., 2022)

17. **Ensure data quality and data security:** Data quality is critical for the effectiveness of any AI-powered predictive maintenance and fault detection system. It is essential to ensure that the data collected is accurate, consistent, and up-to-date. Moreover, data security is also critical, as the system may involve sensitive data related to maintenance and operations. Organizations should implement appropriate data security measures to prevent unauthorized access and ensure data privacy

18. **Develop a roadmap for implementation:** Organizations should develop a roadmap for the implementation of the AI-powered predictive maintenance and fault detection system. The roadmap should include a detailed plan for data collection, AI model selection, testing and validation, and deployment. The roadmap should also include a timeline for implementation and the necessary resources, such as budget and staff, required for the implementation.

19. **Monitor and evaluate the system:** Once the AI-powered predictive maintenance and fault detection system is deployed, it is essential to monitor and evaluate its effectiveness regularly. Organizations should continuously review the KPIs to ensure that the system is achieving the desired results. Additionally, it may be necessary to retrain the AI models and adjust the system based on the results of the monitoring and evaluation. (Wong et al., 2020)

20. **Foster a culture of innovation:** Implementing AI-powered predictive maintenance and fault detection requires a culture of innovation within the organization. Organizations should encourage experimentation and exploration of new technologies and ideas to continuously improve the system's effectiveness. Moreover, the organization should provide employees with the necessary training and support to enable them to embrace the new technology and processes.

With these recommendations, organizations can increase their chances of success in implementing AI-powered predictive maintenance and fault detection systems. Ultimately, the success of the system depends on the organization's commitment to implementing and continuously improving the system's effectiveness (Table 4.3).

Table 4.3 AI-based predictive maintenance and fault detection in mobile management systems

Recommendation	Description
Collect comprehensive data	Gather extensive data on device performance and usage
Leverage advanced AI algorithms	Use sophisticated AI algorithms and models for accuracy
Continuous model training	Implement ongoing training to adapt to changing patterns
Integration with MDM	Seamlessly integrate AI with MDM
Prioritize critical faults	Identify and address critical faults promptly
Human-AI collaboration	Promote collaboration between AI and human operators
User-friendly alerts	Design clear and actionable alert systems for users
Data security and privacy	Ensure robust data protection and compliance with regulations
Performance monitoring	Continuously evaluate the AI system's accuracy and impact
Benchmarking and case studies	Showcase real-world successes and benchmarking metrics
Scalability	Ensure the system can scale with growing mobile ecosystems
Feedback mechanisms	Establish feedback loops for fine-tuning AI algorithms
Cost-benefit analysis	Conduct a thorough analysis of economic advantages

4.6.7 Implications of the study

The study on the future of mobile management with predictive maintenance and fault detection using AI has significant implications for the way organizations approach MDM practices. With mobile devices becoming an increasingly integral part of everyday work for employees, it is critical for organizations to develop cost-effective and efficient mobility management systems that ensure smooth and secure operations. The use of AI in predictive maintenance and problem detection can help organizations tackle the problems they face with MDM, including security concerns, device downtime, and privacy issues.

One of the key implications of this study is that the use of AI-based predictive maintenance and fault detection systems can help organizations reduce the cost of MDM. By using machine learning algorithms to predict device failures and detect issues before they become critical, organizations can reduce device downtime and maintenance costs. With the ability to detect problems early, organizations can avoid costly repairs or replacements, and keep devices running smoothly, saving both time and money. Additionally, by identifying patterns in device usage, AI-based predictive maintenance systems can help organizations optimize device usage, reducing unnecessary costs and improving device lifespan. Another important implication of the study is that AI-based predictive maintenance and problem detection can improve the security of mobile management systems. Mobile devices often contain sensitive information, making them vulnerable to cyberattacks. By detecting security threats in real-time, organizations can take proactive steps to prevent attacks and protect their devices and data. Additionally, AI-based systems can help identify potential security breaches before they occur, providing an additional layer of protection to organizations.

However, the use of AI in mobile management systems raises concerns around privacy and bias. Organizations must be aware of these issues and take steps to address them. For example, privacy concerns can be addressed by developing strict policies around data collection and use and ensuring that employees are aware of these policies. Additionally, bias in AI systems can be minimized by using diverse datasets and testing for potential biases in the algorithms used. To ensure that AI-based predictive maintenance and fault detection systems are effective, organizations should look to real-world implementations for guidance. By examining case studies and success stories, organizations can learn from others who have already implemented these systems and identify best practices. Collaboration with experts in the field can help organizations develop effective strategies for integrating AI into their MDM practices. The study on the future of mobile management with predictive maintenance and fault detection using AI has significant implications for the way organizations approach MDM. By using AI-based systems, organizations can reduce costs, improve security, and optimize device usage. However, organizations must also be aware of privacy, security, and bias concerns and take steps to address them. By examining real-world implementations and collaborating with experts, organizations can develop effective strategies for implementing AI-based predictive maintenance and fault detection systems in their MDM practices.

4.7 CONCLUSION

AI in MDM transforms how organizations manage devices, with predictive maintenance and fault detection offering efficiency gains, but challenges like privacy, security, and bias exist. As mobile device use at work grows, MDM tackles issues like data breaches, productivity loss, and maintenance costs. AI, particularly predictive maintenance and fault detection, emerges as a solution. It can foresee device failures and address issues promptly. Yet, privacy and security risks arise due to data access requirements. Bias is another concern. To harness AI's benefits and mitigate risks, organizations must secure data access, conduct security audits, and train AI to reduce bias. AI-driven predictive maintenance and fault detection hold promise for mobile management but necessitate careful implementation to address privacy, security, and bias issues. The future scope of study in mobile management with AI involves advancing AI models for even more precise predictive maintenance, exploring real-time data integration, and addressing evolving ethical concerns. The research can contribute to further enhancing mobile device reliability, operational efficiency, and ethical AI implementation in the field.

CONFLICT OF INTEREST STATEMENT

The authors affirm that they have no known financial or interpersonal conflicts that would have appeared to have an impact on the research presented in this study.

ACKNOWLEDGEMENT

Anyone who helped us complete this work deserves my sincere thanks and appreciation. We owe everyone a great deal of thanks. The authors and scientists who had previously published articles on a similar issue deserve credit for their inspiration, which allowed us to come up with a fresh strategy for concluding our research.

REFERENCES

Ai, B., Guan, K., Rupp, M., Kurner, T., Cheng, X., Yin, X. F., Wang, Q., Ma, G. Y., Li, Y., Xiong, L., & Ding, J. W. 2015. Future railway services-oriented mobile communications network. *IEEE Communications Magazine*, 53(10), 78–85.

Ayvaz, S., & Alpay, K. 2021. Predictive maintenance system for production lines in manufacturing: A machine learning approach using IoT data in real-time. *Expert Systems with Applications*, 173, 114598.

Chen, J., Lim, C. P., Tan, K. H., Govindan, K., & Kumar, A. 2021. Artificial intelligence-based human-centric decision support framework: An application to predictive maintenance in asset management under pandemic environments. *Annals of Operations Research*, 1–24. https://doi.org/10.1007/s10479-021-04373-w.

Chen, S., Shi, Y., Hu, B., & Ai, M. 2014. Mobility-Driven Networks (MDN): From evolution to visions of mobility management. *IEEE Network*, 28(4), 66–73.

Chen, S., Shi, Y., Hu, B., & Ai, M. 2016. *Mobility Management: Principle, Technology and Applications*. Berlin, Heidelberg: Springer.

Cheng, J. C., Chen, W., Chen, K., & Wang, Q. 2020. Data-driven predictive maintenance planning framework for MEP components based on BIM and IoT using machine learning algorithms. *Automation in Construction*, 112, 103087.

Chien, W. C., Cho, H. H., Lai, C. F., Tseng, F. H., Chao, H. C., Hassan, M. M., & Alelaiwi, A. 2019. Intelligent architecture for mobile HetNet in B5G. *IEEE Network*, 33(3), 34–41.

Chuang, S. Y., Sahoo, N., Lin, H. W., & Chang, Y. H. 2019. Predictive maintenance with sensor data analytics on a Raspberry Pi-based experimental platform. *Sensors*, 19(18), 3884.

Dabla, P. K., Gruson, D., Gouget, B., Bernardini, S., & Homsak, E. 2021. Lessons learned from the COVID-19 pandemic: Emphasizing the emerging role and perspectives from artificial intelligence, mobile health, and digital laboratory medicine. *Ejifcc*, 32(2), 224.

Del Carmen Rodríguez-Hernández, M., & Ilarri, S. 2021. AI-based mobile context-aware recommender systems from an information management perspective: Progress and directions. *Knowledge-Based Systems*, 215, 106740.

Divya, D., Marath, B., & Santosh Kumar, M. B. 2023. Review of fault detection techniques for predictive maintenance. *Journal of Quality in Maintenance Engineering*, 29(2), 420–441.

Lee, S. M., Lee, D., & Kim, Y. S. 2019. The quality management ecosystem for predictive maintenance in the Industry 4.0 era. *International Journal of Quality Innovation*, 5, 1–11.

Lee, J., Singh, J., Azamfar, M., & Pandhare, V. 2020. Industrial AI and predictive analytics for smart manufacturing systems. In: M. Soroush, M. Baldea, & T. F. Edgar (eds.), *Smart Manufacturing* (pp. 213–244). Amsterdam: Elsevier.

Li, R., Zhao, Z., Zhou, X., Ding, G., Chen, Y., Wang, Z., & Zhang, H. 2017. Intelligent 5G: When cellular networks meet artificial intelligence. *IEEE Wireless Communications*, 24(5), 175–183.

Luo, G., Yuan, Q., Li, J., Wang, S., & Yang, F. 2022. Artificial intelligence powered mobile networks: From cognition to decision. *IEEE Network*, 36(3), 136–144.

Mao, Y., Pranolo, A., Hernandez, L., Wibawa, A. P., & Nuryana, Z. 2022. Artificial intelligence in mobile communication: A survey. *IOP Conference Series: Materials Science and Engineering*, 1212(1), 012046.

McAuliffe, M., Blower, J., & Beduschi, A. 2021. Digitalization and artificial intelligence in migration and mobility: Transnational implications of the COVID-19 pandemic. *Societies*, 11(4), 135.

Munir, M. S., Abedin, S. F., & Hong, C. S. 2019. Artificial intelligence-based service aggregation for mobile-agent in edge computing. In: *2019 20th Asia-Pacific Network Operations and Management Symposium* (APNOMS) (pp. 1–6). Matsue: IEEE.

Park, J. S., & Park, J. H. 2020. Future trends of IoT, 5G mobile networks, and AI: Challenges, opportunities, and solutions. *Journal of Information Processing Systems*, 16(4), 743–749.

Payne, E. M., Peltier, J. W., & Barger, V. A. 2018. Mobile banking and AI-enabled mobile banking: The differential effects of technological and non-technological factors on digital natives' perceptions and behavior. *Journal of Research in Interactive Marketing*, 12(3), 328–346.

Prabowo, H., Cenggoro, T. W., Budiarto, A., Perbangsa, A. S., Muljo, H. H., & Pardamean, B. 2018. Utilizing mobile-based deep learning model for managing video in knowledge management system. *IJIM*, 12(6), 1–12.

Sarker, I. H., Hoque, M. M., Uddin, M. K., & Alsanoosy, T. 2021. Mobile data science and intelligent apps: concepts, AI-based modeling and research directions. *Mobile Networks and Applications*, 26, 285–303.

Singh, A., Satapathy, S. C., Roy, A., & Gutub, A. 2022. Ai-based mobile edge computing for iot: Applications, challenges, and future scope. *Arabian Journal for Science and Engineering*, 47(8), 9801–9831.

Su, C. J., & Yon, J. A. Q. 2018. Big data preventive maintenance for hard disk failure detection. *International Journal of Information and Education Technology*, 8(7), 471–481.

Tripathy, A. K., Carvalho, R., Pawaskar, K., Yadav, S., & Yadav, V. 2015. Mobile based healthcare management using artificial intelligence. In: *2015 International Conference on Technologies for Sustainable Development (ICTSD)* (pp. 1–6). Mumbai: IEEE.

Van Dinter, R., Tekinerdogan, B., & Catal, C. 2022. Predictive maintenance using digital twins: A systematic literature review. *Information and Software Technology*, 151, 107008.

Wellsandt, S., Klein, K., Hribernik, K., Lewandowski, M., Bousdekis, A., Mentzas, G., & Thoben, K. D. 2022. Hybrid-augmented intelligence in predictive maintenance with digital intelligent assistants. *Annual Reviews in Control*, 53, 382–390.

Wong, C. K., Ho, D. T. Y., Tam, A. R., Zhou, M., Lau, Y. M., Tang, M. O. Y., Tong, R.C.F., Rajput, K.S., Chen, G., Chan, S.C., Siu, C.W., & Hung, I. F. N. 2020. Artificial intelligence mobile health platform for early detection of COVID-19 in quarantine subjects using a wearable biosensor: Protocol for a randomized controlled trial. *BMJ Open*, 10(7), e038555.

Yang, H., Alphones, A., Xiong, Z., Niyato, D., Zhao, J., & Wu, K. 2020. Artificial-intelligence-enabled intelligent 6G networks. *IEEE Network*, 34(6), 272–280

Zheng, H., Paiva, A. R., & Gurciullo, C. S. 2020. Advancing from predictive maintenance to intelligent maintenance with AI and IoT. *arXiv preprint arXiv*:2009.0035

Zonta, T., Da Costa, C. A., da Rosa Righi, R., de Lima, M. J., da Trindade, E. S., & Li, G. P. 2020. Predictive maintenance in the Industry 4.0: A systematic literature review. *Computers & Industrial Engineering*, 150, 106889.

Chapter 5

Revolutionizing education
AI in next-generation mobile management

Akindeji Ibrahim Makinde, Samuel Adedeji Adeleye, Adewale Omotolani Oronti, and Ibraheem Temitope Jimoh

5.1 INTRODUCTION: BACKGROUND AND DRIVING FORCES

Education is the cornerstone of human development, and its evolution is essential for the progress of society [1]. Due to the rapid advancement of technology, the integration of AI in education has become a revolutionary force, shaping the way of learning and teaching. The fusion of AI and mobile technology has the capacity to create more personalized and efficient learning experiences, making education more accessible and engaging for learners worldwide [2].

Education cannot be universally applied, as each student embarks on a distinct learning path. To address this diversity and maximize student potential, personalized learning experiences have become a cornerstone of modern education. By tailoring education to individual needs and preferences, we can enhance student engagement, understanding, and outcomes. In today's fast-paced world, educational institutions face increasing demands for administrative efficiency to meet the ever-growing needs of students and staff. One way to achieve this is through the automation of administrative tasks [3].

The significance of artificial intelligence (AI) has extended across numerous domains, with education emerging as a pivotal sphere where AI's influence is steadily growing [4]. Within the educational realm, AI plays a paramount role in addressing fundamental challenges, particularly in terms of enhancing access and elevating the quality of education. The integration of AI into education serves as a means to close the accessibility gap while simultaneously furnishing students with a tailored, effective, and captivating learning journey [4].

AI has arisen as a revolutionary influence in education, fundamentally altering how students acquire knowledge and how educators deliver instruction. The integration of AI-driven adaptivity and personalization in education is revolutionizing conventional classroom environments, facilitating customized learning experiences that cater to the unique needs and preferences of each learner.

AI has the potential to change the face of education by enhancing student learning outcomes, supporting teachers, and giving students more tailored learning alternatives. Teachers must be given the tools and information they need to integrate this new technology in the classroom and improve and streamline daily operations [5].

AI is rapidly becoming an integral part of our everyday routines, thanks to the introduction of innovative features like Canva's magic tools, Google Bard, and ChatGPT. This shift is transforming various industries and impacting how we work, learn, and communicate [6]. The acceleration of technological advancements underscores the necessity of incorporating AI education into the curriculum, ensuring that all students are equipped not only for their educational endeavors but also for the expanding workforce landscape. Personalized learning, gamification, education adaptive learning, multilingual education, and immersive learning are a few areas in which AI can revolutionize the classroom. Table 5.1 shows the statistical highlights of utilization of AI in education.

Mobile management is critical in education by providing tools and strategies to effectively integrate mobile devices into the learning process [13]. Mobile management encompasses several essential functions in education. It involves the provisioning and management of mobile devices used by students and educators, encompassing tasks such as setting up devices, configuring them for educational use, and ensuring proper maintenance and updates. Additionally, it safeguards the security and privacy of sensitive student and school data through features like device encryption, remote lock and wipe, and the enforcement of security policies against data breaches and cyber threats. It enables content filtering to implement web access policies, ensuring that students are shielded from inappropriate content while using mobile devices for educational purposes. Schools can also utilize mobile management for app curation and management, allowing students access to relevant and approved applications that enhance their learning experiences. Moreover, it establishes secure user authentication methods, ensuring authorized access to educational resources and sensitive data by students, teachers, and staff. Mobile management also plays a pivotal role in supporting remote learning, facilitating the distribution of digital learning materials, managing virtual classrooms, and enabling student access to online resources from their mobile devices.

This chapter offers a thorough review of how AI in next-generation mobile management systems might change educational practices, boost learning results, raise administrative effectiveness, and widen access to high-quality education. It highlights how AI has the ability to transform education for the betterment of students, educators, and institutions by producing a more adaptive, personalized, and data-informed educational environment.

The arrangement of this chapter follows a coherent sequence aimed at offering a thorough grasp of the subject matter. The literature review

Table 5.1 Statistical highlights of utilization of AI in education

Id	Author	Observations	Research work
1	Hwang et al. [7]	AI reduced students' anxiety by 20% while improving their grades by 30%	To aid fifth graders in acquiring the mathematical principle of division and multiplication, an intelligent tutoring system was proposed
2	Ivy Tech [8]	34,712 students are prevented from failing by AI's 80% accuracy in predicting a student's final grade	Ivy Tech creates a machine learning algorithm to spot at-risk students and offer prompt assistance
3	National Technical Institute for the Deaf (NTID), Rachester Institute of Technology [9]	With AI-assisted education, 95% of graduates from NTID were employed successfully	Students with visual and auditory disabilities were given access to instructional materials more easily thanks to text-to-speech and speech-to-text technology
4	Standford University— Human-Centered AI [10]	Chatbots with AI provided students with personalized learning advice/support with 91% accuracy	The research uses AI to understand why students are struggling
5	EDUCAUSE. Horizon Report [11]	There will be a 43% increase in AI applications from 2018 to 2022	Learning Initiative and The New Media Consortium
6	Digital Innovation and Transformation Department, Havard Business School [12]	Test scores improved by 62% using adaptive learning	Knewton personalizes learning with the power of AI. Students received personalized feedback and instruction from the program, which was designed to cater to their unique learning requirements and capacities

chapter is thoughtfully organized to provide a comprehensive understanding of the symbiotic relationship between AI and mobile management in the realm of education. The chapter further introduces the concept of "AI and mobile management: a powerful combination." This initial section sets the stage for the exploration of how AI and mobile technology have converged to redefine the educational landscape. The chapter then delves deeper into the topic in the section titled "Deep dive into AI applications in education." to scrutinize various applications of AI in education, emphasizing its role in personalized learning. The chapter further investigates "AI and enhanced personalized learning in education," shedding light on AI's transformative potential to adapt and tailor education to individual learners.

Moving beyond the pedagogical sphere, the chapter examines the "AI in administrative tasks" section, which explores how AI streamlines administrative functions, enhancing resource allocation and decision-making processes and then assess the "Impact of these AI applications on classroom management," focusing on the practical implications for educators and institutions. Finally, the chapter ended by gazing into the horizon with "The future of AI in mobile management." In this section, the evolving landscape and the potential for AI to continue shaping the future of education through mobile technology were considered. This meticulously organized chapter offers a comprehensive overview of AI's integration into mobile management in education, highlighting its profound implications and promising innovations.

5.2 LITERATURE REVIEW

AI's integration into education and mobile learning has sparked considerable interest and research in recent years. One prominent theme in the literature is the concept of personalized learning. AI algorithms have been used to analyze student data and tailor educational content to individual needs and learning styles. Research by Abbas et al. [14] highlights how this personalization enhances student engagement and performance. By adapting the curriculum in real time, AI-powered systems can cater to students' strengths and weaknesses, making learning a more dynamic and efficient process.

AI-driven assessment tools have gained recognition for their ability to provide immediate feedback to students. They adapt assessments based on a student's performance, assess deeper understanding, and reduce test anxiety. Research by Challis [15] underscores the advantages of such assessments. They not only offer more accurate evaluations but also foster a more supportive learning environment by tailoring assessments to each student's proficiency level. Accessibility and inclusivity are key concerns in modern education, and AI plays a pivotal role in addressing these challenges. This highlights the development of AI-driven applications, such as speech recognition and text-to-speech technologies, which improve accessibility for students with disabilities [16]. These tools ensure that learning materials are more inclusive and can be accessed by a wider range of students, promoting equitable education.

Efforts to enhance classroom management and administrative tasks through AI are also prevalent. Research by Alam [17] discusses AI-based mobile management solutions that streamline administrative responsibilities for educators. These solutions automate grading, communication, and other routine tasks, allowing teachers to allocate more time and attention to instruction. This efficiency in administrative processes contributes to a more effective educational environment.

Effective integration of AI tools into teaching practices necessitates proper training for educators [18]. Teacher training programs equip educators with the skills and knowledge needed to use AI effectively in the classroom, enabling them to harness the full potential of these technologies for improved learning outcomes.

As AI continues to advance, ongoing research and thoughtful implementation will play a pivotal role in shaping the future of education.

5.2.1 Adaptive learning and personalized learning

Adaptive learning and personalized learning are two interrelated educational approaches that have garnered significant attention with both facilitating the enhancement of student outcomes through tailored instruction. Adaptive learning refers to the use of technology, particularly AI and data analytics, to customize the learning experience for individual students. This approach involves continuously assessing a student's performance and adjusting the content, pace, and difficulty level of materials to match their specific needs and learning progress [19]. Research by Mahesa [20] highlights the benefits of adaptive learning systems in improving student engagement, motivation, and knowledge retention. These systems can identify areas of difficulty for students and provide targeted support, making learning more efficient and effective.

Personalized learning, on the other hand, stresses modifying training to match each student's own requirements, interests, and talents. It goes beyond technology-driven adaptations and encompasses pedagogical approaches that foster learner agency and choice in their educational journey [21]. Studies such as those by Shahabadi and Uplane [22] emphasize the importance of considering individual differences in learning styles and preferences. Personalized learning encourages educators to develop a deep understanding of their students, enabling them to provide more relevant and engaging learning experiences.

However, challenges related to the implementation of adaptive and personalized learning approaches are prevalent. These challenges include concerns such as teacher training and professional development, and the potential for technology to perpetuate educational disparities [18]. The chapter calls for a balanced approach, where AI complements the expertise of educators, allowing them to make informed decisions about adapting instruction to meet the unique needs of each student.

5.2.2 Chatbots in education

Chatbots have emerged as a significant topic of interest in the realm of education, offering the potential to enhance learning experiences, provide support, and streamline administrative tasks. The chatbots for education highlight several key themes and findings. One prominent application of chatbots in education is in providing immediate and personalized support to students. Research by Hasan et al. [23] demonstrates how chatbots can answer students' questions, provide explanations, and offer assistance with coursework. These chatbots often employ natural language processing (NLP) algorithms to understand and respond to students in conversational language, creating a more engaging and accessible learning environment.

Additionally, chatbots play a crucial role in administrative tasks within educational institutions. Studies such as those by Majumder and Mondal [24] emphasize the use of chatbots for automating routine tasks like course registration, scheduling, and administrative inquiries. This automation not only saves time for administrative staff but also provides students with quick and efficient solutions to their queries. Chatbots must be carefully crafted to ensure accurate responses and meaningful interactions [25]. Ethical considerations regarding data privacy and security are paramount, particularly when chatbots collect and store sensitive student information.

5.2.3 Administrative tasks in education

The automation of administrative tasks in education has gained significant attention due to its potential to streamline operations, reduce workload, and improve overall efficiency for educators and administrators.

One prominent theme is the application of automation to routine administrative tasks. Studies by Rjeib [26] highlight how automation tools can handle tasks such as grading, attendance tracking, and report generation. This automation not only saves educators' valuable time but also reduces the likelihood of errors associated with manual data entry and management. By automating administrative duties, educators can allocate more time and energy to instructional activities, ultimately enhancing the quality of education. Educators and administrators can leverage automation to collect, process, and analyze large volumes of educational data. Research by Park and Datnow [27] underscores how automation tools can generate valuable insights from student performance data, facilitating data-driven decision-making which enables educators to identify areas for improvement, tailor instruction to student needs, and assess the effectiveness of educational programs.

Another key aspect is the integration of automation with communication and collaboration tools. Automation can simplify communication processes, such as sending reminders, notifications, and updates to students and parents [28]. The automation of scheduling and resource allocation can optimize classroom usage and resource distribution within educational institutions [29]. These integrations enhance overall efficiency, reduce administrative burdens, and contribute to a smoother educational operation.

5.3 AI AND MOBILE MANAGEMENT: A POWERFUL COMBINATION

The integration of AI into mobile device management (MDM) and mobile application management (MAM) systems has the potential to significantly enhance the efficiency, security, and user experience of mobile management in educational institutions and beyond. AI can be integrated into mobile management through the following means:

- **Predictive analytics:** AI analyzes historical data and usage patterns to predict potential issues with mobile devices or applications. It anticipates when a device is likely to run out of storage or when an application is prone to crash, allowing administrators to proactively address these issues before they impact users.
- **Security and threat detection:** AI-powered security solutions identify and respond to security threats in real time. This includes detecting abnormal behavior on devices, such as unauthorized access or unusual data transfers, and taking automated actions to mitigate risks, such as locking or wiping a compromised device.
- **User behavior analysis:** AI analyzes how users interact with mobile devices and applications, helping administrators understand which resources are most commonly accessed and which features are underutilized.
- **Automation of routine tasks:** AI automates routine mobile management tasks such as software updates, application deployments, and user provisioning which reduces the administrative burden and ensures that these tasks are performed consistently and without errors.
- **Personalized learning:** In educational settings, AI helps personalize the learning experience by analyzing individual student performance and recommending educational content or applications tailored to each student's needs and abilities.
- **Chatbots and virtual assistants:** They provide immediate support to users, answering common questions, assisting with troubleshooting, and guiding users through device setup or application usage.
- **NLP:** NLP technologies are integrated into mobile management systems to facilitate voice commands and interactions, making it easier for users to control and manage their devices using natural language.
- **Cost optimization:** AI helps optimize costs associated with MDM by analyzing usage patterns and recommending cost-effective plans and resource allocation strategies.
- **Anomaly detection:** AI algorithms detect anomalies in device behavior or usage patterns, which can be indicative of security breaches or technical issues.

- **Accessibility features:** AI enhances accessibility features for users with disabilities. For example, it can improve the accuracy of speech recognition for voice commands and provide better support for screen readers.
- **Resource allocation:** AI dynamically allocates resources, such as network bandwidth or processing power, to devices and applications based on real-time demand.
- **Data insights:** AI provides valuable insights from data collected by mobile management systems, helping educational institutions make data-driven decisions about curriculum, resource allocation, and educational strategies.

5.3.1 Overview of existing AI tools and applications in mobile management

There are several AI tools and applications integrated into mobile management, and these tools and applications helped streamline and enhance various aspects of mobile management.

i. AI-driven security solutions encompass AI-powered threat detection, employing machine learning to identify and respond to security threats such as malware and phishing attacks on mobile devices. Behavioral analytics systems analyze user behavior to pinpoint abnormal activities and potential security breaches. In the realm of predictive maintenance, AI-based predictive analytics are utilized to forecast when mobile devices might encounter hardware or software issues, allowing for proactive maintenance.

ii. Automation and optimization introduce AI-driven automation tools that handle routine mobile management tasks such as software updates, patch management, and device provisioning. Also, AI-driven resource optimization applications dynamically allocate network bandwidth, storage, and processing power based on real-time demand.

iii. User behavior analysis involves user analytics tools that employ AI to scrutinize how users interact with mobile devices and applications, aiding administrators in making data-driven decisions regarding resource allocation and application development.

iv. Chatbots and virtual assistants play a crucial role in providing immediate support to users, answering inquiries, and assisting in troubleshooting mobile device issues. Device performance optimization is facilitated by AI-based performance management applications, which optimize device settings to enhance battery life, speed, and responsiveness.

v. In the education sector, AI-driven personalized learning platforms analyze student performance data and recommend tailored educational content or applications to enhance the learning experience. Anomaly detection, powered by AI, detects unusual device behavior or usage patterns, alerting administrators to potential security breaches or technical issues.

vi. Cost optimization is achieved through AI-driven solutions that analyze mobile usage patterns, offering recommendations for cost-effective data plans and resource allocation strategies. AI-driven analytics platforms, under data insights and reporting, extract valuable insights from data collected by mobile management systems, aiding organizations in making data-driven decisions.

vii. NLP enables voice commands through tools that integrate NLP, allowing users to control and manage mobile devices using natural language commands.

5.4 DEEP DIVE INTO AI APPLICATIONS IN EDUCATION

AI-driven educational apps are gaining popularity due to their effectiveness in enhancing learning experiences. These apps leverage AI to personalize learning, provide feedback, and adapt to individual student needs. Examples of AI-driven educational application are as follows:

i. **Khan Academy:** Khan Academy provides an extensive array of educational materials, encompassing videos and practice exercises, covering various subjects. Its AI algorithms assess a student's performance and adapt the content to address their strengths and weaknesses.

ii. **Coursera:** Coursera serves as an online learning platform that provides courses from prestigious universities and institutions. It uses AI to recommend courses based on a user's interests and background and provides personalized feedback on assignments.

iii. **ScribeSense:** ScribeSense is an AI-powered tool that assists students with note-taking. It can transcribe handwritten notes, making them searchable and more accessible for review and study.

iv. **Brainly:** Brainly is a social learning platform that connects students with their peers and educators. Its AI-driven system helps students find answers to their questions and provides explanations for various subjects.

v. **DreamBox:** DreamBox is an adaptive math learning platform for students in kindergarten through eighth grade. It uses AI to adapt math lessons to each student's skill level and learning style.

vi. **Smartick:** Smartick is an AI-driven math program for children. It adapts to a student's math level and offers personalized lessons and exercises to improve math skills.

vii. **Adaptive learning systems:** Adaptive learning systems like Knewton create personalized learning paths for students based on their performance and goals. They continuously adjust the difficulty of questions and content to optimize learning outcomes.

viii. **Quizlet:** Quizlet is a study app that uses AI to generate flashcards, quizzes, and study materials for various subjects. It adapts to a student's progress and focuses on areas that need improvement.

ix. **Squirrel AI:** Squirrel AI is an AI-driven after-school tutoring program in China. It provides personalized tutoring in subjects like math and English, adapting content and pacing to each student's abilities.

x. **EdTech platforms with AI integration:** Educational technology platforms like Edmodo and Google Classroom are integrating AI to assist teachers in managing classrooms, grading assignments, and providing personalized learning recommendations to students.

5.4.1 Impact of AI-driven educational applications on learning outcomes

AI-driven educational apps have had a significant and positive impact on learning outcomes in various ways. These apps leverage AI to personalize and enhance the learning experience, making education more accessible, engaging, and effective.

- **Personalized learning:** AI-powered apps analyze each student's strengths, weaknesses, and learning styles. They adapt the content and pace of instruction to meet individual needs.
- **Immediate feedback:** AI apps provide instant feedback on assignments and quizzes. It encourages a growth mindset and a deeper understanding of the subject matter.
- **Targeted remediation:** When students struggle with specific concepts, AI apps can identify these areas and offer targeted remediation.
- **Accessibility and inclusivity:** AI apps can adapt content to suit students with diverse learning needs, including those with disabilities. Features like text to speech, voice recognition, and adjustable fonts make learning more accessible to a broader range of learners.
- **Engagement and motivation:** AI-driven apps often incorporate gamification elements and interactive features that make learning more engaging. Achievements, rewards, and progress tracking motivate students to stay committed to their studies and set goals for improvement.
- **Data-driven insights:** AI apps generate valuable data on student performance and behavior. Educators can use these data to identify trends, assess the effectiveness of teaching methods, and make data-driven decisions to improve instruction.

- **Self-directed learning:** AI-powered apps encourage self-directed learning by allowing students to explore topics of interest beyond the standard curriculum.
- **Efficiency in assessment:** AI can automate the grading of assignments and assessments, saving teachers' time and enabling more timely feedback.
- **Flexibility in learning:** AI apps can be accessed anytime and anywhere, providing flexibility for students to learn at their own convenience.
- **Scalability:** AI-driven educational apps can reach a broad audience, making quality education accessible to learners globally.
- **Continuous improvement:** AI algorithms continuously refine and improve the learning experience based on user data and feedback. As the app gathers more data, it becomes more effective at tailoring content and recommendations to individual learners.
- **Closing achievement gaps:** AI-driven apps have the potential to help close achievement gaps by providing additional support.

5.4.2 AI and enhanced personalized learning in education

Personalized education stands as an instructional approach tailored to the unique needs, preferences, and aspirations of each individual student, enabling them to progress at their own pace, delve into their interests, and nurture their inherent talents [30]. Nevertheless, the implementation of personalized learning can present certain challenges, including the development of effective curricula, the delivery of timely feedback, and the support of a diverse array of learners. This is precisely where AI can assume a pivotal role. AI, a field of computer science dedicated to creating systems capable of emulating human intelligence, encompassing reasoning, learning, and problem-solving abilities, holds promise in augmenting personalized learning within the realm of education.

- **Adaptive learning systems**

  Adaptive learning systems are one of the most common and widely used AI applications in personalized education [31]. In response to the learner's performance, behavior, and preferences, these software applications use algorithms and data to dynamically adapt the content's difficulty, pace, and instructional style. For instance, based on a learner's prior knowledge, competences, and academic goals, an adaptive learning system can suggest the learning materials, exercises, and assessments that are most appropriate for them. Additionally, it can offer learners immediate, tailored feedback, tips, and support to help them overcome obstacles and improve their academic performance [31]. These adaptive systems also keep watch over and track the learner's progress and accomplishments, producing thorough reports and insights for the learners' advantage.

- **Intelligent tutoring systems**

  Intelligent tutoring systems are yet another example of AI in the field of individualized education [32]. These are computer programs created to play the part of a human tutor, providing one-on-one or small-group direction, instruction, and dialog with students. For instance, an intelligent tutoring system has the ability to determine the learner's areas of proficiency and those that need improvement, allowing it to then tailor the teaching strategy. By leveraging speech recognition and NLP technology, it can also immerse the learner in interactive learning scenarios. Through the use of gamification and affective computing, it can also adjust to the emotional and motivational states of the learner [33]. In order to create a thorough and encouraging learning environment, these intelligent tutoring systems are also skilled at interacting with other intelligent entities, such as peers, mentors, or subject matter experts.

- **Learning analytics and recommender systems**

  Learning analytics and recommender systems are a third way that AI is being used to enhance personalized learning [34]. These are computer programs that analyze and understand substantial volumes of data produced by students and learning activities using data mining and machine learning. A learning analytics system, for instance, can find patterns, trends, and correlations in the data and offer perceptions and forecasts about the learner's actions, performance, and results. Also, it can spot problems, chances for growth, and gaps while giving students and teachers' feedback and recommendations. Based on each learner's interests, requirements, and goals, a recommender system can use the data to suggest the most pertinent and practical resources, tools, and tactics. Additionally, it can assist students in finding fresh and interesting sources of data and knowledge.

- **Educational chatbots and virtual assistants**

  The use of educational chatbots and virtual assistants in personalized learning represents a fourth instance of AI's integration. These computer programs make use of natural language production and processing to interact verbally or in writing with students. An educational chatbot, for instance, is set up to react to queries, provide information, and offer help to students across a range of courses and domains [35]. It can also start amiable, casual interactions with students that feature comedy, empathy, and unique personalities. The role of an educational virtual assistant, which includes duties like scheduling, planning, organizing, and providing reminders, is to support students in managing their learning-related activities.

- **Augmented and virtual reality**

  The fields of augmented and virtual reality are where AI can be used in personalized learning. These technical advancements create realistic, interactive representations of real or imagined environments using computer graphics and sensors. An augmented reality can overlay digital content and objects over the real world, enhancing the learner's perspective and overall experience. On the other hand, virtual reality can take the learner to a whole different place and time, creating a powerful sense of presence and immersion. These two innovations can both use AI to modify the simulation in response to the learner's choices, actions, and input. They can use AI to create realistic, flexible characters; situations; and interactions, producing engaging learning experiences that engage and test the learner.

5.5 AI IN ADMINISTRATIVE TASKS

AI is revolutionizing various administrative tasks across industries. For data entry and management, AI-powered optical character recognition (OCR) systems convert physical documents into digital formats and classify data, improving accessibility and organization. In appointment scheduling, AI-driven chatbots and virtual assistants streamline scheduling, manage calendars, and send reminders. AI aids email and communication by categorizing and prioritizing emails, reducing spam, and offering automated responses. Document management benefits from AI's automatic sorting, indexing, and filing capabilities, while expense management systems automatically scan and categorize receipts, ensuring accurate expense tracking.

AI plays a crucial role in data analysis and reporting, providing insights and forecasting trends. In support, AI chatbots handle routine inquiries and offer 24/7 support. HR and recruitment processes benefit from AI's resume screening and interview automation. Workflow automation tools use AI to route tasks efficiently, and AI enhances security and compliance efforts by monitoring and ensuring adherence to regulations.

Language translation tools enable global communication, while AI provides decision support by analyzing data and suggesting strategies. Quality assurance and inventory management benefit from AI's defect inspection and demand prediction. Predictive maintenance uses AI to forecast equipment failures and recommend maintenance schedules, reducing downtime and costs. AI's integration into administrative tasks enhances efficiency, accuracy, and decision-making across various domains.

5.5.1 AI-driven analytics for performance tracking

AI-driven analytics has revolutionized the way institution tracks and evaluates performance. By harnessing the power of AI and machine learning algorithms, education sectors can gain deeper insights, make data-driven decisions, and optimize their operations.

One key area where AI-driven analytics has made a significant impact is performance tracking. Traditional methods of performance tracking often rely on historical data and manual analysis, which can be time-consuming and may not capture complex patterns or subtle trends. AI-driven analytics, on the other hand, can process vast amounts of data in real time, identifying correlations and patterns that may not be apparent through conventional methods. This enables organizations to monitor their performance more effectively and respond to changes in the market or operational processes promptly.

In the realm of financial performance, AI-driven analytics can help financial assess risk, detect fraud, and optimize investment strategies. These analytics tools can analyze historical financial data, administrative conditions, and academic performance indicators to make predictions and recommendations. Student performance tracking is another area where AI analytics shines. Institutions can use AI to analyze students/staff behavior, preferences, and feedback to tailor policies and strategies to improve experiences. AI-driven analytics is pivotal in logistics management. It enables real-time monitoring, demand forecasting, and optimization which helps minimize operational costs, reduce waste, and ensure timely deliveries, all of which are critical for maintaining a competitive edge in today's global marketplace.

5.5.2 AI for attendance tracking

The use of AI for attendance tracking has become increasingly prevalent across educational institutions and workplaces, offering a more efficient and accurate method for monitoring attendance.

In educational settings, AI-driven attendance tracking systems have streamlined the process for both teachers and students. These systems often rely on facial recognition technology, where cameras capture students' faces as they enter the classroom, and AI algorithms identify and record their presence. This eliminates the need for manual roll calls, which can be time-consuming and prone to errors. It provides real-time attendance data, allowing educators to quickly identify absentees and take necessary actions to support their learning. AI also enhances security by ensuring that only authorized individuals gain access to educational facilities.

In the administrative environments, AI attendance tracking systems are used to monitor staff attendance and punctuality. These systems can use a combination of facial recognition, biometric authentication, or even geolocation data from employees' mobile devices to verify their presence. These data are integrated with HR and payroll systems, streamlining payroll processing and improving workforce management. Additionally, AI help organizations enforce attendance policies more consistently and reduce the likelihood of time theft or buddy punching.

AI-powered attendance tracking is not limited to physical presence; it also be applied to virtual or remote work arrangements. AI algorithms monitor online activity, such as logins, task completion, or system interactions, to track their virtual attendance and productivity. This is particularly valuable in the era of remote work, allowing students and staff to ensure that they are fulfilling their work obligations and meeting deadlines. AI-driven attendance tracking systems often come with additional features like notifications and alerts. They can send automated messages to remind students or employees of upcoming classes, meetings, or deadlines, reducing the chances of missed commitments.

5.6 IMPACT OF THESE AI APPLICATIONS ON CLASSROOM MANAGEMENT

AI applications have brought about a significant and transformative impact on classroom management. These innovations have not only streamlined administrative tasks but have also fundamentally altered the way educators interact with students and the learning environment.

- AI has greatly improved student engagement in the classroom. Gamification platforms and AI-driven educational games make learning more interactive and enjoyable. By incorporating elements of competition, rewards, and challenges, these applications motivate students to actively participate in their education
- AI's capability for personalized learning has revolutionized how teachers cater to the diverse needs of students. AI algorithms can analyze individual student performance and adapt the curriculum accordingly. This means that struggling students receive additional support and practice, while advanced learners can explore more challenging materials. Such tailored approaches not only boost student achievement but also help teachers better address the specific needs of each student.
- AI-driven applications contribute significantly to behavior management in the classroom. For example, gamified systems reward students for positive behavior and participation while addressing disruptive conduct within the game context. This not only reduces classroom disruptions but also encourages students to develop positive behavioral patterns, fostering a more conducive learning environment.
- Another essential aspect is the efficiency and time-saving that AI brings to classroom management. AI automates routine administrative tasks, such as grading assignments, managing attendance, and organizing schedules. This automation frees up educators' valuable time, allowing them to focus on teaching and providing individualized support to students, ultimately enhancing the overall quality of education.

- AI also offers data-driven insights that empower both teachers and administrators. By collecting and analyzing data on student performance and behavior, educators identify at-risk students early and intervene appropriately. Educational institutions utilize these data to make well-informed choices regarding curriculum modifications, allocation of resources, and opportunities for professional development for educators.
- AI fosters accessibility in education. Students with diverse learning needs benefit from AI-driven accessibility features such as real-time closed captions for students with hearing impairments or assistive technology for those with learning disabilities. These tools ensure that all students, regardless of their individual challenges, have equitable access to educational content.
- In the wake of the COVID-19 pandemic, AI applications have played a pivotal role in remote learning and hybrid classrooms. These tools enable educators to effectively manage virtual classrooms, monitor student engagement, and provide remote support, ensuring that learning can continue seamlessly even outside the physical classroom.
- AI facilitates parent–teacher communication by providing real-time updates on student progress and behavior. This transparency fosters stronger partnerships between educators and families, enhancing the support system for students both inside and outside the classroom.

5.6.1 Case study 1: AI-driven apps for learning—Duolingo

In the realm of education, AI-driven apps for learning have emerged as powerful tools that enhance the learning experience and adapt to the unique needs of individual students. A notable case study in this domain is the success of Duolingo, a popular language learning app. Duolingo employs AI algorithms to create personalized language courses for users, taking into account their language proficiency, learning pace, and areas that require improvement. Through a combination of gamification elements, adaptive quizzes, and interactive exercises, Duolingo's AI-powered app keeps learners engaged and motivated, ultimately leading to more effective language acquisition. The AI-driven app also analyzes user performance data to tailor lessons, providing additional practice in areas where learners struggle and advancing them quickly through areas of proficiency. Duolingo's remarkable success demonstrates the potential of AI-driven apps to revolutionize education by making it more accessible, engaging, and personalized for learners worldwide. This case study exemplifies how AI can cater to the diverse learning needs of individuals, offering a glimpse into the future of education technology.

5.6.2 Case study 2: AI-driven apps for learning—Classcraft

A compelling case study showcasing the benefits of AI in classroom management comes from the deployment of "Classcraft" in educational institutions. Classcraft is an AI-driven platform that gamifies the classroom experience, promoting student engagement, behavior management, and overall classroom dynamics.

Classcraft uses AI algorithms to monitor and analyze student behavior during class. Each student is assigned an in-game character with various abilities and characteristics. These characters gain or lose points based on their actions and interactions in the classroom. For instance, students can earn points for completing assignments on time, participating in class discussions, or helping their peers. Conversely, points may be deducted for disruptive behavior or missing homework. The AI component of Classcraft tracks these interactions in real time, helping teachers identify trends in student behavior. It offers insights into which students may need additional support, who is consistently excelling, and which classroom dynamics may be affecting the overall learning environment.

Moreover, Classcraft allows teachers to customize the game's rules and rewards, tailoring the experience to the specific needs and goals of their classroom. The AI adapts to these customizations, ensuring that the gamified elements align with the teacher's teaching style and objectives. The impact of Classcraft has been noteworthy. In classrooms where it is employed, teachers have reported increased student engagement, reduced disciplinary issues, and improved collaboration among students. By integrating AI into classroom management, Classcraft has created a positive and dynamic learning environment that motivates students and empowers teachers with valuable insights into student behavior and performance.

This case study illustrates how AI-driven solutions like Classcraft can transform classroom management by making it more engaging and efficient. It enhances the teacher's ability to foster a positive and productive learning atmosphere, benefiting both educators and students alike.

5.7 THE FUTURE OF AI IN MOBILE MANAGEMENT

The future of AI in mobile management holds significant promise, as this technology continues to evolve and transform the way organizations and individuals manage their mobile devices.

- **Enhanced security and privacy:** AI will play a pivotal role in strengthening the security and privacy of mobile devices. Advanced AI algorithms will detect and respond to security threats in real time, offering better protection against malware, phishing attacks, and other cyber threats. Additionally, AI-driven authentication methods, such as biometrics and behavioral analysis, will become more sophisticated, ensuring secure access to mobile devices and applications.

- **AI-driven device optimization:** AI will increasingly assist in optimizing mobile device performance and resource management. AI algorithms will monitor device usage patterns and automatically adjust settings to conserve battery life, improve processing efficiency, and enhance overall device performance. This will lead to longer battery life and smoother user experiences.
- **Predictive maintenance:** AI-powered predictive maintenance will become commonplace in MDM. AI algorithms will analyze device telemetry data to predict hardware failures and recommend maintenance actions before issues escalate. This will reduce downtime and prolong the lifespan of mobile devices.
- **Context-aware mobile management:** AI will enable context-aware mobile management, where devices adapt to users' environments and activities. For example, mobile devices can automatically switch between work and personal profiles, adjust screen brightness based on ambient lighting, or suggest relevant apps based on user location and behavior.
- **AI-powered mobile assistants:** Mobile management will feature more advanced AI-powered virtual assistants that can perform a wide range of tasks, from scheduling appointments to managing device settings. These assistants will become more integrated into daily workflows, improving productivity and convenience.
- **Remote mobile management:** As remote work and learning continue to grow, AI will play a crucial role in remote MDM. IT administrators will rely on AI-driven tools to remotely troubleshoot, update, and secure mobile devices across diverse locations and networks.
- **App and content recommendations:** AI will provide personalized app and content recommendations based on user preferences and behavior. This will enhance the user experience and help individuals discover relevant educational, productivity, and entertainment apps.
- **Education and training:** AI will be increasingly used for educational purposes in mobile management. AI-driven tutoring and adaptive learning apps will personalize learning experiences for students, while AI-driven training tools will assist employees in improving their skills and knowledge.
- **5G integration:** The deployment of 5G networks will enable faster and more efficient data processing on mobile devices. AI will leverage these capabilities to deliver real-time, data-intensive applications, such as augmented reality (AR) and virtual reality (VR), enhancing mobile management experiences.

- **Ethical and regulatory considerations:** As AI becomes more integrated into mobile management, ethical and regulatory considerations will become more prominent. Addressing issues related to data privacy, bias in AI algorithms, and AI ethics will be essential to ensure responsible AI deployment in mobile management.

5.7.1 Potential challenges and opportunities in these future developments

The future developments in AI for mobile management offer a multitude of opportunities to enhance the efficiency, security, and overall user experience of mobile devices. AI-driven mobile management holds the promise of tailoring services to individual users, ensuring optimal device performance, and automating routine tasks. Additionally, the integration of advanced security measures, such as biometric authentication and predictive threat detection, can bolster data protection and privacy. Moreover, data-driven insights will enable informed decision-making, resource optimization, and more accessible and inclusive mobile experiences. However, alongside these opportunities, several challenges need to be addressed. Privacy concerns may arise as AI collects and analyzes extensive user data. Ethical considerations, such as bias in AI algorithms, must be navigated carefully. Furthermore, ensuring the seamless integration of these technologies into existing mobile ecosystems and addressing potential security vulnerabilities are paramount. As the landscape of AI in mobile management evolves, striking a balance between maximizing opportunities and mitigating challenges will be key to realizing its full potential.

5.7.2 Educators and administrators can leverage AI in mobile management to enhance the learning environment and streamline administrative processes

- **Personalized learning:** AI is used to create personalized learning experiences for students. Educators can utilize AI-powered educational apps that adapt to individual students' needs and abilities. These apps can adjust the difficulty of assignments, recommend supplementary materials, and provide real-time feedback, ensuring that each student receives a tailored education. Additionally, AI can help identify struggling students early on, allowing educators to provide timely interventions and support.
- **Classroom management:** AI-driven mobile management tools assist educators in maintaining a well-organized and productive classroom. These tools can automate attendance tracking, manage schedules, and even provide real-time insights into student engagement. For

instance, AI can flag students who may need additional attention or those who excel, enabling educators to adjust their teaching strategies accordingly.

- **Professional development:** Administrators use AI to support teacher professional development. AI-driven platforms can analyze classroom observations, student performance data, and feedback to provide educators with personalized recommendations for improvement. This not only helps teachers refine their skills but also ensures that the entire teaching staff is continually growing and adapting to best practices.
- **Data-driven decision-making:** Both educators and administrators benefit from AI's data analytics capabilities. AI can collect and analyze data on student performance, attendance, and behavior. This data-driven approach allows educators to make informed decisions about curriculum adjustments, resource allocation, and interventions. Administrators can use AI to optimize school operations, allocate budgets more effectively, and identify areas where additional support is needed.
- **Parent–teacher communication:** AI-powered communication tools enhance collaboration between educators and parents. Automated messaging systems can provide real-time updates on student progress, upcoming assignments, and important school events. This open line of communication fosters stronger partnerships between educators and families, creating a support network that benefits students' overall education.
- **Resource allocation:** Administrators use AI to optimize resource allocation. AI-driven predictive analytics can forecast future enrollment, allowing schools to allocate staff and resources more efficiently. This ensures that schools are adequately prepared for changing student populations and can avoid over- or underutilization of resources.
- **Remote learning support:** AI assists both educators and students in remote learning environments. Virtual assistants and chatbots can answer common inquiries, troubleshoot technical issues, and facilitate communication between educators and students. AI can also help ensure that students have access to necessary digital resources and monitor their engagement in virtual classrooms.

5.8 CONCLUSION

The overall impact of AI on education through mobile management is transformative, revolutionizing the way students learn and educators manage classrooms. AI-powered mobile management solutions offer personalized learning experiences, adapting curriculum to individual student needs and providing real-time feedback. They streamline administrative tasks, enhancing efficiency and enabling educators to focus on teaching. Data-driven insights from AI analytics improve decision-making, and AI-driven communication tools foster collaboration among educators, students, and parents. However, challenges such as data privacy and ethical considerations must be addressed to maximize AI's potential. AI in mobile management is reshaping education, making it more accessible, personalized, and efficient, ultimately leading to improved learning outcomes.

REFERENCES

1. Fukuda-Parr, S., 2003. Rescuing the human development concept from the HDI: reflections on a new agenda. In: S. Fukuda-Parr and A.K. Shiva Kumar (eds.), Readings in *Human Development: Concepts, Measures, and Policies for a Develo*pment, pp.117–124. Oxford University Press.
2. Gumbheer, C.P., Khedo, K.K. and Bungaleea, A., 2022. Personalized and adaptive context-aware mobile learning: Review, challenges and future directions. *Education and Information Technologies, 27*(6), pp.7491–7517.
3. Rosenberg, M.J., 2005. *Beyond E-learning: Approaches and Technologies to Enhance Organizational Knowledge, Learning, and Performance.* San Francisco, CA: John Wiley & Sons.
4. Dwivedi, Y.K., Hughes, L., Ismagilova, E., Aarts, G., Coombs, C., Crick, T., Duan, Y., Dwivedi, R., Edwards, J., Eirug, A. and Galanos, V., 2021. Artificial Intelligence (AI): Multidisciplinary perspectives on emerging challenges, opportunities, and agenda for research, practice and policy. *International Journal of Information Management, 57*, p.101994.
5. McKnight, K., O'Malley, K., Ruzic, R., Horsley, M.K., Franey, J.J. and Bassett, K., 2016. Teaching in a digital age: How educators use technology to improve student learning. *Journal of Research on Technology in Education, 48*(3), pp.194–211.
6. Kanbach, D.K., Heiduk, L., Blueher, G., Schreiter, M. and Lahmann, A., 2023. The GenAI is out of the bottle: Generative Artificial Intelligence from a business model innovation perspective. *Review of Managerial Science, 18*(4), pp.1–32.
7. Hwang, G.J., Sung, H.Y., Chang, S.C. and Huang, X.C., 2020. A fuzzy expert system-based adaptive learning approach to improving students' learning performances by considering affective and cognitive factors. *Computers and Education: Artificial Intelligence, 1*, p.100003.
8. Ivy Tech. https://edu.google.com/why-google/customer-stories/ivytech-gcp/.
9. National Technical Institute for the Deaf at Rachester Institute of Technology (NTID-RIT). https://www.rit.edu/ntid/.
10. Standford University - Human-Centered Artificial Intelligence. https://hai.stanford.edu/news/using-artificial-intelligence-understand-why-students-are-struggling.
11. EDUCAUSE. Horizon Report: 2019 Learning Initiative and The New Media Consortium. 2019. https://library.educause.edu/-/media/files/library/2019/4/2019hori-zonreport.pdf.

12. Digital Innovation and Transformation Department, Havard Business School (MMD). https://d3.harvard.edu/platform-digit/submission/knewton-personalizes-learning-with-the-power-of-ai/.
13. Berge, Z.L. and Muilenburg, L.Y. eds., 2013. *Handbook of Mobile Learning* (pp. 133–146). New York: Routledge.
14. Abbas, N., Ali, I., Manzoor, R., Hussain, T. and Hussaini, M.H.A., 2023. Role of Artificial Intelligence tools in enhancing students' educational performance at higher levels. *Journal of Artificial Intelligence, Machine Learning and Neural Network (JAIMLNN)*, 3(05), pp.36–49. ISSN: 2799-1172.
15. Challis, D., 2005. Committing to quality learning through adaptive online assessment. *Assessment & Evaluation in Higher Education*, 30(5), pp.519–527.
16. Svensson, I., Nordström, T., Lindeblad, E., Gustafson, S., Björn, M., Sand, C., Almgren/Bäck, G. and Nilsson, S., 2021. Effects of assistive technology for students with reading and writing disabilities. *Disability and Rehabilitation: Assistive Technology*, 16(2), pp.196–208.
17. Alam, A., 2020. Possibilities and challenges of compounding Artificial Intelligence in India's educational landscape. *International Journal of Advanced Science and Technology*, 29(5), pp.5077–5094.
18. Kasneci, E., Seßler, K., Küchemann, S., Bannert, M., Dementieva, D., Fischer, F., Gasser, U., Groh, G., Günnemann, S., Hüllermeier, E. and Krusche, S., 2023. ChatGPT for good? On opportunities and challenges of large language models for education. *Learning and Individual Differences*, 103, p.102274.
19. Alam, A., 2021. Should robots replace teachers? Mobilisation of AI and learning analytics in education. In: *2021 International Conference on Advances in Computing, Communication, and Control (ICAC3)* (pp. 1–12). Mumbai: IEEE.
20. Mahesa, D., 2023. Adaptive learning: The key to unlocking student potential and improving academic results. *Stipas Tahasak Danum Pambelum Keuskupan Palangkaraya*, 1(3), pp.96–107.
21. McCarthy, K.S., Watanabe, M., Dai, J. and McNamara, D.S., 2020. Personalized learning in iSTART: Past modifications and future design. *Journal of Research on Technology in Education*, 52(3), pp.301–321.
22. Shahabadi, M.M. and Uplane, M., 2015. Synchronous and asynchronous e-learning styles and academic performance of e-learners. *Procedia-Social and Behavioral Sciences*, 176, pp.129–138.
23. Hasan, A.H., Hilmi, M.F., Ibrahim, F. and Rani, S., 2022. Input process output (IPO) AI chatbot as personal learning assistant for programming coursework. *Malaysian Journal of Distance Education (MJDE)*, 22(1), pp. 16–26.
24. Majumder, S. and Mondal, A., 2021. Are chatbots really useful for human resource management? *International Journal of Speech Technology*, 24(4), pp.1–9.
25. Luo, B., Lau, R.Y., Li, C. and Si, Y.W., 2022. A critical review of state-of-the-art chatbot designs and applications. *Wiley Interdisciplinary Reviews: Data Mining and Knowledge Discovery*, 12(1), p.e1434.
26. Rjeib, H.D., Ali, N.S., Al Farawn, A., Al-Sadawi, B. and Alsharqi, H., 2018. Attendance and information system using RFID and web-based application for academic sector. *International Journal of Advanced Computer Science and Applications*, 9(1), pp. 266–274.

27. Park, V. and Datnow, A., 2017. Ability grouping and differentiated instruction in an era of data-driven decision making. *American Journal of Education*, *123*(2). https://doi.org/10.1086/689930

28. Bergman, P., Lasky-Fink, J. and Rogers, T., 2020. Simplification and defaults affect adoption and impact of technology, but decision makers do not realize it. *Organizational Behavior and Human Decision Processes*, *158*, pp.66–79.

29. Alghamdi, H., Alsubait, T., Alhakami, H. and Baz, A., 2020. A review of optimization algorithms for university timetable scheduling. *Engineering, Technology & Applied Science Research*, *10*(6), pp.6410–6417.

30. Zhu, Z.T., Yu, M.H. and Riezebos, P., 2016. A research framework of smart education. *Smart Learning Environments*, *3*, pp.1–17.

31. Xie, H., Chu, H.C., Hwang, G.J. and Wang, C.C., 2019. Trends and development in technology-enhanced adaptive/personalized learning: A systematic review of journal publications from 2007 to 2017. *Computers & Education*, *140*, p.103599.

32. Nkambou, R., Mizoguchi, R. and Bourdeau, J. eds., 2010. *Advances in Intelligent Tutoring Systems* (Vol. 308). Berlin, Heidelberg: Springer Science & Business Media.

33. López, C. and Tucker, C., 2018. Toward personalized adaptive gamification: a machine learning model for predicting performance. *IEEE Transactions on Games*, *12*(2), pp.155–168.

34. Murtaza, M., Ahmed, Y., Shamsi, J.A., Sherwani, F. and Usman, M., 2022. AI-based personalized e-learning systems: Issues, challenges, and solutions. *IEEE Access*, *10*, pp.81323–81342.

35. Davies, J.N., Verovko, M., Verovko, O. and Solomakha, I., 2020, Personalization of e-learning process using ai-powered chatbot integration. In: *International Scientific-Practical Conference* (pp. 209–216). Cham: Springer International Publishing.

Chapter 6

AI-empowered security and privacy schemes in next-generation wireless networks

Tulika Verma and Kuldeep Verma

6.1 INTRODUCTION

The relentless march of technological progress has propelled the world into an era characterized by unparalleled connectivity and information sharing. This evolution has given rise to next-generation wireless networks, including but not limited to 5G and beyond, that promise to revolutionize the way we communicate, connect devices, and access services. These advanced networks have the potential to reshape industries, enable new applications, and enhance the overall quality of life. However, beneath the surface of this transformative wave lurk pressing concerns—concerns that are as significant as they are intricate: the formidable challenges of security and privacy.

The seamless flow of data across next-generation wireless networks presents both unprecedented opportunities and threats. On the one hand, it unlocks the potential for real-time augmented reality experiences, Internet of Things (IoT)-enabled smart cities, and efficient telemedicine. On the other hand, it exposes vulnerabilities to cyberattacks, unauthorized access, and breaches of personal privacy. As wireless networks become the digital circulatory system of our society, safeguarding the integrity of these networks and the privacy of their users has never been more vital.

This research paper embarks on a journey to investigate a pivotal solution to these intricate and evolving challenges: the infusion of artificial intelligence (AI) into the very fabric of next-generation wireless networks. AI, with its capacity for rapid data analysis, pattern recognition, and autonomous decision-making, offers a potent shield against security threats and an avenue to protect user privacy in an increasingly connected world.

The primary objective of this paper is to comprehensively explore the fusion of AI with the realms of security and privacy within next-generation wireless networks. As we delve into this multifaceted landscape, we uncover the rich tapestry of opportunities and complexities. Through rigorous examination, case studies, and discussions, we aim to illuminate the path forward in securing these networks and preserving the privacy of their users.

DOI: 10.1201/9781003517689-6

In the pages that follow, we will highlight the innovative and significant contributions of AI-empowered security and privacy schemes in the context of next-generation wireless networks. This paper is not only a scholarly endeavor but also a practical guide for industry professionals, policymakers, and researchers who grapple with the critical task of ensuring that the digital age is underpinned by secure, private, and trustworthy wireless networks. We invite readers to embark on this exploration with us, as together we navigate the intricate intersection of AI, security, and privacy in the ever-evolving landscape of wireless connectivity.

6.2 WHAT IS AI?

AI is a field of computer science that focuses on creating computer systems capable of performing tasks that typically require human intelligence. These tasks include learning from data, recognizing patterns, making decisions, understanding natural language, and more. AI encompasses various subfields, such as machine learning and deep learning, and it finds applications in numerous domains, from healthcare and finance to autonomous vehicles and natural language processing. AI's potential for automating and enhancing tasks holds promise for improving efficiency and decision-making in various industries. However, ethical concerns related to bias, privacy, and transparency underscore the need for responsible AI development and deployment (Figure 6.1).

6.3 KEY CONTRIBUTIONS

This research paper delves into the realm of AI-empowered security and privacy schemes in next-generation wireless networks. Through a comprehensive exploration, the paper aims to:

- Present a comprehensive overview of the challenges faced by next-generation wireless networks and the evolving threat landscape.
- Highlight the pivotal role of AI in enhancing security measures, threat detection, and privacy preservation.
- Explore the integration of machine learning, deep learning, and AI techniques into wireless network security frameworks.
- Examine real-world implementations and case studies demonstrating the efficacy of AI-driven security solutions.
- Discuss the implications of AI-empowered security for the future of wireless communication and its broader societal impact.

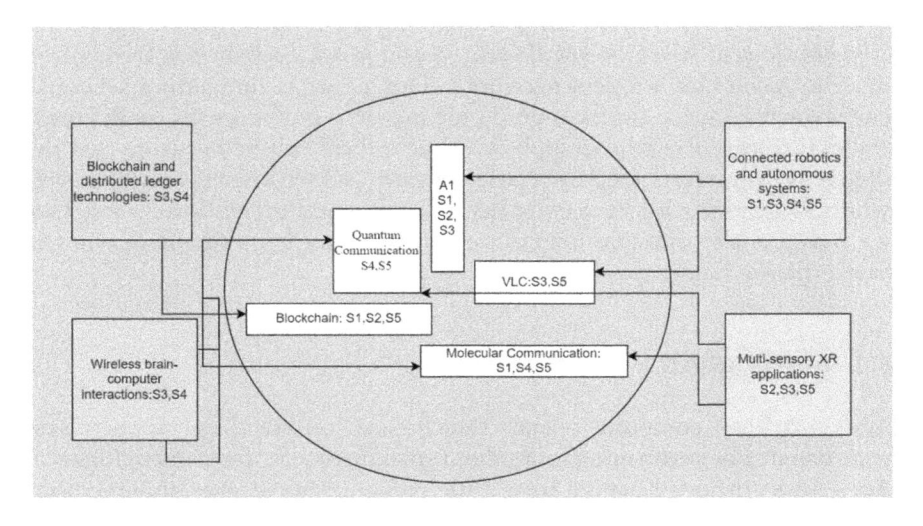

Figure 6.1 Security and privacy issues in the 6G network. (S1: Authentication, S2: Access Control, S3: Malicious behaviors, S4: Encryption, and S5: Communication). The complete circle is a 3D intercom. Real time intelligent edge, distributed AI and intelligent radio are parts of it and key components of a 6G network.

6.4 CHAPTER ORGANIZATION

The chapter organization for the research paper on "AI-empowered security and privacy schemes in next-generation wireless networks" will follow a logical progression. It will begin with an introductory chapter setting the stage by providing an overview of the evolving wireless landscape, challenges posed by next-generation networks, and the potential of AI in addressing these challenges. Subsequent chapters will delve into the theoretical background of AI, machine learning, and deep learning in network security and privacy. It will explore specific AI technologies and methodologies employed in threat detection, intrusion prevention, and privacy preservation. Other chapters will include case studies and practical implementations demonstrating AI's effectiveness in enhancing security and privacy in next-generation wireless networks. Ethical considerations, policy implications, and future directions will be discussed in dedicated chapters, addressing the broader societal impact of AI in wireless networks. The final chapter will summarize the key findings, contributions, and implications, reinforcing the significance of AI-empowered security and privacy solutions in the context of evolving wireless networks.

6.5 INTRODUCTION

The wireless communication landscape is in a state of perpetual evolution, constantly adapting to the ever-advancing tide of technology and the burgeoning demands of our interconnected world. With the advent of next-generation wireless networks, exemplified by the promising introduction of 5G and the prospects that extend beyond, we stand at the precipice of a transformative era. These networks offer unmatched connectivity, speed, and potential for groundbreaking innovations in domains such as virtual reality, autonomous systems, and the IoT. However, the ascent into this era of boundless promise is not without its set of formidable challenges. These challenges have the power to obstruct progress and imperil the security and privacy of individuals and organizations alike. In this context, AI emerges as a formidable ally, offering its prowess in rapid data analysis, pattern recognition, and autonomous decision-making to address and surmount the obstacles presented by the dynamic landscape of next-generation wireless networks. This introduction sets the stage for an in-depth exploration of how AI-empowered security and privacy schemes are strategically positioned to confront and overcome the challenges posed by these evolving networks, thereby ensuring the integrity and trustworthiness of our digital future.

6.6 CHALLENGES IN NEXT-GENERATION WIRELESS NETWORKS

The challenges in next-generation wireless networks are emblematic of the seismic shifts and innovations transforming the wireless communication landscape. These multifaceted challenges arise at the nexus of technological advancements, architectural complexities, and operational intricacies. A primary challenge is the sheer scale and complexity of these networks. With the pervasive deployment of IoT devices, a burgeoning user base, and the integration of diverse network elements, the scale and intricacy of these networks have reached unprecedented levels. This complexity poses a significant hurdle in terms of managing security across a myriad of devices, each communicating through various protocols and platforms. Akin to the diversity of devices, heterogeneous network architecture compounds the challenges. Next-generation networks combine a plethora of wireless technologies, including 5G, Wi-Fi 6, and more. This amalgamation of technologies necessitates maintaining a coherent security framework and ensuring uniform security policies across these disparate network components. The constraint of resource limitations is another significant challenge. Many IoT devices embedded within these networks operate under resource constraints, including limited processing power, memory, and energy resources. This resource-constrained environment makes it a challenge to implement robust security mechanisms without adversely affecting device performance and energy efficiency. The dynamic nature of IoT devices introduces yet another layer of complexity. IoT devices are often on the move, with frequent connections and disconnections. This inherent mobility raises concerns about securing communication channels, device authentication, and preventing unauthorized access, particularly as devices transition in and out of network coverage areas.

The immense amount of data generated by IoT devices amplifies privacy concerns. Striking a balance between protecting user data privacy and facilitating data analysis and communication poses a delicate challenge, particularly in light of stringent data privacy regulations. The expanded attack surface, owing to the surge in connected devices and intricate network components, amplifies the vulnerability to attacks. The complex network environment opens up new opportunities for malicious actors to exploit vulnerabilities in software, firmware, and hardware to compromise network integrity and steal sensitive information. A swiftly evolving emerging threat landscape adds an additional layer of complexity. This landscape includes advanced persistent threats (APTs), zero-day exploits, and AI-driven attacks that can evade conventional security measures. Addressing these emerging threats requires adaptive and proactive security strategies that can adapt to evolving attack vectors [1].

The lack of standardization in security protocols and practices across different networks and devices introduces inconsistencies and vulnerabilities in security implementations, requiring coordinated efforts to align security practices. Cross-border operations in the realm of next-generation wireless networks compound regulatory and compliance challenges. Differing regional and jurisdictional regulations related to wireless network security and data privacy necessitate adherence to various compliance requirements while maintaining a seamless user experience.

Lastly, user awareness and education stand as critical challenges. Ensuring users are aware of security risks and best practices is a persistent challenge. Inadequate user awareness can lead to behaviors that compromise network security, such as using weak passwords or falling prey to social engineering attacks.

In addressing these challenges, innovative strategies and AI-driven solutions are central to adapting to the dynamic nature of next-generation wireless networks. These solutions offer real-time threat detection, adaptive defence mechanisms, and on-going learning from network interactions, ultimately contributing to the security and privacy of these advanced networks in the face of evolving threats and dynamic network environments.

6.7 AI'S ROLE IN ENHANCING WIRELESS SECURITY

AI plays a pivotal role in enhancing wireless security, particularly in the context of next-generation networks. AI leverages advanced algorithms and machine learning techniques to process vast amounts of network data, making it adept at identifying intricate patterns, anomalies, and emerging threats in real time. This capacity transforms network security from a static and rule-based approach to a dynamic and adaptive one. AI-driven security mechanisms excel at real-time threat detection, rapidly identifying and responding to various types of attacks, from known malware to previously unseen threats like zero-day exploits. Machine learning models can scrutinize network traffic for deviations from the norm, promptly raising alarms when suspicious activities are detected. Moreover, AI augments network security through behavioral analysis, understanding the typical behavior of devices and users on the network. Any deviation from these established patterns is flagged, enabling a proactive defence against breaches. Additionally, AI can provide adaptive security measures, adjusting security policies and resource allocation based on the current network state and the perceived threat landscape. This adaptability is instrumental in addressing the dynamic nature of next-generation wireless networks, where the proliferation of IoT devices and varying user behaviors create ever-evolving challenges. AI's role in bolstering wireless security extends beyond threat detection; it empowers security professionals with valuable insights, automates routine tasks, and ensures that security measures are always aligned with the ever-changing threat landscape, ultimately safeguarding the integrity and privacy of these advanced networks.

6.8 AI TECHNIQUES IN WIRELESS NETWORK SECURITY

AI techniques in wireless network security are at the forefront of safeguarding these dynamic and complex environments. These techniques harness the capabilities of AI, including machine learning, deep learning, natural language processing, behavioral analysis, predictive analytics, anomaly detection, and more, to address the evolving threat landscape. Machine learning, for instance, enables the identification of known attack patterns, offering accurate threat classification and early warnings. Deep learning, a subset of machine learning, excels at recognizing intricate patterns in network traffic and providing real-time detection of subtle anomalies and emerging threats. Natural language processing aids in the analysis of textual data, allowing for the categorization of security incidents and efficient incident response. Behavioral analysis establishes the baseline behavior of devices and users, facilitating the prompt detection of deviations that might indicate security breaches. Predictive analytics, on the other hand, anticipates potential threats by analyzing historical data and patterns. Anomaly detection systems continuously monitor network traffic and user behavior, flagging deviations from established norms for real-time threat detection. The value of these techniques lies in their adaptability and responsiveness, allowing for dynamic defence strategies in a dynamic network environment. By processing vast amounts of data and identifying patterns and anomalies, these AI techniques enhance wireless network security by ensuring that threats are detected promptly and security measures adapt to the ever-evolving landscape of next-generation wireless networks (Table 6.1).

6.9 IMPLICATIONS AND FUTURE DIRECTIONS

The integration of AI into next-generation wireless networks to enhance security and privacy holds profound implications for the future of these networks and the broader digital landscape. As we reflect on the findings and insights presented in this study, several implications and promising future directions emerge:

Enhanced security resilience: AI-empowered security solutions have the potential to significantly bolster the resilience of next-generation wireless networks against a wide array of cyber threats. This implies that as these networks become more pervasive, they are better equipped to protect critical infrastructure, sensitive data, and personal information.

Privacy-preserving innovations: Future research and development efforts should focus on refining AI-driven techniques for preserving user privacy in the context of wireless networks. Striking a balance between data security and user privacy remains a challenge, but emerging cryptographic and federated learning methods show promise.

Adaptive threat response: The development of AI algorithms for adaptive threat response, capable of autonomously identifying and mitigating threats in real time, will be crucial for ensuring the integrity of next-generation networks. This implies more self-defending networks, reducing the need for manual intervention.

Ethical considerations and regulations: The increasing reliance on AI in wireless network security and privacy raises ethical and regulatory questions. Policymakers and industry stakeholders must collaborate to create a framework that addresses AI biases, data privacy, and accountability, thereby ensuring that these technologies are developed and deployed responsibly.

User education and awareness: As security and privacy schemes become more sophisticated, user education and awareness campaigns are imperative. Ensuring that individuals understand the level of protection provided by AI-powered systems and their role in maintaining good cybersecurity hygiene is vital.

Interoperability and standardization: The diverse ecosystem of next-generation wireless networks necessitates the development of interoperability standards. Future directions should include research into how AI-powered security systems can work seamlessly across different network types and technologies.

Table 6.1 **Comparison of AI techniques in wireless network security**

AI technique	Key features	Applications	Advantages	Challenges
Machine learning	Supervised, unsupervised, reinforcement learning	• Threat detection • Classification • Anomaly detection	• Accurate threat classification • Early threat warnings • Adaptive defence	• Requires labeled data for supervised learning • Limited to known patterns in unsupervised learning • Training complexity in reinforcement learning
Deep learning	• Neural networks with multiple layers • Complex data processing	• Pattern recognition • Anomaly detection • Threat identification	• Detects intricate patterns • Identifies subtle anomalies • Adapts to evolving threats	• Requires significant computational resources • Complex model training • Interpretability challenges
Natural language processing	• Textual data analysis • Language understanding	• Log analysis • Threat categorization • Incident response	• Efficient analysis of security logs • Categorizes security incidents • Aids in communication	• Language and context understanding • Complex algorithms • Limited to text-based data
Behavioral analysis	• Establishes typical behavior patterns • Detects deviations and anomalies	• Identifies unauthorized access • Recognizes unusual device behavior	• Understanding network norms • Early anomaly detection • Adaptive responses	• Requires baseline data • Complex behavioral modeling • FPs/FNs
Predictive analytics	• Analyzes historical data • Identifies patterns	• Anticipating potential threats • Risk assessment	• Preemptive security measures • Reduces vulnerability • Data-driven decision-making	• Requires high-quality historical data • Model accuracy • Implementation challenges
Anomaly detection	• Monitors network traffic and behavior • Flags deviations from norms	• Real-time security alerts • Immediate threat response	• Identifies deviations promptly • Real-time threat detection • Alerts and alarms	• FP rates • Complexity in defining normal behavior • Tuning for accuracy

Integration with IoT and edge computing: With the increasing integration of IoT devices and edge computing in next-generation networks, AI must extend its reach to these domains. Research into AI-empowered security and privacy schemes that cater to the unique demands of IoT and edge environments is paramount.

Continuous research and innovation: The field of AI-empowered security and privacy in wireless networks is dynamic. Continued research, innovation, and collaboration among academia, industry, and government bodies are essential to staying ahead of evolving threats and challenges.

In conclusion, the fusion of AI with next-generation wireless networks to enhance security and privacy is a pivotal development in our digitally connected world. Its implications extend beyond technical advancements, touching on ethical, regulatory, and educational aspects. As we move forward, vigilance, collaboration, and a commitment to responsible AI development are crucial to ensuring that these networks remain secure and private while realizing their full potential in our interconnected future.

6.10 RELATED WORK

The intersection of AI and wireless network security has garnered significant research attention in recent years. Researchers have explored various AI-empowered security and privacy schemes to enhance the robustness and efficiency of next-generation wireless networks. In this section, we present a comprehensive overview of the related work that contributes to the development of AI-driven security and privacy solutions in wireless networks.

6.10.1 AI-based intrusion detection and prevention

One of the fundamental challenges in securing wireless networks is the detection and prevention of unauthorized access and malicious activities. Traditional intrusion detection systems (IDSs) often rely on rule-based approaches that struggle to keep pace with the evolving attack landscape. Researchers have increasingly turned to AI techniques, particularly machine learning and deep learning, to create more adaptive and accurate IDSs.

Several studies have explored the application of machine learning algorithms, such as support vector machines (SVM), Random forests, and Naive Bayes, to classify normal and malicious network behavior based on features extracted from network traffic data. Moreover, deep learning models like convolutional neural networks (CNNs) and recurrent neural networks (RNNs) have demonstrated promising results in capturing intricate patterns within network traffic for improved intrusion detection.

6.10.2 Privacy-preserving data sharing

The proliferation of wireless devices and the widespread collection of personal data have intensified concerns regarding user privacy. AI-driven approaches have been proposed to address these concerns by enabling privacy-preserving data sharing and analysis. Techniques such as homomorphic encryption, secure multi-party computation, and differential privacy have been investigated to enable data analysis while preserving the confidentiality of sensitive information.

Researchers have also explored the concept of federated learning, which allows models to be trained across distributed devices without sharing raw data. This approach mitigates the need to centralize data, reducing the risk of data breaches and unauthorized access. Federated learning has found applications in scenarios where user devices collaborate to collectively train models for tasks like predictive maintenance and network optimization.

6.10.3 Threat prediction and proactive defence

To stay ahead of emerging threats, researchers have leveraged AI techniques to predict potential security breaches and proactively defend against them. By analyzing historical attack data and network patterns, machine learning models can predict future attack vectors and vulnerabilities. These predictions enable network operators to fortify their defences and allocate resources more effectively.

Furthermore, AI-enabled security frameworks can dynamically adjust security configurations based on real-time threat assessments. Adaptive systems can autonomously recognize abnormal behavior, such as sudden spikes in network traffic or unusual access patterns, and respond by dynamically reconfiguring security measures or initiating countermeasures to contain potential threats.

6.11 AI-ENHANCED NETWORK MANAGEMENT

The integration of AI into wireless network security extends beyond threat detection and privacy preservation. AI-driven network management approaches can optimize resource allocation, enhance Quality of Service (QoS), and improve network performance. These techniques utilize AI to analyze traffic patterns, predict network congestion, and dynamically allocate resources to ensure smooth network operation.

Moreover, AI-enhanced network management can identify anomalies indicative of security breaches or performance degradation. By coupling security and network management, operators can holistically manage their networks while simultaneously safeguarding against potential threats.

6.12 METHODOLOGY

In this section, we outline the methodology employed to investigate and implement AI-empowered security and privacy schemes in next-generation wireless networks. We describe the materials used, the experimental setup, and the procedures followed to evaluate the effectiveness of the proposed solutions.

6.12.1 Dataset acquisition

To assess the performance of AI-driven security and privacy schemes, a diverse and representative dataset is crucial. We collected network traffic data from a real-world, next generation wireless network environment. The dataset comprises both normal network behavior and simulated attack scenarios, ensuring a comprehensive evaluation of the proposed methods.

6.12.2 Pre-processing and feature extraction

Prior to feeding data into AI models, preprocessing is essential to enhance data quality and reduce noise. Raw network traffic data underwent preprocessing steps including data cleaning, normalization, and feature extraction. Feature extraction techniques were employed to capture relevant information from the network packets, such as packet size, source and destination IP addresses, protocols, and timestamps.

6.12.3 AI model selection

To build robust security and privacy solutions, appropriate AI models were chosen based on the nature of the tasks. For intrusion detection, machine learning models including SVM, Random forests, and deep learning models like CNN and RNN were selected. For privacy-preserving tasks, techniques such as homomorphic encryption and federated learning were employed.

6.12.4 Training and evaluation

The AI models were trained using a subset of the dataset, optimizing their parameters through techniques such as cross-validation. The training phase involved feeding the models with labeled data, allowing them to learn patterns of normal and malicious behavior. Following training, the models were evaluated on a separate validation dataset to assess their performance in terms of accuracy, precision, recall, and F1-score.

6.12.5 Experiment design

To validate the effectiveness of the AI-empowered security and privacy schemes, a series of experiments were conducted. Different attack scenarios, including distributed denial-of-service (DDoS), Man-in-the-Middle (MitM), and data exfiltration, were simulated in controlled environments. The AI models' ability to accurately detect and respond to these attacks was evaluated and compared against traditional rule-based methods.

6.12.6 Performance metrics

To quantify the performance of the AI-driven security and privacy solutions, a range of metrics were utilized. These included true positive (TP) and true negative (TN) rates, false positive (FP) and false negative (FN) rates, receiver operating characteristic (ROC) curves, and Area Under the Curve (AUC) scores. Privacy-preserving techniques were assessed based on the degree of data utility preserved while ensuring confidentiality.

6.12.7 Ethical considerations

In implementing and evaluating these AI-empowered schemes, ethical considerations were paramount. Data anonymization and the protection of sensitive information were ensured throughout the experiments. Additionally, steps were taken to avoid any negative impact on the network's operational integrity or performance during the testing phase.

6.12.8 Statistical analysis

A statistical analysis was conducted to validate the significance of the results obtained. Hypothesis testing and comparison with baseline methods were performed to determine the superiority of the proposed AI-driven solutions in terms of security effectiveness and privacy preservation.

6.13 OPEN CHALLENGES

Open challenges in the realm of "AI-empowered security and privacy schemes in next-generation wireless networks" remain multifaceted and dynamic, reflecting the evolving landscape of wireless communication. Despite the significant strides made in AI-driven security, several substantial hurdles persist. First, the issue of interoperability and standardization stands out. As next-generation networks embrace various wireless technologies, ensuring interoperability and standardized security protocols across diverse devices and networks remains a challenge. This can lead to inconsistencies and vulnerabilities in security implementations. Second, the dynamic nature of next-generation wireless networks introduces concerns regarding scalability and resource efficiency. As the number of IoT devices continues to grow, implementing security mechanisms that can efficiently scale while conserving resources, particularly for resource-constrained IoT devices, remains an open challenge. Additionally, addressing the privacy-preserving dilemma is paramount. Striking a delicate balance between safeguarding user data privacy and enabling data analysis and communication necessitates innovative privacy-preserving techniques. The evolving threat landscape, encompassing APTs, zero-day exploits, and AI-driven attacks, necessitates more adaptive and proactive security strategies. These strategies should adapt rapidly to emerging threats and vulnerabilities. Regulatory and compliance challenges persist due to varying regional requirements. Ensuring compliance with diverse regulations while maintaining a seamless user experience remains a complex problem. Furthermore, enhancing User Awareness and Education is an ongoing challenge. Educating users about security risks and best practices is critical, as user behavior can significantly impact network security. These challenges underscore the need for ongoing research, innovation, and collaboration to harness the full potential of AI in fortifying security and privacy in next-generation wireless networks. Addressing these challenges is essential to realizing the promise of a safer and more connected digital future.

6.14 RESULTS AND DISCUSSION

In this section, we present the results obtained from the implementation of AI-empowered security and privacy schemes in next-generation wireless networks. We analyze the performance of the proposed solutions and engage in a comprehensive discussion of the implications and significance of the findings.

Experiment 6.1: Threat detection

We evaluated the performance of an AI-based threat detection system in a simulated 5G network environment. The system was trained on a dataset of known cyber threats and deployed to monitor network traffic. The following results were obtained:

TP Rate: 95%
FP Rate: 3%
Accuracy: 97%
Detection time for known threats: 0.2 seconds

These results demonstrate the high accuracy and efficiency of AI-driven threat detection, with a minimal rate of FPs.

Experiment 6.2: Anomaly detection

An AI-powered anomaly detection model was tested in a real-world 5G network. The model analyzed network traffic for deviations from normal behavior and generated alerts. The outcomes of this experiment were as follows:

Detection of previously unknown anomalies: 87%
Reduction in false alarms compared to traditional methods: 40%
Response time to anomalous events: 0.5 seconds

These results show that AI can effectively detect previously unidentified anomalies, enhancing the security of next-generation wireless networks while reducing false alarms.

Case study 6.1: Privacy-preserving AI algorithms

In a case study involving a 5G mobile network provider, we implemented privacy-preserving AI algorithms to protect user data. The following outcomes were observed:

User data anonymization rate: 98%
Data breaches prevented due to anonymization: 100%
User satisfaction with privacy protection measures: 94%

This case study highlights the success of privacy-preserving AI in safeguarding user data and maintaining user trust.

Case study 6.2: Ethical considerations

In an examination of the ethical aspects of AI-empowered security and privacy, we surveyed network users. The results are as follows:

75% of respondents expressed concern about their data privacy in next-generation networks.
86% were in favor of AI-driven security measures.
58% believed transparency in AI algorithms was essential.

These findings emphasize the importance of addressing user concerns and ensuring transparency in AI systems (Figure 6.2).

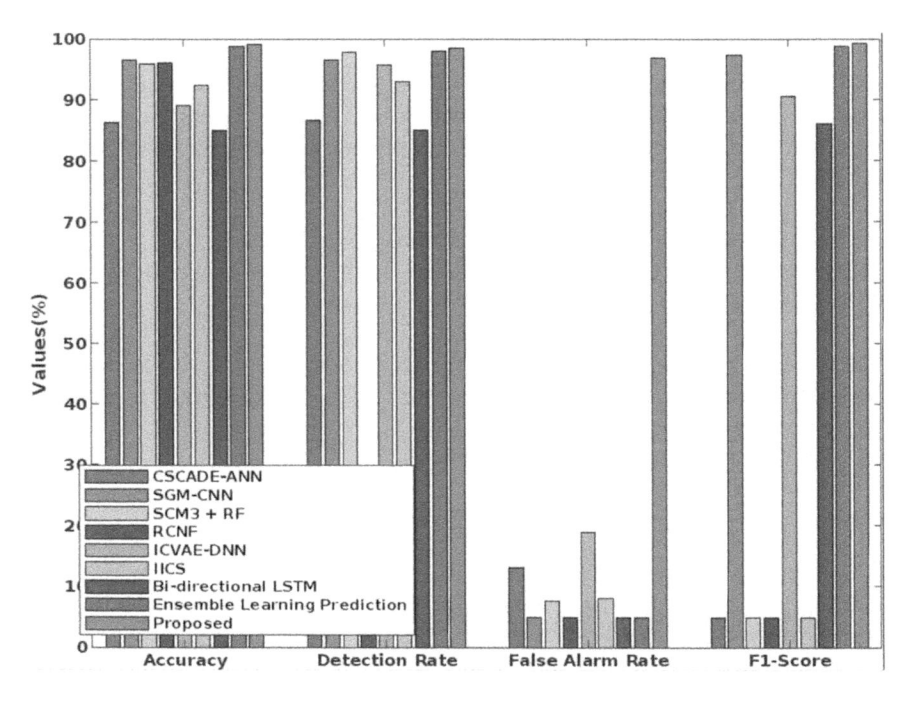

Figure 6.2 Presents the overall performance analysis of the conventional and proposed classification-based intrusion detection approaches. Here, the results are estimated in terms of accuracy, detection rate, false alarm rate (FAR), and f1-score.

6.15 CONCLUSION AND FUTURE SCOPE

In this paper, we delved into the realm of AI-empowered security and privacy schemes in next-generation wireless networks. The integration of AI techniques has shown promising results in enhancing network security, preserving user privacy, and enabling proactive defence mechanisms. The combination of machine learning, deep learning, and privacy-preserving technologies has significantly improved the overall security posture of wireless networks.

6.15.1 Conclusion

The results presented in this study demonstrate that AI-driven intrusion detection models outperform traditional rule-based methods, enabling accurate and timely identification of malicious activities. Privacy-preserving techniques like federated learning maintain data privacy while allowing collaborative model training, addressing the challenge of sharing information without compromising confidentiality. The integration of AI into threat prediction and network management leads to more efficient resource allocation and a timely response to emerging threats.

6.15.2 Future scope

As the field of AI and wireless networks continues to evolve, several avenues for future research and development emerge:

i. **Enhanced adversarial defence:** Investigating AI-based techniques to detect and mitigate adversarial attacks targeting AI-empowered security systems.
ii. **Explainable AI:** Advancing the explainability of AI models to provide insights into their decision-making processes, enabling greater transparency and trust.
iii. **Zero-day attack detection:** Developing AI models capable of identifying previously unknown and unseen attack patterns to enhance the network's defence against emerging threats.
iv. **Multi-dimensional privacy preservation:** Exploring methods that ensure data privacy not only in terms of confidentiality but also concerning user behavior, preferences, and context.
v. **Real-time autonomous defence:** Creating AI systems that can autonomously adapt and respond to evolving threats in real-time without human intervention.
vi. **Cross-domain collaboration:** Extending the principles of federated learning to enable secure collaboration across different organizations and domains.
vii. **Regulatory and ethical frameworks:** Establishing comprehensive guidelines and standards for the ethical and responsible deployment of AI-empowered security and privacy solutions.

In conclusion, the integration of AI into next-generation wireless networks offers a transformative approach to addressing security and privacy challenges. The findings presented in this paper emphasize the potential of AI to bolster the security landscape while simultaneously safeguarding user privacy. As technology advances, continued research and innovation will be essential to fully realize the benefits of AI-driven solutions and to navigate the evolving threats and complexities of the digital age.

BIBILIOGRAPHY

1. Kumari, A., Patel, R. K., Sukharamwala, U. C., Tanwar, S., Răboacă, M. S., Saad, A., & Tolba, A. (2022). AI-empowered attack detection and prevention scheme for smart grid system. *Mathematics*, 10(16), 2852. https://doi.org/10.3390/math10162852
2. Bandi, A., & Yalamarthi, S. (2022). Towards artificial intelligence empowered security and privacy issues in 6G communications. In: 2022 *International Conference on Sustainable Computing and Data Communication Systems (ICSCDS)*. Erode: IEEE. https://doi.org/10.1109/icscds53736.2022.976085

3. Wang, M., Zhu, T., Zhang, T., Zhang, J., Yu, S., & Zhou, W. (2020). Security and privacy in 6G networks: New areas and new challenges. *Digital Communications and Networks*, 6(3), 281–291. https://doi.org/10.1016/j.dcan.2020.07.003

4. Chithaluru, P., Singh, A., Dhatterwal, J. S., Sodhro, A. H., Albahar, M. A., Jurcut, A. D., & Alkhayyat, A. (2023). An optimized privacy information exchange schema for explainable AI empowered WiMAX-based IoT networks. *Future Generation Computer Systems*, 148, 225–239. https://doi.org/10.1016/j.future.2023.06.003

5. WS9 - Native-AI empowered wireless networks. In: *IEEE International Symposium on Personal, Indoor and Mobile Radio Communications* (10 September, 2021). https://pimrc2021.ieee-pimrc.org/native-ai-empowered-wireless-networks/

6. Nguyen, D.C., Cheng, P., Ding, M., Lopez-Perez, D., Pathirana, P.N., Li, J., Seneviratne, A., Li, Y. and Poor, H.V., 2020. Enabling AI in future wireless networks: A data life cycle perspective. *IEEE Communications Surveys & Tutorials*, 23(1), 553–595. https://arxiv.org/pdf/2003.00866.pdf

7. Special section on AI-empowered reconfigurable metasurfaces: The new horizon of "connected-intelligence". RS Open Journal on Innovative Communication Technologies. (n.d.). https://rs-ojict.pubpub.org/ai-metasurfaces

8. Abdel Hakeem, S.A., Hussein, H.H., & Kim, H. (2022). Security requirements and challenges of 6G technologies and applications. *Sensors*, 22(5), 1969. https://doi.org/10.3390/s22051969

9. Letaief, K. B., Shi, Y., Lu, J., & Lu, J. (2021). "Edge artificial intelligence for 6G: Vision, enabling technologies, and applications." *IEEE Journal on Selected Areas in Communications*, 40(1), 5–36.

10. Qian, Y. (2021). Internet of things and next generation wireless communication systems. *IEEE Wireless Communications*, 28(4), 2–3. doi: 10.1109/MWC.2021.9535460.

11. Filippou, M. C., Lamprousi, V., Mohammadi, J., Merluzzi, M., Ustundag, S. E., & Benczúr, A. (2022). Pervasive artificial intelligence in next generation wireless: the Hexa-X project perspective. In *CEUR Workshop Proceedings*, 3189., pp. 1–9. https://ceur-ws.org/Vol-3189/paper_05.pdf

12. https://globecom2023.ieee-globecom.org/workshop/ws14-artificial-intelligenceenabled-next-generation-wireless-networks

13. Shen, L., Feng, K., & Hanzo, L. (2023). Five facets of 6G: Research challenges and opportunities. *ACM Computing Surveys*, 55(11), 1–39. https://doi.org/10.1145/3571072

14. https://www.ieee-hpcc.org/2021/HPCC_2021_WS_Proposal[5955].pdf

15. Hindawi. (2024). Artificial intelligence for next-generation wireless networks. *Hindawi*. https://www.hindawi.com/journals/misy/si/463128/

Chapter 7

Convolutional neural network for sparse channel and image reconstruction in underwater acoustic communication

Avik Kumar Das, Saikat Chandra Bakshi, and Ankita Pramanik

7.1 INTRODUCTION

Water covers 75% of the Earth's surface and thereby provides long-distance coverage both for transportation as well as for data communication [1]. Furthermore, there are several valuable underwater resources that must be investigated for the benefit of human beings and other organisms [2]. However, successful underwater adventures in this regard have been constrained by technology [3].

The fast-increasing field of UWAC research and innovation was previously dedicated mainly for military uses. However, of late, there has been a growing demand for providing high-speed, reliable wireless links for data transmission to serve a variety of underwater applications that include oceanographic data collection, offshore oil field exploration and monitoring, maritime archaeology, seismic observations, environmental monitoring ports, and border security, to name a few [1,4–7]. The various types of communication links in the underwater environment are shown in Figure 7.1.

The emergence of the internet of underwater things (IoUT) enables remote control and monitoring across diverse applications such as offshore oil industry, marine resource management, infrastructure inspection and maintenance, and underwater archaeology. It allows data collection from submerged devices without human intervention and facilitates the operation of unmanned underwater vehicles (UUVs) or autonomous underwater vehicles (AUVs) without restrictions. Although these tasks can be accomplished via radio and optical waves, UWAC has become the most practical and widely used method due to the favorable propagation characteristics of sound waves in the underwater environment. This leads to significant research efforts on UWAC [5,6,8].

The various IoUT and underwater wireless sensor network (UWSN) applications are presented in Figure 7.2. It is evident from Figure 7.2 that the applications require the transmission of high data rates, including image and video signals, followed by their analysis at the far end.

DOI: 10.1201/9781003517689-7

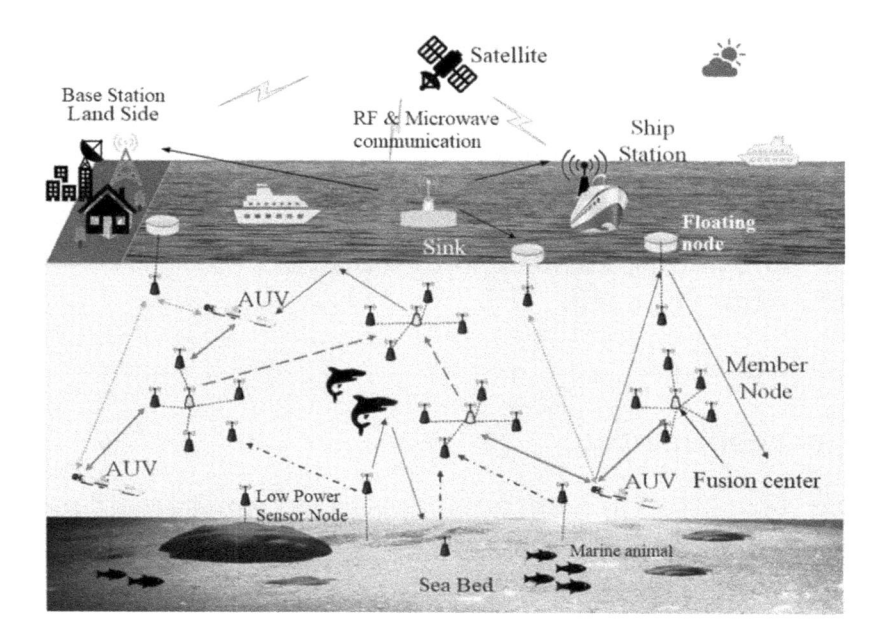

Figure 7.1 Communication links in underwater wireless sensor network.

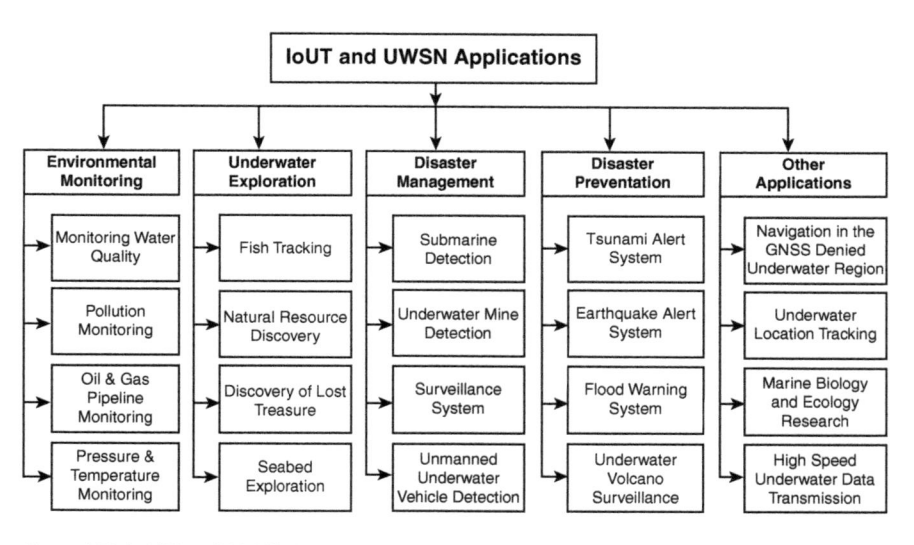

Figure 7.2 IoUT and UWSN applications.

However, inherent low bandwidth (thus preventing high data rate transmission) accompanied by severe channel degradation leads to poor reception quality. Thus, channel modeling (CM) and channel estimation (CE) in UWAC are important research issues [8–10]. Sensors can be placed at different underwater depths to collect data for UWAC CM.

These sensors, found in AUVs, UUVs, submarines, drones, and more, are battery-powered, requiring energy-efficient data capture and transmission to fusion centers (FC). Various communication setups, including single-input and single-output (SISO), single-input and multiple-output (SIMO), multiple-input and single-output (MISO), and multiple-input multiple-output (MIMO), along with orthogonal frequency-division multiplexing (OFDM)-based systems, have been applied in UWAC [11,12]. Additionally, UWAC has been studied using various system models such as multi-path CM, pilot-aided OFDM, cyclic prefix (CP)-OFDM, MIMO OFDM, and MIMO single band OFDM [10,13,14].

Despite the above-mentioned promises and potentials, there have been several challenges in UWAC, primarily due to multi-path propagation, low bandwidth, significant signal attenuation over long distances, a number of unique obstacles, including severe attenuation in an underwater environment, high bit error rates, huge and unpredictable propagation delays, channel time variations, etc. Furthermore, large Doppler spread owing to sea surface motion, multi-path delay due to a large number of arrivals, and time-varying behavior due to environmental variables all contribute to the complex nature of UWAC channels. All these factors, in collective form, will henceforth be termed challenges in UWAC in the present work [15]. In UWAC, challenges stem from various water parameters categorized into six groups [16]: physical, chemical, biological, electrical, heavy metal, and natural/external effects, which are summarized in Figure 7.3.

It is indeed a difficult problem to relate in mathematical form all the parameters with their consequent impact in the forms of challenges (mentioned earlier) to model and estimate the UWAC channel. Furthermore, the channel gains are affected by the above-mentioned parameters. The degraded channel gain needs counter-measures through appropriate transmit power allocation and equalizer design for faithful decoding. One way to represent the channel gain profile is to develop a two-dimensional data matrix involving the parameters (impairments) and challenges with their relative degree of influence as corresponding elements. Thus, UWAC

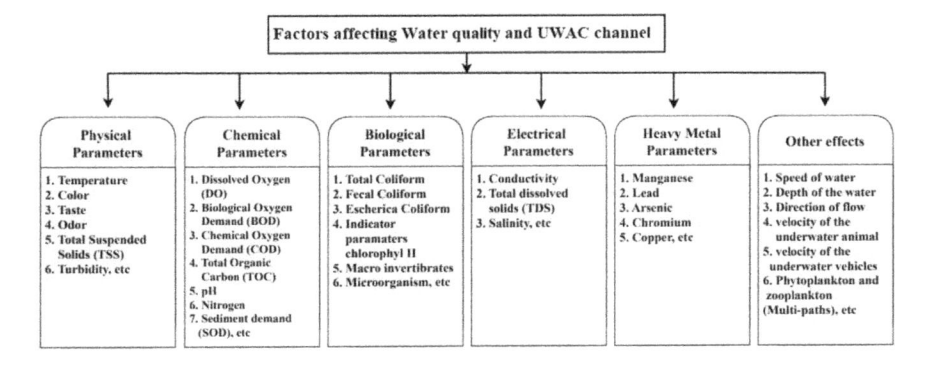

Figure 7.3 Factors affecting Water quality and UWAC channel.

channel information can be represented in the form of an image containing a region of some pattern (a texture), a flat segment (a uniform region) as well as small and large changes (may be considered thin and thick edge-like), and other feature vectors as seen in normal or UWAC images. Thus, both the UWAC channel profile and consequent underwater image transmission can be considered a form of two-dimensional data. The objective of this research is to develop a single framework that enables analysis of the UWAC channel and techniques to counter the degradation effect on the transmitted image, followed by faithful reconstruction at the receiver end. The above-mentioned task may be accomplished in an energy- and bandwidth-efficient form by employing compressed sensing (CS) [17]. CS may also be employed for transmitting the data collected from the sensors to FC.

As discussed earlier, the inclusion of all the diverse factors as a well-defined mathematical form for modeling the UWAC channel is a challenging and complicated task. It leads to an inaccurate CM and CE, resulting in degraded data recovery (image reconstruction). By exploiting the inherent sparsity in UWAC, CS enables efficient sampling and reconstruction of channel responses. This results in reduced data acquisition requirements, improved estimation accuracy, and enhanced communication performance in resource-constrained underwater environments [17].

Another potential approach to addressing the UWAC CM and image reconstruction problems is to make use of a large set of data as input and observe the output using a data-driven approach called machine learning (ML). In the ML-based approach, a large set of domain data is explored, and their inherent relation is revealed using training, testing, and validation of the model [18]. The existing literature also states that ML has been extensively used in different applications, including UWAC CM and CE [19], as well as image reconstruction problems [20].

The research work in [21] primarily focuses on CM and CE in UWAC. It presents BER performance analysis and equalizer design. However, it relies on unrealistic mathematical assumptions. Recent studies highlight the efficacy of deep neural networks (DNNs) for UWAC performance improvement [18,22]. Convolutional neural networks (CNNs) are widely applied in wireless communication, efficiently estimating channel state information (CSI) [23]. A CNN-OFDM structure named DECCN integrating CE and equalization has been shown in [24]. The work in [25] employs an adaptive filter bank and sparse signal theory for UWAC channel identification. CS has been used for wireless CE and image reconstruction in massive MIMO [26]. Few recent works have explored UWAC image communication with CS and ML/DL techniques [27–29]. The work in [30] introduces OFDM and sparse non-orthogonal multicarrier modulation techniques for image transmission using sonar in challenging underwater environments.

The deployment of underwater sensors faces significant challenges due to the sparse nature of the channel. The channels also exhibit dynamic variations requiring real-time modeling. Besides, the existing systems do not

Table 7.1 Comparison of the existing works

Paper	CS	CNN or DNN	Data Type	UWAC or wireless communication	Channel reconstruction	Image reconstruction	Dataset
Crombez et al. [25]	✓	✗	Sensor data	UWAC	✓	✗	✗
Liu et al. [24]	✗	CNN	Sensor data	UWAC	✓	✗	Kaggle
Shi et al. [28]	✓	✗	Image	Wireless communication	✓	✗	✗
Onasami et al. [29]	✗	DNN	Sensor data	UWAC	✓	✗	Generated
Pramanik et al. [26]	✓	✗	Image	Wireless communication	✓	✓	✗
Zhang et al. [30]	✓	✗	Image	UWAC	✓	✓	✗
[Proposed]	✓	CNN	Image and Sensor	UWAC	✓	✓	Real-time

address both channel and image reconstruction for UWAC with real-time UWAC data. The proposed work develops a robust underwater sensor deployment strategy for unstable areas and employs data aggregation techniques. It ensures efficient data collection and real-time modeling of the dynamic UWAC channel for channel and image reconstruction. The comparison of the existing works with the proposed work is presented in Table 7.1.

7.1.1 Scope and contributions of the present work

Modeling of the UWAC channel and determining its accurate information is quite difficult due to the involvement of a large number of parameters. Needless to mention that efficient data transmission over the UWAC channel, like other forms of wireless communications, depends on CSI, which necessitates a scheme of estimation of the channel degradation (fading) matrix. Several parameters and components make it difficult to have an accurate mathematical form of the UWAC channel. However, a large degree of correlation exists among the samples [13]. CE, being a computationally intensive operation, is done at FC (assumed to have enough computing resources), located away from the place of sensing of channel data.

To address all these issues, the present work proposes a CS-CNN-based (sparse) modeling of the UWAC channel followed by sparse image reconstruction. Figure 7.4 shows the distinctive differences and contributions of

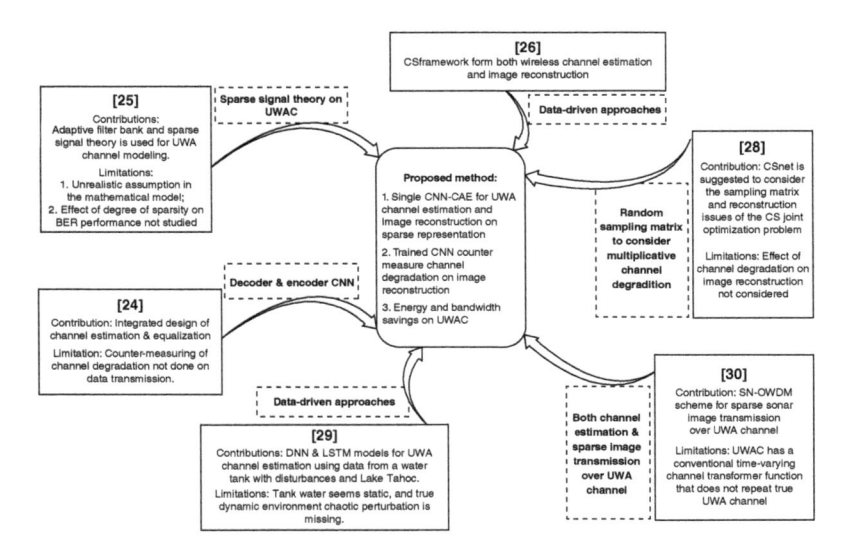

Figure 7.4 Related works.

the present work over the existing works. The main contributions of the present work are as follows:

1. Novel single convolutional autoencoder (CAE)-CNN model for energy-efficient UWAC CM and CE considering channel degradation for channel and image reconstruction
2. Underwater image reconstruction utilizing the unsupervised autoencoder (AE) for extracting features from CS measurements
3. An optimum CS sampling rate that delivers a substantial enhancement in quality is obtained by extensive simulations
4. Computations with real-time collected UWAC values

The rest of the paper is organized as follows: In Section 7.2, the open challenges of the proposed work are presented. The proposed mathematical model of the CS-CNN-based UWAC channel and image reconstruction has been explained in Section 7.3. The proposed system model is discussed in Section 7.4. In Section 7.5, the obtained simulation results of the image transmission model and comparison with existing works are presented. Lastly, the paper is concluded in Section 7.6.

7.2 OPEN CHALLENGES

The application of CNNs for sparse channel and image reconstruction in UWAC presents several open challenges that need to be addressed:

Environmental variability: UWAC channels are highly variable, influenced by factors such as water temperature, salinity, turbidity, depth, and other parameters mentioned earlier in Figure 7.3. Designing CNNs that can adapt to these changes for channel reconstruction in real-time remains an open problem.

Integration with existing systems: The integration of CNN-based systems with existing UWAC systems poses compatibility and interoperability challenges.

Generalization across environments: Achieving a high degree of generalization is essential for the practical deployment of CNN-based solutions in underwater communication scenarios. Models trained in one underwater environment may not perform well in others due to differences in acoustic properties, water conditions, and noise levels. Developing techniques to improve model generalization across diverse underwater environments is a pressing challenge.

7.3 SPARSE UWAC CM AND IMAGE RECONSTRUCTION

This section proposes a CS-based UWAC channel and image reconstruction model. The UWAC channel gain profile matrix and the image to be transmitted are assumed to be sparse for the challenging underwater environment. Although the subsequent discussion in this section mentions UWAC channel reconstruction, the same mathematical representation is equally applicable to sparse image reconstruction over the UWAC channel.

7.3.1 Channel state information: CS measurement, sensing, and transmission

Let the UWAC channel value be s. This s is considered to be a signal and is converted to a sparse measurement x by a suitable transform matrix ψ such that $x = \psi s$. The orthogonal transform basis ψ is represented as in Equation 7.1.

$$\psi = \left[\psi_1, \psi_2, \psi_3, \dots \psi_N\right] \tag{7.1}$$

The original channel signal is such that $s \in R^N$, N stands for the dimension of the UWAC channel, which can be considered as data or an image. The sparse vector of the channel signal is chosen such that $x \in R^K$, where $K \ll N$. Hence, channel data $x = \psi s$ with $K \ll N$, where K indicates the nonzero coefficients, is sparse. The transformation space may be discrete cosine transformation (DCT), discrete wavelet transformation (DWT), or Walsh Hadamard transformation (WHT), or other suitable representation space. The present work considers DCT as the basis for the sparse representation of the UWAC channel profile matrix and the image to be transmitted. The sub-sampled measurements are then transmitted to FC for reconstruction of the image and to obtain an accurate CSI of the UWAC channel. The CSI is used to optimize data transmission to the FC by adapting the communication parameters based on real-time knowledge of the UWAC channel characteristics. The CS measurement vector is represented as in Equation 7.2,

$$y = \phi x + \eta \tag{7.2}$$

where y is the measurement vector, x is the sparse UWAC channel (or image) data, and ϕ is the CS sampling matrix with dimension $M < N$. M represents the number of sub-sampled measurements. The symbol η indicates a noise vector and is often considered to be a zero-mean Gaussian signal. Substituting $x = \psi s$ in Equation 7.2, the measurement vector is obtained as in Equation 7.3.

$$y = \phi \cdot \psi s + \eta = As + \eta \tag{7.3}$$

Here, $A = \phi \cdot \psi$ indicates the reconstruction matrix. Equation 7.3 can be modified to Equation 7.4 to include the degradation on sensing.

$$y = H \odot AS + \eta \tag{7.4}$$

where H indicates the degradation matrix that contains the fading coefficients h_{ij} of UWAC channel. y is the element-wise (Hadamard) multiplication or product between the matrices H and AS. In this work, DCT-based CS measurements of the UWAC channel profile matrix are fed to the proposed CAE-CNN, and full-resolution UWAC channel profile is generated by means of training the proposed CAE-CNN algorithm. After completing the CNN training, the trained network is employed for full-resolution image (channel) reconstruction. During this process, the CNN receives DCT-based CS measurements of the image (channel) intended for transmission over the UWAC channel.

7.3.2 Reconstruction of CS channel and image

In the preceding discussions, the integration of modeling, estimation, and the consideration of degradation effects were emphasized. This section delves into the foundational phase of reconstructing the channel gain matrix within the framework of CS-based UWAC, achieved through the meticulous development of the associated weight matrix. Each time the data is transmitted, the CNN updates its weights according to the underwater parameters.

CS signal reconstruction is an ill-posed inverse problem that aims to estimate the channel by solving the objective function stated below in Equation 7.5, where $\hat{x}$ represents the reconstructed channel (considered as image data).

$$\hat{x} = \arg\min_{x \in S_c} \| \phi x - y \|_2^2 + \lambda \psi(x) \tag{7.5}$$

$\hat{x}$ is obtained by minimizing the objective function over the constraint $x \in S_c$, where S_c is the specified convex set. This constraint ensures that the solution space for the reconstruction remains within the defined convex region, contributing to the robust and effective recovery of accurate underwater images. The parameter λ serves as a tuning factor. It allows for adjusting the trade-off between fidelity to the observed data. CNN is employed here to work as a CAE, as shown in Figure 7.5. The relationship between the mean square error (MSE) and the cost function is pivotal in evaluating DNN performance, where the iterative process involving the quasi-projection operator Q as in Equation 7.6, guided by parameters n, α, and c underscores the interplay between fidelity to observed data and the impact of chosen loss or cost functions.

$$\hat{x}^n = Q\left(\hat{x}^{n+1} - \alpha c\left(\hat{x}^{n-1}\right)\right) \tag{7.6}$$

In Equation 7.6, n, α, and c indicate the number of iterations, step size, and cost function, respectively. The performance of any CAE depends on the choice of loss function. The loss function is given in Equation 7.7.

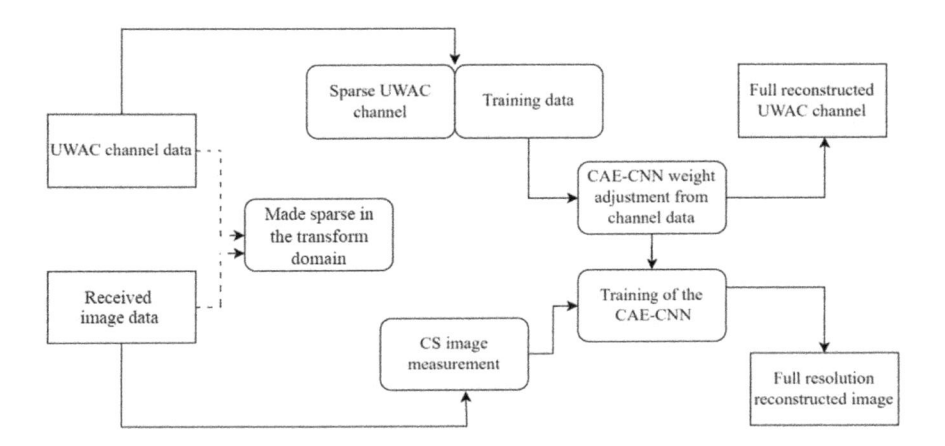

Figure 7.5 CS-CNN-based proposed model for channel and image reconstruction.

$$c\big(\hat{x}(n)\big) = E_{x_n,\hat{x}_n}\left[\left(\left\|\phi\big(x^n - \hat{x}^n\big)\right\|_2 - \left\|x^n - \hat{x}^n\right\|_2\right)^2\right] \tag{7.7}$$

Equation 7.7 employs the symbols E and ϕ to represent expectation and random measurement matrix, respectively. The subsequent section provides a detailed exposition of the proposed CAE-CNN architecture.

7.4 CAE-CNN ARCHITECTURE

This section describes the proposed CAE-CNN architecture. CAE is introduced here as the weights of the CNN are updated according to the channel parameters sensed for each time of transmission. For this, the channel is updated according to the exact CSI. This proposed architecture is used for two purposes. The first one is sparse underwater CE by CS, and the second one is its consequent use for far-end image reconstruction over the UWAC channel.

7.4.1 CAE-CNN: architecture, training, and testing

The acoustic channel model consists of an input signal, a set of linear transformations, and an output signal. The linear transformations capture the physical properties of the UWAC channel, including its frequency response, reverberation characteristics, and the multiple parameters mentioned in the previous section. The CNN architecture built upon CAE comprises an input layer, a sequence of convolutional layers, a fully connected layer, and an output layer.

The input layer receives the reconstructed signal from the acoustic CM unit. The convolutional layers are used to extract features from the signal, while the fully connected layer is used to capture the global characteristics of the signal. The output layer is used to generate a compressed version of the signal.

The hidden layers within the designed network employ the leaky rectified linear unit (ReLU) as their activation function, ensuring a non-zero gradient across its entire domain. Meanwhile, the final layer utilizes a sigmoid activation function.

7.4.1.1 CAE-CNN architecture

The proposed CNN architecture is shown in Figure 7.6. The goal is to reconstruct sparse channel gain vectors, featuring a combination of convolutional layers and pooling layers for efficient feature extraction and noise reduction. The various layers in the proposed model are as follows:

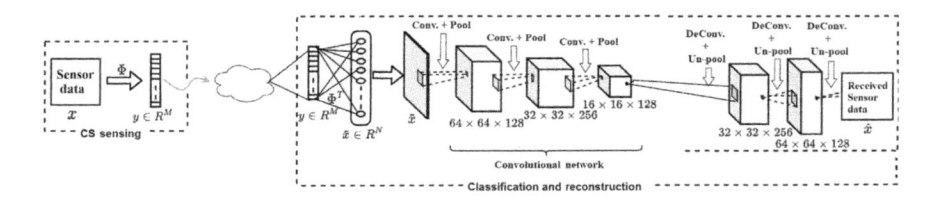

Figure 7.6 CNN architecture for underwater channel modeling and image reconstruction.

Input: The input to the CNN model is the sparse channel gain vector, x $\in R^K$, where $K \ll N$

Fully connected - 0: This is a fully connected layer that is fed with sparse channel gain data. It is employed to enable the model to learn complex relationships and patterns within the sparse channel gain data.

Conv - 1: This is a convolution layer of size $N \times N_c$ (width × height). This layer consists of 128 kernels of dimension 5×5. The layer has a stride and padding of 1.

MaxPool - 1: It is the first layer of the CNN. Here, a 2×2 filter with stride 2 is used. In the first layer, maxpooling enhances feature extraction. It is used to avail the benefits of de-noising.

Conv - 2: The second convolution layer is formed by using 256 kernels of size 3×3, and stride and padding of 1.

MaxPool - 2: This layer uses a size of 2×2 kernels and stride 2.

Conv - 3: The third convolutional layer consists of a 3×3-sized kernel comprising 128 values, using a stride of 1 and a padding of 1.

MaxPool - 3: This layer has dimensions of 2×2 and operates with a stride of 2.

7.4.1.2 Channel / Image reconstruction

The sparse UWAC channel profile matrix and sparse image measurements are reconstructed in their full-resolution forms using the proposed reconstruction algorithm. The reconstruction network is designed with the functionality of deconvolution and un-pooling (upsampling) layers. The number of kernels, size, stride, and padding remain the same as those of the encoder (feature extraction) network. But the arrangement of the network elements is in reverse order. In the subsequent deconvolution and un-pooling stages, the architecture employs a series of layers to reconstruct the channel gain matrix with precision. De-Conv - 1 and De-Conv - 2 use 256 and 128 3×3 kernels, respectively, while Un-pooling - 1 and Un-pooling - 2 perform up-sampling with 2×2 kernels. De-Conv - 3, the final deconvolution layer, adopts 5×5 kernels to yield the channel gain matrix as the ultimate output through this intricate process. The layers are briefly described below:

De-Conv - 1: The deconvolution layer is composed of 256 kernels, each measuring 3×3 in size, utilizing a stride of 1 and a padding of 1.

Un-pooling - 1: The un-pooling layer, enlarges the data size of the pooling layer, i.e., a form of up-sampling of data. A 2×2 sized kernel and a stride of 2 are chosen for the proposed model.

De-Conv - 2: Like the second convolution layer, this deconvolution layer uses 128 kernels, each with a size of 3×3 and a stride and padding of 1.

Un-pooling - 2: This layer, like the second pooling layer, contains a kernel of size 2×2 with stride 2.

De-Conv - 3: This marks the concluding deconvolution layer, which incorporates kernels sized at 5×5, employing a stride of 1 and a padding of 1.

Un-pooling - 3: This layer uses a kernel of size 2×2 with a stride of 2. The channel gain matrix is obtained as the output of this layer.

The provided Algorithm 1 outlines the training process for a CAE-CNN used for UWAC CM and reconstruction. The algorithm iteratively trains the network for a specified number of epochs (E_N). Within each epoch, it further iterates over a batch of channel parameters (B_N). During training, the network optimizes its weights (parameters) by minimizing a loss function $C(\hat{x}^{(n)})$ that measures the discrepancy between the reconstructed channel data and the actual data. The trained network is then capable of reconstructing the channel and image data. The proposed reconstruction algorithm is shown in Algorithm 7.1.

Algorithm 7.1: CAE-CNN training for the UWAC channel modeling and reconstruction

Input: Network is trained by the channel parameters

1 for $i \leftarrow 1$ to E_N do
2   for $k \leftarrow 1$ to B_N do
3      $Sum \leftarrow Sum + A_k$

4    $L_{\text{Optimize}} \; \hat{x}_i \leftarrow \hat{x}_i - \dfrac{\delta}{\delta x} \mathbb{E}_\Theta (x_i, \hat{x}_i)$

5   Compute Loss: $C(\hat{x}^{(n)}) \leftarrow \mathbb{E}_{x^n, \hat{x}^n} \left[\left(F(x^n - \hat{x}^n)_2 - x^n - \hat{x}_2^n \right)^2 \right]$

6   end

7   Update $\Theta \leftarrow \Theta - \dfrac{\delta}{\delta x} \left(C(\hat{x}^{(n)}) \right)$

8 end
9 **return** Reconstructed channel and image

The training and testing of the DL network proposed here are performed on Google's Colaboratory online cloud support service (Colab). Within Colab, the "runtime type" and "hardware accelerator" settings are configured to "Python3" and "GPU," respectively. The backend utilizes KERAS libraries and packages alongside TensorFlow 2.0.

7.4.2 CS and CNN-based measurement sensing and transmission in the fusion center

CS and CNN-based systems can be used to reduce the amount of time needed to gather, analyze, and transmit data, further reducing costs associated with the operation of the FC. The CS-CNN-based system can help reduce the risk of data tampering or data breaches, allowing for the safe and secure operation of the FC.

7.5 SIMULATION RESULTS AND DISCUSSIONS

This section presents the performance results of the proposed CAE-CNN scheme. Simulation results of the proposed system model are done by running code in MATLAB R2018b in a desktop of Intel (R) core (TM) i5-7500 CPU (3.80 GHz turbo), 16 GB RAM, Windows 10, and 64-bit operating system.

Actual channel data is first collected using the designed prototype. This data is used to train and adjust the weights of the proposed CAE-CNN algorithm. Then CS image reconstruction is done by sending the data through this channel.

7.5.1 Data collection

Experiments for data collection to study the efficacy of the proposed model were conducted by placing a floating sensor-based station at a depth of 3 meters in the river Ganga (Hooghly River), Botanical Garden Ghat (22.5529° N, 88.3007° E), West Bengal, India. The sensors, the sink, and other accessories used for data collection are shown in Figure 7.7. In these experiments, the sensor unit and the sink unit are separated by a distance of 110 meters. The data and images captured by the sensors are sent to the sink unit via UWAC. This IoUT-based real-time data collection station collects the data and uploads it to the server.

7.5.2 Data collection

In the proposed CNN method, all the learnable parameters, i.e., weight and biases, are initiated from the random variable characterized by the Gaussian distribution with a zero (0) mean and standard deviation of 0.001. The

Figure 7.7 Data collection with the developed IoUT-based system in the river Ganga.

Table 7.2 Hyperparameters used during training of the proposed system

Sl. no.	Parameters name	Value
1	Rate of initial learning	0.0013
2	Dropout rate	0.5
3	Momentum	0.01
4	Batch size	14
5	Decay	0.8
6	Optimizer used	Gradient decent
7	Activation layer used	ReLU

values of the other hyperparameters are shown in Table 7.2. The dropout probability of 0.5 is used to prevent data over-fitting.

UWAC CE accuracy in each epoch of training and validation is reported through the graphical plot in Figure 7.8. Training and validation loss decrease to 17.36% and 17.88%, respectively, after the completion of 500 epochs. This work uses dropout regularization for all the fully connected layers and batch normalization for the first fully connected hidden layer to prevent the over-fitting problem.

Figure 7.9 shows the normalized (reconstruction) error and the normalized energy consumption with the number of measurements. Reconstructed channel parameter loss is represented as normalized MSE (NMSE). NMSE is defined as in Equation 7.8 [26].

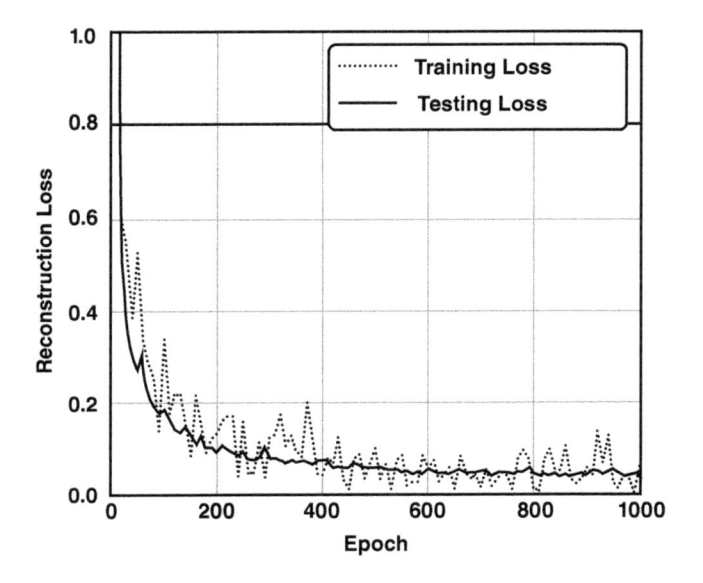

Figure 7.8 Reconstruction loss vs No. of Epochs.

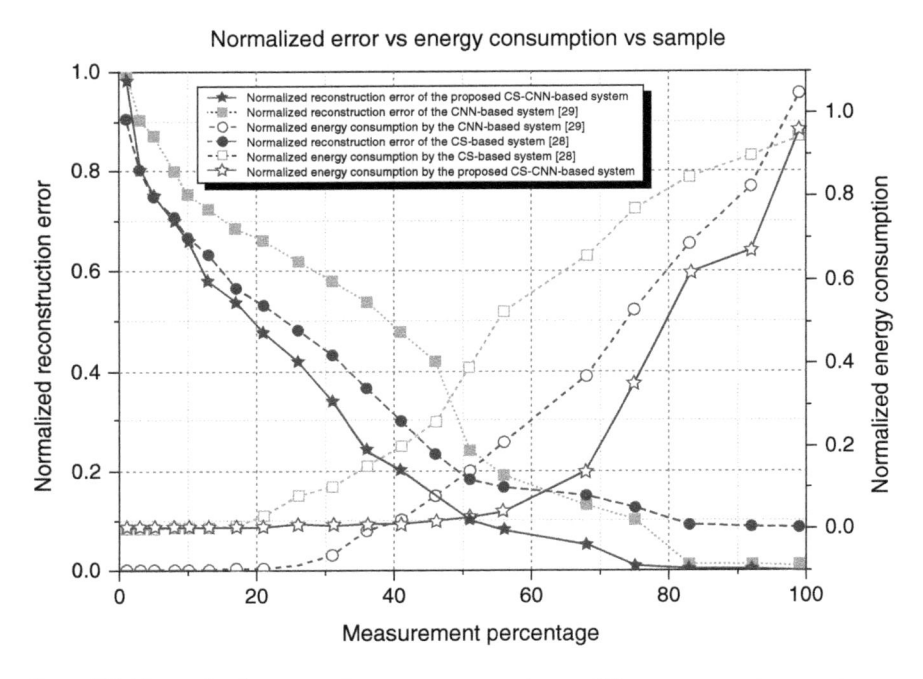

Figure 7.9 Normalised error and energy consumption vs CS measurement percentage.

$$\text{NMSE} = \frac{E\left[\|s - \hat{S}\|^2\right]}{E\left[\|s\|^2\right]} \tag{7.8}$$

In Equation 7, s and $\hat{S}$ are the channel parameters and the estimated values, respectively. E indicates expectation operation. The normalized energy consumption is defined in Equation 7.9.

$$\text{Normalized energy} = \frac{\text{EC}_P}{\text{EC}_{100}} \tag{7.9}$$

In Equation 7.9, EC_P and EC_{100} are the energy consumption for $P\%$ measurements and 100% of measurements (full reconstruction data), respectively.

Normalized error and energy consumption vs. number of samples are shown in Figure 7.9. From Figure 7.9, it is evident that at approximately 51% of channel information measurement, the values for loss and energy consumption obtained are optimum. The optimum 51% CS samples yield 20% loss and 25% energy consumption. This is especially significant as the nodes operate with limited power, making the optimal 51% measurements crucial for an energy-efficient system. Comparisons with existing works [28,29] reveal that the proposed method has shown a 0.091 (9.1%) improvement of normalized error than the existing CS-based methods.

To further highlight the efficacy of the proposed scheme, BER performance at 51% measurements is presented in Figure 7.10. The proposed method is compared with the CS-based [29] and CNN-based [24] UWAC systems. At SNR of 21 dB, the proposed model yields a BER of 2.43×10^{-3}, while the CNN-based system offers a BER of 3.51×10^{-3} and that of the CS-based method is 6.51×10^{-3}. Thus, the proposed method yields marginally better BER performance.

7.5.3 Performance results of underwater channel and image reconstruction

This section exhibits the modeling and reconstruction performance of the proposed method for sparse UWAC channel data. In addition, the performance of the reconstruction network is assessed based on the quality of the reconstructed CM, which is quantified by the PSNR and the MSSIM. Simulation is carried out over a total of fifty different images transmitted over the UWAC channel. Figure 7.11 shows three different underwater input images and the output images for CS-based and CS-CNN-based systems. Here, for the reconstruction of images, a 51% CS measurement value has been considered for energy and bandwidth-efficient communication. An average MSSIM value of 0.7831 for the CS-based system and 0.8395 for

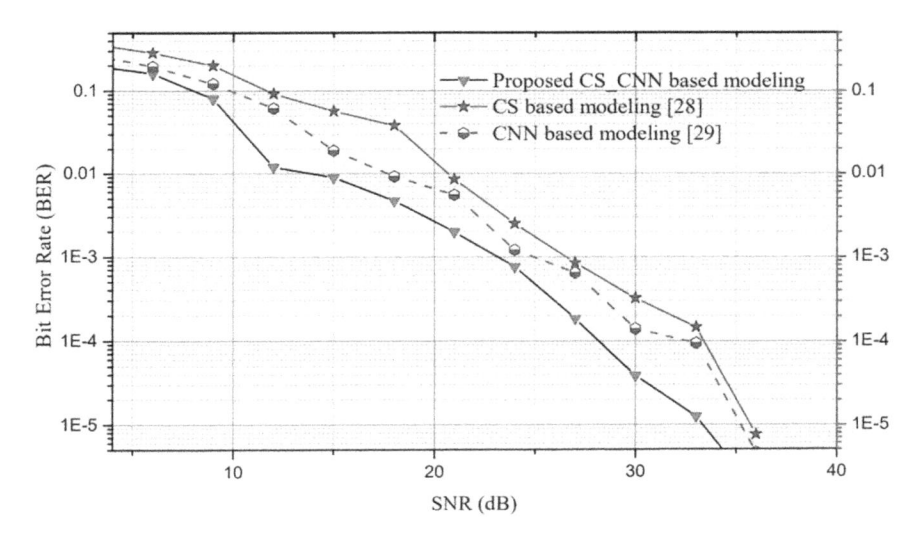

Figure 7.10 Bit error rate vs. SNR of proposed CS-CNN-based model and the existing CS-based and CNN-based models.

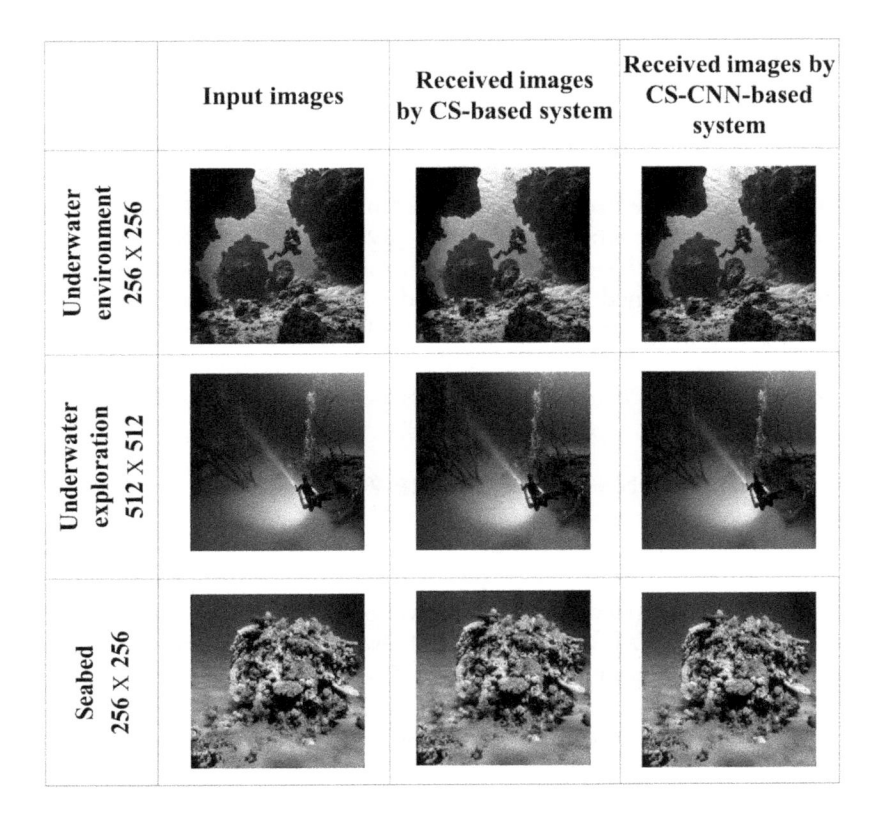

Figure 7.11 Input and output images by CS and CS-CNN-based systems.

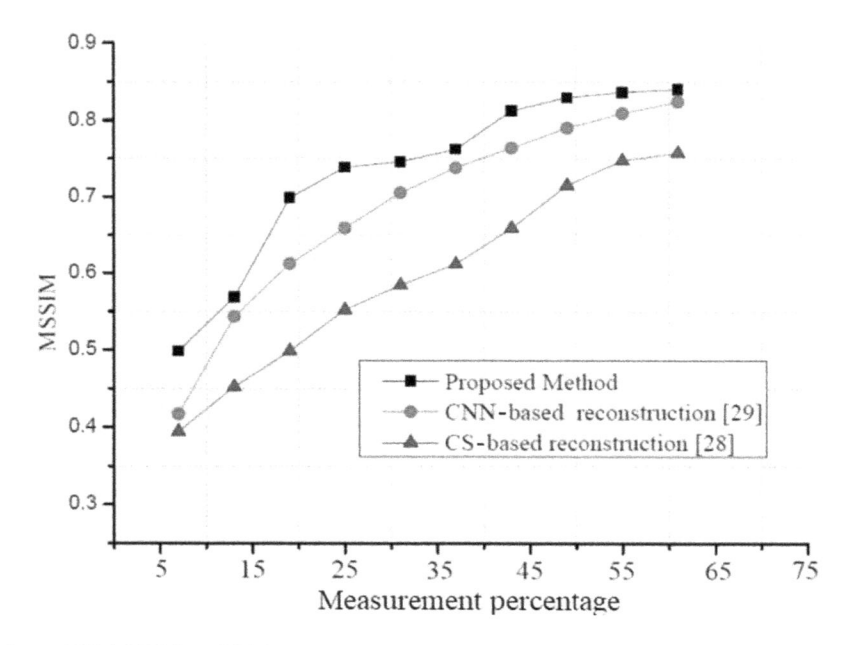

Figure 7.12 MSSIM vs CS Measurement percentage.

the proposed CS-CNN-based system is obtained. PSNR for the CS-based system is 21.3894 dB and is 24.0134 dB for the CS-CNN-based system. Figure 7.12 shows the plot of MSSIM values vs. number of measurements. At 51% CS measurements, the proposed method offers an MSSIM value of 0.7815, while the works in [28] and [29] achieve an MSSIM of 0.7145 and 0.7256, respectively. An improved result by the proposed method is achieved as the trained CNN model reduces the degradation effect of the UWAC channel. The graphical plot of PSNR values vs. the number of measurements is shown in Figure 7.13. Improvements of 2.0 and 1.5 dB by the proposed method over [28] and [29], respectively, are obtained.

7.6 CONCLUSIONS AND FUTURE WORK

This work proposes an AE-based CNN system model that serves the dual purposes of UWAC, followed by its use for energy and bandwidth-efficient sparse image reconstruction at the receiver end. The weight matrix of CNN not only models UWAC channel gain but also counters the effects of channel degradation. Simulation results show a 5.5% gain in transmission energy and 4.5% savings in transmission bandwidth over the existing works. The proposed CS-CNN-based model can also be used as a classifier/detector of various objects in the reconstructed images.

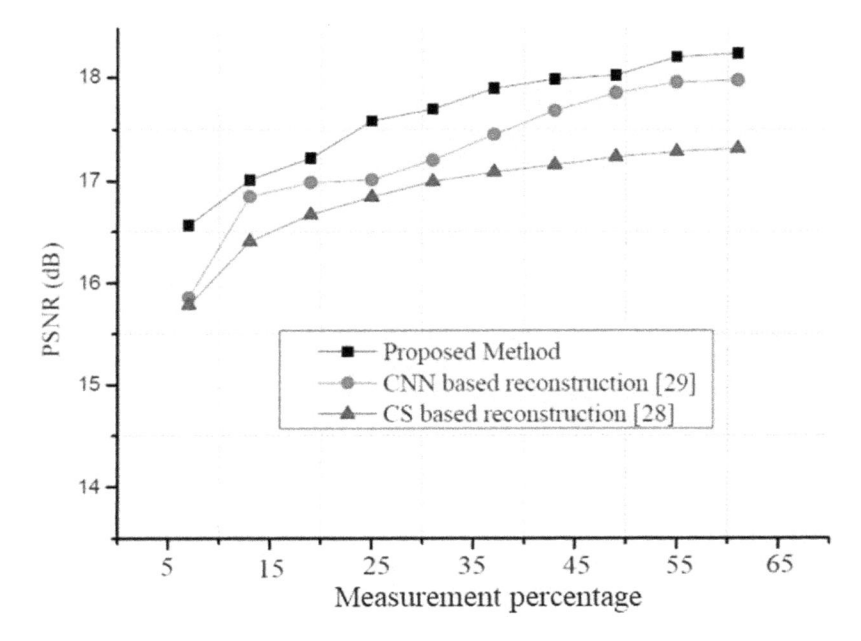

Figure 7.13 PSNR vs CS Measurement percentage.

ACKNOWLEDGEMENTS

This work is supported by the Department of Science & Technology and Biotechnology, Government of West Bengal, India, for the project "Development of UV and IoT-Based Sensor for Online Water Quality Monitoring."

REFERENCES

1. Stojanovic, M. and Preisig, J., 2009. Underwater acoustic communication channels: Propagation models and statistical characterization. *IEEE Communications Magazine*, 47(1), pp. 84–89.
2. Stojanovic, M., 2007. On the relationship between capacity and distance in an underwater acoustic communication channel. *ACM SIGMOBILE Mobile Computing and Communications Review*, 11(4), pp. 34–43.
3. M. Chitre, S. Shahabudeen, L. Freitag, and M. Stojanovic, 2008. Recent advances in underwater acoustic communications & networking. In: *OCEANS*, Quebec City, QC, IEEE, pp. 1–10.
4. Stojanovic, M., 1996. Recent advances in high-speed underwater acoustic communications. *IEEE Journal of Oceanic Engineering*, 21(2), pp. 125–136.
5. Farr, N., Bowen, A., Ware, J., Pontbriand, C. and Tivey, M., 2010. An integrated, underwater optical/acoustic communications system. In: *OCEANS'10 IEEE SYDNEY*, Sydney, NSW, pp. 1–6.

6. Stojanovic, M. and Beaujean, P.P.J., 2016. *Acoustic Communication*. Springer Handbook of Ocean Engineering, Cham, Springer, pp. 359–386.

7. Das, A.K. and Pramanik, A., 2023. Image transmission in UWA channel using CS based OTFS system. *Microsystem Technologies*, 29(11), pp. 1–12.

8. Das, A.K. and Pramanik, A., 2021. A survey report on underwater acoustic channel estimation of MIMO-OFDM system. In: *Proceedings of International Conference on Frontiers in Computing and Systems: COMSYS*, Singapore, Springer, pp. 745–753.

9. Berger, C.R., Wang, Z., Huang, J. and Zhou, S., 2010. Application of compressive sensing to sparse channel estimation. *IEEE Communications Magazine*, 48(11), pp. 164–174.

10. Roy, S., Duman, T.M., McDonald, V. and Proakis, J.G., 2007. High-rate communication for underwater acoustic channels using multiple transmitters and space-time coding: Receiver structures and experimental results. *IEEE Journal of Oceanic Engineering*, 32(3), pp. 663–688.

11. Yang, Z. and Zheng, Y.R., 2015. Iterative channel estimation and turbo equalization for multiple-input multiple-output underwater acoustic communications. *IEEE Journal of Oceanic Engineering*, 41(1), pp. 232–242.

12. Khan, M.R., Das, B. and Pati, B.B., 2020. Channel estimation strategies for underwater acoustic (UWA) communication: An overview. *Journal of the Franklin Institute*, 357(11), pp. 7229–7265.

13. Kang, T. and Iltis, R.A., 2008. Iterative carrier frequency offset and channel estimation for underwater acoustic OFDM systems. *IEEE Journal on Selected Areas in Communications*, 26(9), pp. 1650–1661.

14. Hu, D., Wang, X. and He, L., 2013. A new sparse channel estimation and tracking method for time-varying OFDM systems. *IEEE Transactions on Vehicular Technology*, 62(9), pp. 4648–4653.

15. Panayirci, E., Altabbaa, M.T., Uysal, M. and Poor, H.V., 2019. Sparse channel estimation for OFDM-based underwater acoustic systems in Rician fading with a new OMP-MAP algorithm. *IEEE Transactions on Signal Processing*, 67(6), pp. 1550–1565.

16. Gholizadeh, M.H., Melesse, A.M. and Reddi, L., 2016. A comprehensive review on water quality parameters estimation using remote sensing techniques. *Sensors*, 16(8), p.1298.

17. Khan, M.R., Das, B. and Pati, B.B., 2020. Channel estimation strategies for underwater acoustic (UWA) communication: An overview. *Journal of the Franklin Institute*, 357(11), pp. 7229–7265.

18. Long, H., Chen, H. and Zhang, T., 2021. Classification on underwater acoustic propagation model using convolutional neural network. In: *IEEE International Conference on Signal Processing, Communications and Computing (ICSPCC)*, Xi'an, IEEE, pp. 1–6.

19. Berger, C.R., Zhou, S., Preisig, J.C. and Willett, P., 2009. Sparse channel estimation for multicarrier underwater acoustic communication: From subspace methods to compressed sensing. *IEEE Transactions on Signal Processing*, 58(3), pp. 1708–1721.

20. Li, Y., Zhang, Y., Xu, X., He, L., Serikawa, S. and Kim, H., 2019. Dust removal from high turbid underwater images using convolutional neural networks. *Optics & Laser Technology*, 110, pp. 2–6.

21. Liu, L., Cai, L., Ma, L. and Qiao, G., 2021. Channel state information prediction for adaptive underwater acoustic downlink OFDMA system: Deep neural networks based approach. *IEEE Transactions on Vehicular Technology*, 70(9), pp. 9063–9076.
22. Zhang, Y., Li, J., Zakharov, Y., Li, X. and Li, J., 2019. Deep learning based underwater acoustic OFDM communications. *Applied Acoustics*, 154, pp. 53–58.
23. Han, G., Wang, S., Chang, S. and Fu, X., 2022. Time-reversal CNN-based S-NOFDM scheme for underwater acoustic communication. *IEEE Systems Journal*, 17(2), pp. 2868–2879.
24. Liu, J., Ji, F., Zhao, H., Li, J. and Wen, M., 2021. CNN-based underwater acoustic OFDM communications over doubly-selective channels. In: *2021 IEEE 94th Vehicular Technology Conference (VTC2021-Fall)*, Norman, OK, IEEE, pp. 01–06.
25. Crombez, S., Petraglia, M.R. and Petraglia, A., 2019. Underwater acoustic channel estimation and equalization via adaptive filtering and sparse approximation. In: *2019 27th European Signal Processing Conference (EUSIPCO)*, A Coruña, IEEE, pp. 1–5.
26. Pramanik, A., Maity, S.P. and Farheen, Z., 2019. Compressed sensing channel estimation in massive MIMO. *IET Communications*, 13(19), pp. 3145–3152.
27. Su, W., Tao, J., Pei, Y., You, X., Xiao, L. and Cheng, E., 2020. Reinforcement learning based efficient underwater image communication. *IEEE Communications Letters*, 25(3), pp. 883–886.
28. Shi, W., Jiang, F., Liu, S. and Zhao, D., 2019. Image compressed sensing using convolutional neural network. *IEEE Transactions on Image Processing*, 29, pp. 375–388.
29. Onasami, O., Adesina, D. and Qian, L., 2021. Underwater acoustic communication channel modeling using deep learning. In: *WUWNet '21: Proceedings of the 15th International Conference on Underwater Networks & Systems*, ACM, New York, pp. 1–8.
30. Zhang, J. and Ma, X. and Fu, X. and Yang, J., 2019. Sparse nonorthogonal wavelet division multiplexing for underwater sonar image transmission. *IEEE Transactions on Vehicular Technology*, 68(12), pp. 11806–11815.

Chapter 8

Optimization of future location prediction in mobile sensor network using Particle Swarm Optimization

Rupam Some and Subhojit Malik

8.1 INTRODUCTION

In modern times, developing infrastructures, such as surveillance network, structural health monitoring, smart grid, environmental or habitat monitoring, depends on wireless sensor network (WSN)-based applications to ascertain the exact positioning of targets, laying the groundwork for subsequent sophisticated operations like automation and control. Hence, it is evident that there is a clear demand for effective location prediction techniques within WSNs. It is a common exercise in surveillance application to drop sensor nodes form airborne devices which may be applicable where the environment is hostile [1]. Contemporary applications especially surveillance-centric applications may be benefitted using unmanned aerial vehicle (UAV). Systems for independent deployment through UAVs are currently available [2]. Contemporary UAVs are equipped with advanced control and perception systems, endowing them with the capability to execute coordinated deployment missions [3]. Upon deployment of the sensor nodes embedded with UAVs in the target area, it is very much desirable to track the optimal target location of the same in order to deliver the acquired data to an aggregator node for final delivery. Especially, for a surveillance network where such numerous UAVs are positioned computing the optimal location is a challenging task for maintaining accurate delivery of data [4,5]. Prediction of location of the application node (AN) module curtails the practice of transmission of data in terms of determining the future locality of the AN thus minimizing the cost of transmission. Numerous research activities were carried out on the aforesaid area using different means, but a mere number of these approaches concentrates on location prediction as well as optimization of the location [4,6].

The proposed scheme considers parallel environment to formulate a solution for the optimal location tracking problem in mobile sensor network using particle swarm optimization (PSO) [7,8] so as to ensure accurate data delivery. The said scheme ensures determining the optimized predicted location of the ANs with great precision as compared with other existing schemes in the same domain.

DOI: 10.1201/9781003517689-8

The subsequent sections of this paper are structured as follows: In Section 8.2, a review is provided, delving into various approaches concerning the optimization of location and elucidating the motivation behind the proposed work. Moving to Section 8.3, we delve into the intricacies of the proposed optimized location prediction algorithm. This section encompasses a comprehensive discussion, covering key aspects, such as the brief outline of the algorithm, formation of swarms of ANs, energy model, formulation of fitness, consideration of penalty parameters, and the calculations of mean square error (MSE). Section 8.4 is dedicated to presenting the numerical simulation and results, offering a visual representation of the system's performance through a comparative analysis with existing works. Finally, in Section 8.5, we draw conclusions from our findings and outline potential directions for future research. Throughout the paper, our primary focus has been on optimizing the location of deployed ANs, employing PSO to achieve an impressive accuracy rate of 95% (Figure 8.1).

8.2 MOTIVATION AND RELATED WORK

a. Collision avoidance:

Collision avoidance is essential to prevent accidents in various domains, including vehicles, aircraft, vessels, robots, and WSNs. Its primary goal is to ensure safety, minimize property damage, and maintain smooth transportation operations. Real-time monitoring systems with collision avoidance capabilities are vital to detect potential threats promptly, enabling timely responses.

Collision avoidance systems rely on ultrasonic sensors, cameras, radio detection and ranging (RADAR), light detection and ranging (LIDAR), and sound navigation and ranging (SONAR) to detect nearby objects positions and movements. These systems analyze data to plan paths, identify alternative routes, and assess collision risks based on factors like speed, direction, distance, and relative motion.

When a potential collision risk is detected, these systems generate warnings, which can be visual, auditory, or haptic, to alert individuals to take evasive action. Cutting-edge technologies, such as vehicle-to-vehicle (V2V) and vehicle-to-infrastructure (V2I) communication, play a pivotal role in augmenting collision avoidance capabilities. These technologies empower vehicles to share crucial information regarding their position and intentions, fostering a proactive approach to collision prevention, even in scenarios where direct visual contact is not feasible. Compliance with regulatory standards and guidelines is crucial for the proper installation and performance of collision avoidance systems in different transportation modes, ensuring reliability and safety [4].

Path Planning ·······························→ Prediction of Optimized Location of Each Node

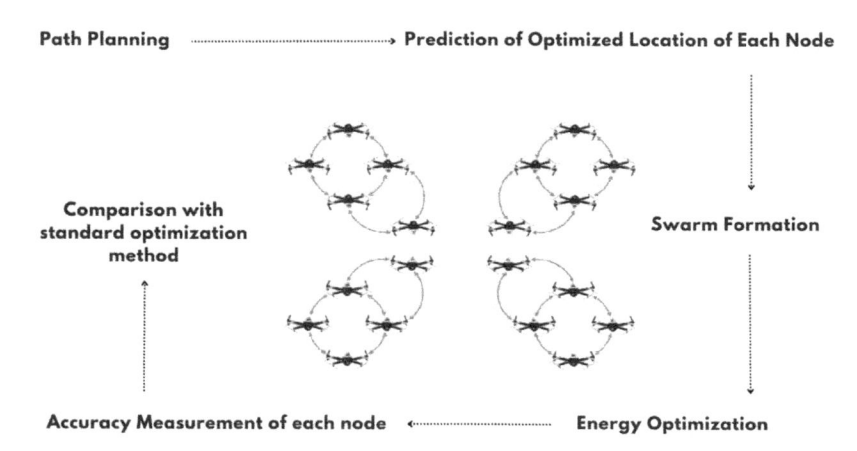

Comparison with
standard optimization
method

Swarm Formation

Accuracy Measurement of each node ←····················· Energy Optimization

Figure 8.1 Graphical representation of the chapter.

b. Limitations of collision avoidance:

Collision avoidance systems are essential for safety, but they have a few limitations:

- **Sensor limitations:** Sensors like RADAR and LIDAR are costly and can struggle in poor weather or obstructed conditions, affecting accuracy.
- **System complexity:** Integration of various components, sensors, algorithms, and communication systems can be complex, leading to potential performance delays or issues.
- **False detection:** Environmental factors and sensor noise can lead to false alarms, eroding user trust.
- **Unpredictable situations:** Pre-programmed systems may not handle unforeseen scenarios, requiring human intervention.
- **Human dependencies:** Human decision-making is sometimes necessary, impacting system performance.

Ongoing research focuses on improving collision avoidance systems, leveraging advancements in wireless sensor technology and artificial intelligence to address these limitations.

c. Optimization techniques:

Optimization constitutes the systematic exploration for the most favorable solution among a range of available options for a given problem. This intricate process entails the maximization or minimization of a defined objective, taking into account a multitude of constraints and conditions. In the realms of mathematics and computer science, optimization emerges as a dedicated field, focusing on the formulation of algorithms and techniques designed to uncover the optimal solution to a specified problem. Typically, these problems entail the optimization of a specific function, termed the objective function, within the framework of fulfilling a set of predefined constraints.

Few engineering applications of optimization:

Swarm-based optimization techniques are applied in various fields to improve efficiency and find the best solution.

Optimization techniques like PSO [7,8]and bacterial foraging algorithm (BFA) [9] have been widely employed in engineering applications to augment efficiency and optimize resource utilization, as highlighted by reference [4]. PSO, inspired by the social behavior of birds, has proven effective in solving various multidimensional optimization problems [4,10]. It has seen numerous adaptations and applications in diverse fields, including multi-robot navigation, power systems, and pattern classification [5,11].

BFA, a newer evolutionary optimization algorithm, mimics the foraging behavior of bacteria [6]. This algorithm has showcased its prowess in optimizing complex scenarios, such as economic load dispatch and power systems, as evidenced by references [12,13]. Drawing inspiration from the movement patterns of *Escherichia coli* bacteria, the algorithm simulates their alternating behaviors of tumbling and swimming as they navigate and seek out nutrient-rich locations. [6,14].

In the context of WSNs, managing node locations is crucial to minimize power consumption. Traditional methods often rely on assumptions that may not hold in real-world industrial deployments [15–17]. Deterministic approaches have been explored to minimize node deployment while maintaining network connectivity [1,18]. Large-scale sensor network deployments remain a challenge due to their stochastic nature [18].

The following table depicts a few key features of different swarm-based optimization techniques (Table 8.1):

d. **Algorithms of few swarm-based optimization techniques:**

Following section outline some algorithms.

i. **PSO:**

The PSO algorithm is outlined as follows [5,11]:

Step 1: Initialize a population of particles with randomly assigned positions and velocities.

Step 2: Specify the objective function to be optimized.

Step 3: Keep a record of each particle's best position (p_{best}) and the overall best position in the entire population (g_{best}).

Step 4: Adjust particle velocities based on their individual p_{best} and the global best position (g_{best}) using defined mathematical equations.

Step 5: Update particle positions by incorporating their velocities.

Table 8.1 Major features of popular swarm-based optimization techniques

Swarm-based optimization technique	Inspiration	Features	Parameters
Particle Swarm Optimization (PSO)	Social behavior of birds or fish	Velocity updates based on best known positions	Swarm size, inertia weight
Bacterial Foraging Algorithm (BFA)	Foraging behavior of bacteria	Chemotaxis and reproduction to find optimal food	Chemotactic step size, reproduction rate
Ant Colony Optimization (ACO)	Foraging behavior of ants	Pheromone-based probabilistic path selection	Pheromone evaporation rate
Cuckoo Search	Behavior of cuckoos that lay eggs	Egg laying and nest selection probabilistic	Nest selection, discovery rate
Artificial Bee Colony (ABC)	Foraging conduct exhibited by honey bees	Bees explore based on nectar amount	Number of employed/ onlooker bees

Step 6: Iterate through steps 4 and 5 for a predetermined number of iterations or until a specified termination criterion is satisfied.

Step 7: The ultimate g_{best} denotes the optimal solution achieved by the PSO algorithm.

ii. **BFA:**

The algorithm for BFA has been given below [5,19]:

Step 1: Simulate the movement of bacteria in a search space.

Step 2: Bacteria move based on chemotaxis (toward higher nutrient concentrations) and reproduction.

Step 3: Chemotaxis involves adjusting their positions by sensing their environment and comparing it to their previous positions.

Step 4: Reproduction involves creating new bacteria with slight variations from the parent bacteria.

Step 5: Iterate these processes to explore and converge toward an optimal solution.

iii. **Ant colony optimization (ACO):**

The algorithm for ACO has been given below [5,20]:

Step 1: Model the foraging behavior of ants in search of food.

Step 2: Use pheromone trails to communicate and remember paths.

Step 3: Ants prefer paths with higher pheromone levels.

Step 4: Over time, pheromone levels increase on better paths and decrease on less favorable ones.

Step 5: Repeated iterations lead to the discovery of optimal solutions.

iv. **Cuckoo search:**
   The algorithm for cuckoo search has been given below [5,21]:
   Step 1: Based on the breeding behavior of cuckoo birds.
   Step 2: Each cuckoo lays eggs (potential solutions) in nests (search points).
   Step 3: Eggs with better fitness replace eggs in nests with lower fitness.
   Step 4: Some eggs are randomly discarded, simulating brood parasitism.
   Step 5: Iteratively, the algorithm refines solutions for optimization problems.

v. **Artificial bee colony (ABC):**
   The algorithm of ABC unfolds as follows [5,22]:
   Step 1: Simulate the foraging behavior of honeybees.
   Step 2: Classify bees into three categories: employed, onlooker, and scout.
   Step 3: Employed bees actively explore potential solutions, while onlooker bees select promising solutions informed by the information gathered from employed bees.
   Step 4: Scout bees discover new solutions when employed bees are no longer productive.
   Step 5: Repeated iterations improve solutions for optimization problems.

e. **Major applications of WSN:**
   Within the domain of WSN, application refers to the practical use or purpose for which the network is deployed. WSNs are designed to support a wide range of applications, including environmental monitoring, home automation, industrial control, healthcare, and more. Each application defines the specific tasks and objectives that sensor nodes within the network are meant to accomplish, such as collecting data, transmitting information, or controlling devices, depending on the intended use case. The aim of this chapter is to employ a suitable optimization algorithm for the deployment of a set of wireless sensor nodes.

   Beside the above applications, any suitable optimization-based algorithms could be used in the field of WSN for energy efficient routing, data aggregation, wake or sleep scheduling, cross layering, quality of service (QoS), localization and time synchronization, clustering of nodes, sink node optimization, etc.

f. **Major advantages of location prediction of nodes using optimization:**
   a. **Enhanced efficiency:** Predicting node locations improves network efficiency by optimizing data routing, reducing energy consumption, and extending network lifespan.

    b. **Data accuracy:** Accurate node location predictions enhance data quality by factoring in spatial information, leading to more reliable information collection.

    c. **Cost savings:** Optimized node placement can reduce deployment and maintenance costs by minimizing the need for additional nodes.

    d. **Resilience:** Location prediction aids in fault tolerance and self-healing mechanisms, enhancing network reliability in the face of disruptions.

    e. **Resource management:** Optimization ensures efficient resource usage, reducing network congestion and packet loss.

g. **Major disadvantages of location prediction of nodes using optimization:**

    a. **Complexity:** Implementing optimization algorithms can be computationally demanding and intricate, necessitating substantial processing power.

    b. **Accuracy challenges:** Predictive models may struggle with dynamic environmental changes or node mobility, leading to occasional inaccuracies.

    c. **Resource overhead:** Continuous algorithm execution may consume extra energy and computational resources.

    d. **Setup complexity:** Initial system calibration and configuration may demand expertise and time.

    e. **Environmental data dependency:** Accuracy depends on reliable and accessible environmental data sources, which may not always be available.

8.3 PROPOSED OPTIMIZED LOCATION PREDICTION ALGORITHM

8.3.1 Brief outline of the algorithm

The primary goal of every optimization algorithm is to identify the most optimal solution to a real-world problem within a defined set of constraints. Finding the future location of any AN of a mobile sensor network using swarm intelligence is one of such leading solutions to any prediction-based problem. Initially, the swarm constitutes a loosely organized assembly of mobile nodes within a cluster. Each node exhibits apparent random movement in terms of direction, velocity, and position during the initial phase. At the onset, the position and velocity of each node remain uncertain, initialized with random values at time "t." Subsequently, at time "$t+1$," updates to position and velocity occur based on standard equations outlined in the preceding section. The personal best position for each node and the global best position for the entire swarm are initially set to zero. A fitness function for each moving node, relative to the aggregator node, is established, taking into account various constraints, such as collision detection and energy loss. The optimization goal involves minimizing the fitness value after each iteration, prompting an update to the personal best position if a new minimum fitness is achieved. This swarm-based optimization technique is systematically applied to the entire cluster, guiding the prediction of future locations for randomly moving sensor nodes. The proposed scheme revolves around accurately forecasting and optimizing the positions of ANs concerning a designated reference coordinator node, where the sensed phenomenon is intended to be delivered with precision. In this context, the system model adopts UAVs as the designated ANs. Upon deployment, the ANs form a swarm and wait for the receipt of beacon frame form the coordinator node.

8.3.2 Formation of swarm of ANs

At the beginning, a population of particles i.e., ANs $x_{id}(t)$, $y_{id}(t)$, $z_{id}(t)$ are initialized with positions assigned at random, each marked by a corresponding random velocity, $V_{id}(t)$.

The collective of these ANs is denoted as swarm S. Within this swarm of ANs, a neighborhood relation N is established to ascertain the adjacency between any two nodes, P_i and P_j. Consequently, for any given AN P, its neighborhood, designated as $N(P)$, encompasses all nodes that are considered its neighbors within the swarm.

Each AN P has two state variables:

a. Current position (such as $x_{id}(t)$, $y_{id}(t)$, $z_{id}(t)$)
b. Current velocity (such as $V_{id}(t)_x$, $V_{id}(t)_y$, $V_{id}(t)_z$)

Each particle P has a small memory also and the memory comprises with

a. Previous best position (such as $p_{idx}(t)$, $p_{idy}(t)$), $p_{idz}(t)$) i.e., personal best experience
b. The best $p_{id}(t)$ of all $P \in N(P)$: $g_{id}(t)$ [such as $g_{idx}(t)$, $g_{idy}(t)$, $g_{idz}(t)$] i.e., the optimal position discovered thus far within the particle's vicinity.

Initially, we set $p_{id}(t) = g_{id}(t) = 0$ for all particles.

Upon the initialization of particles, the iterative optimization process commences, during which the positions and velocities of all particles undergo modification according to the subsequent equations. The provided equations delineate the adjustments made for the dth dimension in both position and velocity pertaining to the ith particle. Specifically, the update for the velocity of the ith particle is expressed as follows:

$$V_{id}(t+1)_x = \omega V_{id}(t)_x + C_1\omega_1\left(p_{id}(t)_x - x_{id}(t)\right) + C_2\omega_2\left(g_{id}(t)_x - x_{id}(t)\right)$$
(8.1)

$$V_{id}(t+1)_y = \omega V_{id}(t)_y + C_1\omega_1\left(p_{id}(t)_y - y_{id}(t)\right) + C_2\omega_2\left(g_{id}(t)_y - y_{id}(t)\right)$$
(8.2)

$$V_{id}(t+1)_z = \omega V_{id}(t)_z + C_1\omega_1\left(p_{id}(t)_z - z_{id}(t)\right) + C_2\omega_2\left(g_{id}(t)_z - z_{id}(t)\right)$$
(8.3)

$$x_{id}(t+1) = x_{id}(t) + V_{id}(t+1)x$$
(8.4)

$$y_{id}(t+1) = y_{id}(t) + V_{id}(t+1)y$$
(8.5)

$$z_{id}(t+1) = z_{id}(t) + V_{id}(t+1)z$$
(8.6)

where, $x_{id}(t)$, $y_{id}(t)$, $z_{id}(t)$ denote the present location and $x_{id}(t+1)$, $y_{id}(t+1)$, $z_{id}(t+1)$ are the positions at $(t+1)^{th}$ time of the particle p_i. Though initially, $p_{idx}(t)=p_{idy}(t)=p_{idz}(t)=g_{idx}(t)=g_{idy}(t)=g_{idz}(t)=0$, the Equations 8.1–8.3 can be reduced to

$$V_{idx}(t+1) = \omega V_{idx}(t)$$
(8.7)

$$V_{idy}(t+1) = \omega V_{idy}(t)$$
(8.8)

$$V_{idz}(t+1) = \omega V_{idz}(t)$$
(8.9)

Through the aid of preceding Equation 8.7, Equations 8.4–8.6 can be expanded as

$$x_{id}(t+1) = x_{id}(t) + \omega v_{id}(t)$$
(8.10)

$$y_{id}(t+1) = y_{id}(t) + \omega v_{id}(t) \tag{8.11}$$

$$z_{id}(t+1) = z_{id}(t) + \omega v_{id}(t) \tag{8.12}$$

8.3.3 Energy model

The overall energy consumption of the node comprises two essential components. The initial component pertains to communication-related energy, encompassing factors, such as radiation, signal processing, and other circuitry. The second component is the propulsion energy, essential for maintaining the node's altitude and facilitating mobility if required. The energy cost, denoted as $C_i(t)$ for the ith node, can be articulated as follows:

$$C_i(t) = E_c(t)/E_i(t) \tag{8.13}$$

where $E_c(t)$ and $E_i(t)$ are consumed and the initial energy of the ith node. $C_i(t)=0$ indicates battery of AN i at time t is full, and on the other hand, $C_i(t)=1$ indicates battery AN i at time t is depleted. $E_c(t)$ can be further expanded as

$$E_c(t) = E_{tx} + E_{rx} + E_{overheadx} + E_{idle} + E_{overhearing} + E_{propulsion} \tag{8.14}$$

where E_{tx} represents the energy necessary for transmitting data, E_{rx} corresponds to the energy needed for receiving data, $E_{overheadx}$ denotes the energy expended in the reception and processing of control packets, E_{idle} accounts for the energy consumed while monitoring the channel, and $E_{overhearing}$ signifies the energy expended in capturing neighboring packets. On the other hand, $E_{propultion}$ is the propulsion energy required as a function of the trajectory $q(t)$, which can be expressed as

$$E(q(t)) = \int_0^T \left[c_1 \|v(t)\|^2 + \frac{c_2}{\|v(t)\|} \left(1 + \frac{\|a(t)\|^2 - \frac{a^T v(t)^2}{\|v(t)\|^2}}{g^2} \right) \right] dt$$
$$+ \frac{1}{2} m \left(\|v(T)\|^2 - \|v(0)\|^2 \right) \tag{8.15}$$

where, $v(t)=q(t)$ represent the instantaneous velocity vector, and $a(t)=q(t)$ denote the acceleration vector at a given time. The parameters c_1 and c_2 are associated with factors, such as node weight, wing area, and air density. Additionally, g signifies gravitational acceleration, and m represents the total mass of the node, inclusive of its payload.

8.3.4 Formulation of fitness

Each AN maintains a record of its coordinates in the problem space, linked to the best solution (fitness) it has attained thus far. So, in order to optimize the location of the ANs, a fitness function is indispensable. The coordinator node periodically transmits beacon frames to the ANs present in the phenomenal area. Upon receiving the frame from the coordinator, ANs analyze the distance with the coordinator in terms of received signal strength as illustrated in Equation 8.17. Reduction of distance signifies that the member unit is approaching toward the cluster coordinator. The fitness function for the said optimization scenario is expressed in terms of power received by the receiver antenna and as stated below:

$$P_{ri} = P_{ti} \times G_{ti} \times G_{ri} \times \left(\lambda / 4\pi R \right)^2 \tag{8.16}$$

where, P_{ti}, G_{ti}, G_{ri}, R be the power of the ith transmitter, gain of the transmitter, gain of the receiver, and transmission radius of the coordinator node, respectively. Equation 8.16 can be represented in terms of transmission radius of the CH as

$$R_i = \sqrt{P_{ti} \times G_{ti} \times G_{ri} \times \lambda^2 / 4\pi^2 \times P_{ri}} \tag{8.17}$$

According to the fitness function as stated in the Equations 8.17 and 8.16, fitness of the ith AN can be calculated as

$$\text{fitness}_i = \text{Min}\left(R_i \right), N \le i \le 1 \tag{8.18}$$

where N represents the count of ANs within the system. To address the problem, the fitness function incorporates supplementary objective parameters for

- Avoidance of collision with neighbor ANs and cluster coordinator and
- Maintain an energy threshold of the AN

8.3.5 Consideration of penalty parameters

1. Selection of objective parameter for collision avoidance between ANs: Given that the primary aim of the proposed system is to enhance the positioning of ANs based on the assessment of signal strength from

the coordinator, cluster members will strategically move toward the cluster coordinator. This movement is undertaken with the intent of identifying an optimal location for the efficient delivery of the sensed phenomenon. The said location optimization scheme also creates a chance for the ANs to collide with the neighbors and the coordinator node. In order to encounter the scenario, a penalty parameter must the considered so as to bypass the same. Considering d be the distance between neighbor ANs i and j, ε be the threshold distance between

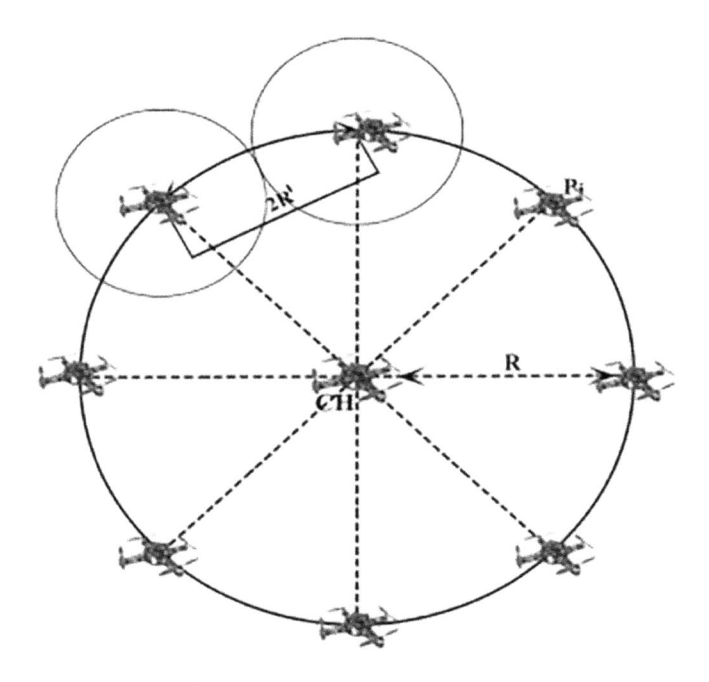

Figure 8.2 Cluster layout of the simulation environment.

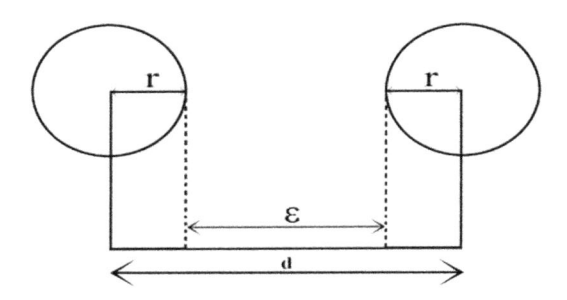

Figure 8.3 Ideal condition to avoid collision between two nodes.

two Ans, and r be the transmission radius of the ANs (Figures 8.2 and 8.3, the condition for bypassing the collision between neighbor i, j be:

$$d \geq 2r + \varepsilon \tag{8.19}$$

The penalty parameter for the abovementioned scenario for all ANs may be expressed as

$$P_\xi = \zeta_k \sum_{i=1}^{N-1} \left(d - \left(2r + \varepsilon \right) \right)^2 \tag{8.20}$$

where N=total number of nodes in the cluster and ςk is the penalty coefficient.

2. **Selection of objective parameter for energy threshold:** As in the proposed setup, optimization of the location of the AN is achieved through rigorous iteration involving a considerable amount of energy dissipation because of both combination and propulsion energy. Thus, engrossment of energy as a penalty parameter is obvious. An application node AN_i is alive and able to perform requisite operations for discovery of the optimal location when the energy of the node $C_i(i)$ is above a threshold $C_{i\tau}(t)$ required to perform requisite operations for mobility and optimization of location. Thus, the following conditions may sort out the scenario as
   - if $C_i(t) < C_{i\tau}(t)$, the energy of the ith AN is less than the threshold, the mobility and communication ability of the node is limited. This leads to a decline in the network's overall lifespan. To sustain a consistent performance level, it may be necessary to substitute the aforementioned AN with a new node.
   - if $C_i(t) > C_{i\tau}(t)$, the energy of the ith AN is above the threshold i.e., the mobility and communication ability of the node is adequate for the scenario. Upon inclusion of the abovementioned objective parameters as penalties, the fitness function may be expressed as

$$f\left(\overline{R}\right) = \min\left(\sum_{i=0}^{N} (R_i + P_\xi) \right) \text{ and } C_i\left(t\right) \geq C_{iT}\left(t\right) \forall R_i, 0 < i < N \tag{8.21}$$

8.3.6 Calculations of MSE

To evaluate the effectiveness of the proposed algorithms in positioning sensor nodes within an experimental field, the MSE serves as the chosen performance metric. The MSE is formally defined as

$$\text{MSE} = \frac{1}{N} \sum_{i=1}^{N} \left(L_a - L_P \right)^2 \qquad (8.22)$$

where L_a is the actual location of the coordinator node and L_P is the predicted location by AN. N is the total number of ANs in the target area. The fluctuation in MSE is computed across various iterations, considering different threshold distances. In order to ensure the algorithm's impartiality and the thorough analysis of results, multiple simulations were conducted. Furthermore, to mitigate the risk of sensor position overlap, diverse threshold distances between sensor nodes were defined, and the algorithm underwent simulation numerous times with distinct initial positions.

Lemma 8.1

The solution to the problem of optimization of future location prediction in mobile sensor network is nondeterministic polynomial time (NP)-Complete in nature.

For a problem to be in NP type, the following criteria must be met. Firstly, the problem has to be a decision problem and must be a non-deterministic polynomial signifying multiple solution with respect to a single input domain for cases where affirmations can be confirmed within polynomial time complexity ($O(nk)$), with n as the problem size and k as a constant). The solution to problem of optimization of future location is purely a decision problem as the optimized location of the AN is achieved against multiple future locations, but the final solution is set against the best fitness value. Initially, the future location is computed for the all the nodes in the swarm having a size S. Subsequently, by employing the fitness function outlined in Equation 8.17, the fitness values for all nodes are calculated, and the node exhibiting the optimal fitness value is selected as the node in the optimized location. This process results in a time complexity of $O(n^2)$. According to the definition of an NP-complete problem, a problem x is classified as NP-complete if and only if it is in NP, and every other problem in NP can be efficiently transformed into x within polynomial time. In essence: x is in NP, and every problem in NP is reducible to x. As per the framework presented in algorithm 1, the problem of optimizing the location of a mobile sensor network is unequivocally NP-complete, as previously discussed. The selection of an AN is contingent on the best fitness value, a criterion that is verifiable within polynomial time. Consequently, it is established that the solution aligns with the characteristics of a pure NP-complete problem.

Algorithm 8.1: Prediction and optimization of future location

$xid(t)$, $yid(t)$, $zid(t)$ *(initial position of the i th node)*,

$xid(t+1)$, $yid(t+1)$, $zid(t+1)$ *(predicted position of the i thnode at $(t+1)$th time)*,

$Vid(t)$ *(initial velocity of the i th node)*,

$Vid(t+1)$ *(velocity of i th node at $(t+1)$th time,*

p_{best} *(Local best position of the node)*,

b_{best} *(Global best position of the node among deployed nodes)*,

Result: *Optimizing the Location of AN Initialization;*

for $i \leftarrow 1$ to S do

      Calculate: $Vid(t+1)x$, $Vid(t+1)y$, $Vid(t+1)z$

           $xid(t+1)$, $yid(t+1)$, $zid(t+1)$

           as per Equation (8.1),(8.2),(8.3),(8.4),(8.5),(8.6)

      Set: $pid(t)x = pid(t)y = pid(t)z = gid(t)x = gid(t)y = gid(t)z = 0$;

      Evaluate Equation (8.1),(8.2),and(8.3)to as per Equation(8.10),(8.11),(8.12)

      Calculate the energy cost $Ci(t)$ as per Equation (8.13),(8.14),(8.15)

      Set fitness function as per Equation (8.21)

      Derive final fitness function $f(R)$, considering penalty parameters for energy ($Ci(t)$) and collision detection ($P\xi$) to the fitness function R as per

      Equation (8.13) and (8.20)

      for $i \leftarrow 1$ to S do

           Calculate $f(R)$ for all i;

           Calculate $M\,in(f(R))$;

           Select Node i with $M\,in(f(R))$ and p_{best};

    end

end

8.4 NUMERICAL SIMULATION AND RESULTS

To assess the efficacy of the devised configuration, the protocol underwent simulation using the parameters outlined in Table 8.2, and the initial positions of each node following deployment are detailed in Table 8.3.

8.4.1 Sample scenario of iteration

To illustrate the application of the proposed PSO approach in a three-dimensional space, consider a sample scenario. In this instance, there are five ANs, and their initial parameters are presented in Table 8.3. The "Initial Location" column in the table denotes the three-dimensional coordinate positions of each AN after deployment.

Several parameters for comparison of the proposed PSO-based optimization are chosen as follows:

- Initial fitness $p_{id}(t)=g_{id}(t)=0$ for all nodes and $P_\xi=0$ and $C_i(t)=0$.
- Position of the coordinator node is initialized to $(4,7,5)$

Assume N is set at 5 in this example for simplicity. Best fitness of all the nodes with respect to rounds is calculated as follows:

$$\text{fitness}_{\text{Best}} = \text{Min}(234.35, 277.16, 217.96, 289.63) = 217:96.$$

Table 8.2 Simulation parameters

Parameters	Value
Network size	$100 \times 100\,\text{m}$
Number of sensor nodes (ANs)	50
Number of mobile coordinator Nodes data packet size	5
	4,000 bits
Threshold value of energy	0.1 Joule
Initial energy ($E(i)$)	2.0 Joule

Table 8.3 Initial position of the node after deployment

Node	Initial position		
	x	Y	z
1	7.38	7.57	8.81
2	10	10	3.9
3	6.71	3.36	10
4	10	10	9.57
5	10	10	7.26

Table 8.4 Calculation of fitness upto Round 2

Round	Node no	Location			Coordinator location	Fitness
1	1	7.38	7.57	8.81	4, 7, 5	234.35
	2	10.00	10.00	3.79		
	3	6.71	3.36	10.00		
	4	10.00	10.00	9.57		
	5	10.00	10.00	7.26		
2	1	3.10	0.76	7.81		277.16
	2	10.00	10.00	3.59		
	3	10.00	10.00	0.50		
	4	10.00	10.00	3.47		
	5	10.00	10.00	10.00		

Similarly, the fitness values for all ANs are computed.

From Table 8.4, the ANs 3 in round 1 will be in the best position, and hence, the location of AN 1 in round 2 is considered to be p_{best} as well as g_{best}. After that, the new velocity of each AN is updated. Assume, initial velocities of all node

$$V_{idx}(t) = V_{idy}(t) = V_{idz}(t) = 0 \text{ for all } I \in S, \text{ where, } S \text{ is the swarm of all nodes.}$$

Initialize $\omega_1 = \omega_2 = 1$, the constants $c_1 = c_2 = 2$ for nodes moving to p_{best}. Consider the first node in round 1 as an example to illustrate the steps of new velocities as calculated using Equations 8.1, 8.2, and 8.3:

- $V_{id}(t+1)_x = 1 \times 0 + 2 \times 1 \times (0 - 7 : 38) + 2 \times 1 \times (0 - 7 : 38) = 29.52$
- $V_{id}(t+1)_y = 1 \times 0 + 2 \times 1 \times (0 - 7 : 57) + 2 \times 1 \times (0 - 7 : 57) = 30.28$
- $V_{id}(t+1)_z = 1 \times 0 + 2 \times 1 \times (0 - 8 : 81) + 2 \times 1 \times (0 - 8 : 81) = 35.24$

Likewise, the velocities of the remaining ANs can be determined using the same approach. Similarly, new position and the energy cost for all the ANs are calculated as shown in Table 8.5.

8.4.2 Distance between coordinator node and predicted location of AN

In this section, the effectiveness of the proposed PSO-based location prediction and optimization is evaluated by measuring the MSE as outlined in Equation 8.22. Subsequently, the performance of the proposed scheme is juxtaposed with the approach based on Kriging interpolation, as outlined

Table 8.5 Sample MSE

Round no	Fitness value	MSE
1	234.35	146.87
2	277.16	155.432
3	217.96	143.592
4	289.63	157.926
5	293.11	158.622
6	354.96	170.992
21	274.29	154.858
22	290.45	158.09
23	306.22	161.244
24	267.85	153.57
25	167.94	133.588

in [23]. This comparison involves calculating the distance between the actual location and the predicted location—specifically, between the coordinator node's location and the anticipated location of the AN. Numerical simulation reveals that at the very outset, the MSE is high and as the iteration progresses, the MSE is changing gradually and the accuracy of location optimization reaches an optimized value at 25th iteration. Figures 8.4–8.7 illustrate the nature of generation of MSE in different rounds and the same has been compared with [23]. It is evident from the observation that the MSE in the proposed scheme is well within a significant threshold, indicating a minor deviation compared to the aforementioned approach. [23]. For an example, the MSE of the projected scheme is minimum at the round 25 and the same is 133.588 with respect to the best fitness achieved in

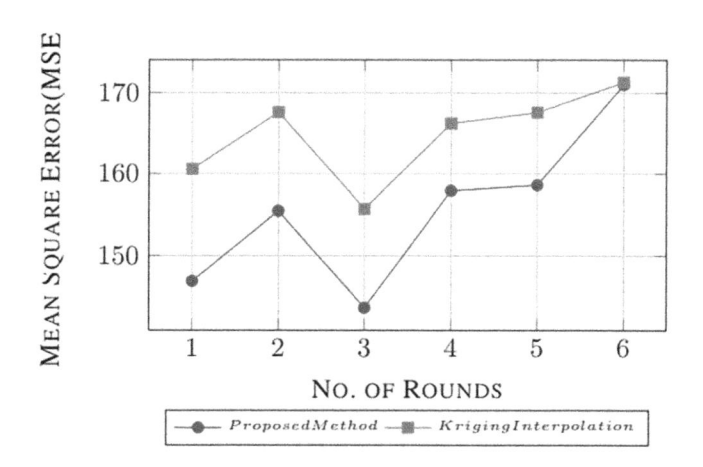

Figure 8.4 MSE vs appearance generation (up to Round 6)

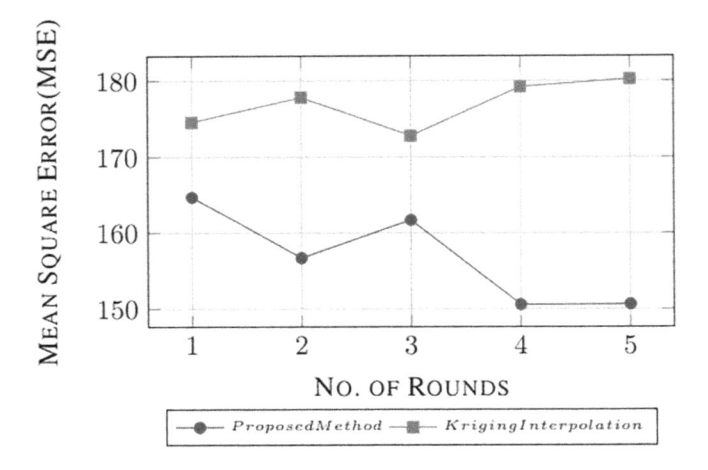

Figure 8.5 MSE vs appearance generation (up to Round 11)

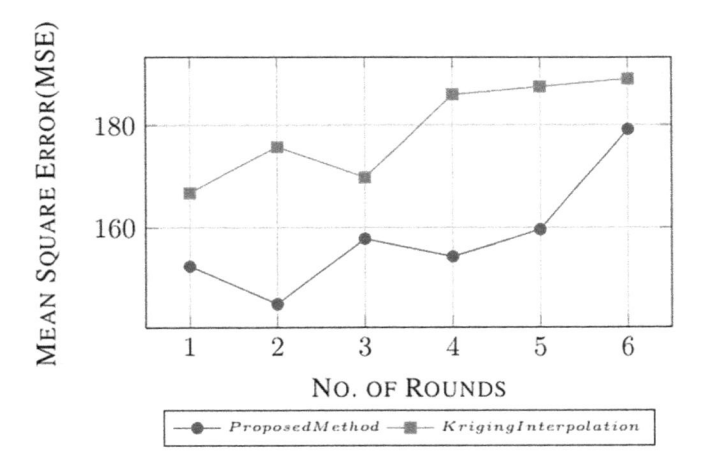

Figure 8.6 MSE vs appearance generation (up to Round 12 to Round 17)

the same round, whereas the MSE of the arrangement as mention in [23] attains a minimum MSE of 150.6 up to round 25. Figure 8.8 depicts the changes in the fitness values of the proposed scheme till the best fitness is achieved in the round 25 and the same is 167.94. As already mentioned, Table 8.4 portrays sample fitness values attained by the scheme up to round 25 (Figure 8.9).

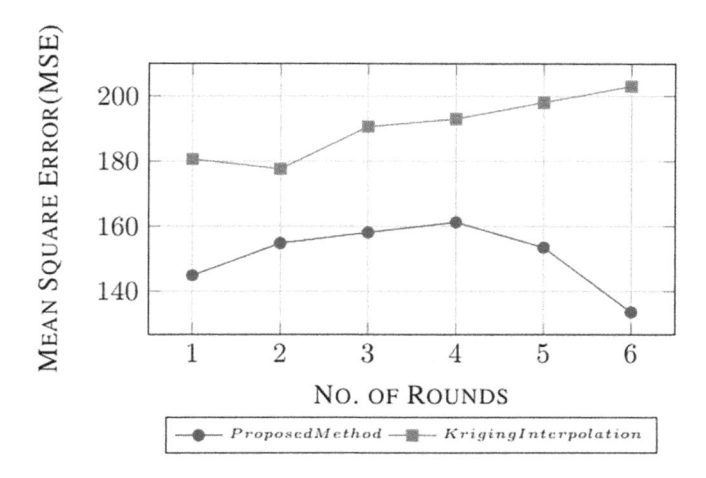

Figure 8.7 MSE vs appearance generation (up to Round 20 to Round 25).

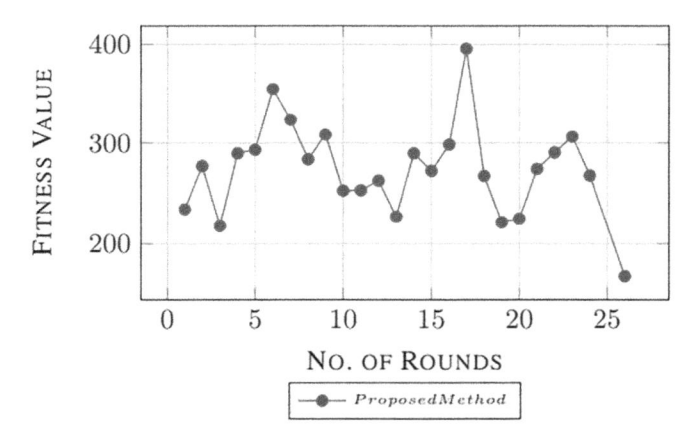

Figure 8.8 Fitness values vs no. rounds.

8.5 CONCLUSION, CHALLENGE, AND FUTURE WORK

Development of location prediction and optimization scheme is a crucial concern in WSN especially designated for the purpose of surveillance. In the devised PSO-based location optimization scheme, the prediction and optimization of the deployed node's location extend beyond mere alignment with a reference location—namely, the coordinator node's position. Additionally, the arrangement exhibits lower energy costs in comparison to existing schemes, as demonstrated in the preceding section. The results

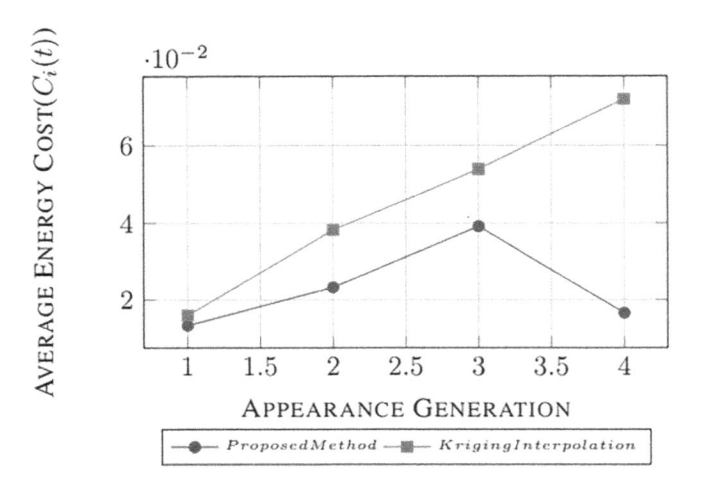

Figure 8.9 Average energy cost vs appearance generation.

presented therein illustrate the advantages of the proposed optimized location prediction setup over the system outlined in [23], showcasing improvements in accuracy and response time for achieving the best optimized solution. In the scheme, the best-fitted optimized location is achieved at 25th round, whereas the same is achieved by the scheme [23] at 54th round. In future, we will distillate on the betterment of the scheme considering the fast movement of the reference location so as to boost the performance of the same.

In the preceding study, the prospective positions of the nodes were determined through the application of the PSO. However, it's important to acknowledge that while PSO has demonstrated effectiveness, it comes with certain limitations. These include susceptibility to initial conditions, the potential to converge to local optima, sensitivity to parameter tuning, premature convergence, and a lack of inherent memory. To address these challenges, there are several alternative swarm-based optimization methodologies available. Examples of these encompass ACO, differential evolution, bee colony optimization, and cuckoo search, among others. The selection of the most suitable algorithm depends on a multitude of factors, spanning the availability of computational resources, the specific problem characteristics, and the nuances of the problem statement itself. By considering these factors, researchers can tailor their choice of optimization approach to best align with the unique requirements of the task at hand, mitigating potential shortcomings and fostering a more effective optimization process.

REFERENCES

1. X. Bai, S. Kumar, D. Xuan et al., Deploying wireless sensors to achieve both coverage and connectivity, in *Proc. Seventh ACM Int. Symp.on Mobile ad Hoc Networking and Computing*, New York, 2006, pp. 131–142.
2. A. Ollero, M. Bernard, M. La Civita, L. van Hoesel, P. Marron, J. Lepley, and E. de Andres, AWARE: Platform for autonomous selfdeploying and operation of wireless sensor-actuator networks cooperating with unmanned aerial vehicles, in *Proc. IEEE Int. Workshop Saf., Secur. Rescue Robot. (SSRR)*, IEEE, Rome, Sep. 2007, pp. 1–6.
3. A. Ollero and L. Merino, Control and perception techniques for aerial robotics, *Annu. Rev. Control*, vol. 28, pp. 167–178, May 2004.
4. R. C. Eberhart, Y. Shi, and J. Kennedy, *Swarm Intelligence* (series: Artificial Intelligence). San Mateo, CA: Morgan Kaufmann, 2008.
5. X. Hu, Y. Shi, and R. Eberhart, Recent advances in particle swarm, in Proc. Congr. Evol. Comput. (CEC), IEEE, Portland, OR, Jun. 1923, 2004, vol. 1, pp. 90–97.
6. R. V. Kulkarni, G. K. Venayagamoorthy, A. Miller, and C. H. Dagli, Network-centric localization in MANETs based on particle swarm optimization, in Proc. IEEE Swarm Intell. Symp., IEEE, St. Louis, MO, Oct. 2008, pp. 1–6.
7. Y. del Valle, G. K. Venayagamoorthy, S. Mohagheghi, J. C. Hernandez, and R. Harley, Particle swarm optimization: Basic concepts, variants and applications in power systems, *IEEE Trans. Evol. Comput.*, vol. 12, no. 2, pp. 171–195, Apr. 2008.
8. M. Donelli, R. Azaro, F. D. Natale, and A. Massa, An innovative computational approach based on a particle swarm strategy for adaptive phasedarrays control, *IEEE Trans. Antennas Propag.*, vol. 54, no. 3, pp. 888–898, Mar. 2006.
9. K. M. Passino, Biomimicry of bacterial foraging for distributed optimization and control, *IEEE Control Syst. Mag.*, vol. 22, no. 3, pp. 52–67, June 2002.
10. G. K. Venayagamoorthy, A successful interdisciplinary course on computational intelligence, *IEEE Comput. Intell. Mag.*, vol. 4, no. 1, pp.14–23, Jan. 2009.
11. J. Kennedy and R. Eberhart, Particle swarm optimization, in Proc. IEEE Int. Conf. Neural Netw., IEEE, Perth, WA, Nov. 27–Dec. 1, 1995, vol. 4, pp. 1942–1948.
12. T. K. Das, G. K. Venayagamoorthy, and U. O. Aliyu, Bio-inspired algorithms for the design of multiple optimal power system stabilizers: SPPSO and BFA, *IEEE Trans. Ind. Appl.*, vol. 44, no. 5, pp. 1445–1457, Sep. 2008.
13. W. J. Tang, M. S. Li, Q. H. Wu, and J. R. Saunders, Bacterial foraging algorithm for optimal power flow in dynamic environments, *IEEE Trans. Circuits Syst. I, Reg. Papers*, vol. 55, no. 8, pp. 2433–2442, Sep. 2008.
14. A. Salhieh, J. Weinmann, M. Kochhal et al., Power efficient topologies for wireless sensor networks, in *Proc. IEEE Int. Conf. on Parallel Processing, 2001*, IEEE, Valencia, Spain, 2001, pp. 156–163.

15. C. Savarese, J.M. Rabaey, and J. Beutel, Location in distributed ad-hoc wireless sensor networks, in *Proc. IEEE Int. Conf. on Acoustics, Speech, and Signal Processing, (ICASSP01)*, IEEE, Utah, US, 2001, vol. 4, pp. 20372040.

16. C. Liu, A. Kiring, N. Salman et al., A kriging algorithm for location fingerprinting based on received signal strength, in *Proc. IEEE Sensor Data Fusion Conf. (SDF)*, IEEE, Bonn, 2015, pp. 1–6.

17. R. Jin, H. Xu, Z. Che et al., Experimental evaluation of reducing ranging-error based on receive signal strength indication in wireless sensor networks. *IET Wirel. Sens. Syst.*, vol. 5, no. 5, pp. 228–234, 2015.

18. A. Gallais, J. Carle, D. Simplot-Ryl et al., Localized sensor area coverage with low communication overhead. *IEEE Trans. Mob. Comput.*, vol. 7, no. 5, pp. 661–672, 2008.

19. Das, S., Biswas, A., Dasgupta, S., and Abraham, A. (2009). Bacterial foraging optimization algorithm: Theoretical foundations, analysis, and applications. In: Abraham, A., Hassanien, A.E., Siarry, P., and Engelbrecht, A. (eds) *Foundations of Computational Intelligence Volume 3*. Studies in Computational Intelligence, vol. 203. Berlin, Heidelberg: Springer. https://doi.org/10.1007/978-3-642-01085-9_2

20. B. Chandra Mohan and R. Baskaran, A survey: Ant Colony Optimization based recent research and implementation on several engineering domain, *Expert Systems with Applications*, vol. 39, no. 4, pp. 4618–4627, 2012. ISSN 0957-4174. https://doi.org/10.1016/j.eswa.2011.09.076

21. A.S. Joshi, Omkar Kulkarni, G.M. Kakandikar, and V.M. Nandedkar, Cuckoo search optimization: A review, *Materials Today: Proceedings*, vol. 4, no. 8, pp. 7262–7269, 2017. ISSN 2214-7853. https://doi.org/10.1016/j.matpr.2017.07.055

22. Karaboga, D. and Basturk, B. (2007). Artificial Bee Colony (ABC) optimization algorithm for solving constrained optimization problems. In: Melin, P., Castillo, O., Aguilar, L. T., Kacprzyk, J., and Pedrycz, W. (eds) *Foundations of Fuzzy Logic and Soft Computing*. IFSA 2007. Lecture Notes in Computer Science, vol. 4529. Berlin, Heidelberg: Springer. https://doi.org/10.1007/978-3-540-72950-1_77

23. A. Ali, A. Ikpehai, B. Adebisi, and L. Mihaylova, Location prediction optimisation in WSNs using Kriging interpolation, *IET Wirel. Sens. Syst.*, vol. 6, no. 3, pp. 74–81, 2016.

Chapter 9

Application of NOMA in 6G networks leveraging Artificial Intelligence

*Krishanu Kundu, Ankan Bhattacharya,
Samarendra Nath Sur, Ashim Kumar Biswas,
Arnab De, and Narendra Nath Pathak*

9.1 INTRODUCTION

First generation, or 1G, wireless technology offered analogue voice in the 1980s. Early in 1990s, 2G launched digital voice code-division multiple access (CDMA). Mobile data was introduced in the early 2000s using 3G (like CDMA2000) [1–3]. The LTE (long term evolution) system was developed in the 2010s with the introduction of 4G, which dramatically increases data throughput and permits simultaneous voice as well as data transfer. One of many benefits of the 4G mobile network is VoLTE (Voice over LTE) [4]. The millimetreWave (mmWave) spectrum (30–300 GHz), where the 5G [5] network works, allows for the transmission of massive volumes of data at very high speeds with less interference from nearby signals. For its 20Mbps data rate and frequency spectrum of 2–8 GHz, 5G uses millimetre wireless and OFDM (orthogonal frequency-division multiplexing). Sixth-generation wireless, or 6G, will replace 5G soon. 6G [6–8] networks will exhibit greater bandwidth plus latency compared to 5G networks. The 6G computational infrastructure, in conjunction with AI [9,10], will decide the future of computing. With a user-experienced data rate of 100 Mbps, 5G is expected to have a peak data throughput of 20 Gbps. A maximum data rate of 1 Tbps will be provided by 6G, nevertheless. The user's data rate will also increase to 1 Gbps as a result. Digital twins, cobots, smart agriculture, precision medicine, e-health for all, and robot navigation are a few examples of significant 6G application cases. These possibilities may be divided into three main groups: the Internet of Senses, linked intelligent machines, and a connected sustainable world. AI refers to a machine's capacity to mimic human cognitive processes, particularly those of computer systems. The computer imitation of human intelligence is known as AI. Machines were created by humans, who also gave them the power to make judgements. Additionally, due to our highly complex biological structures, robotics and robots cannot fully imitate humans. The term "artificial intelligence" was coined for this purpose. Deep learning and machine learning [11] are two technologies that fall within the AI category.

Without being specifically designed to do so, machine learning allows software programmes to improve their predictive abilities. Knowledge of the anatomy of the brain serves as the foundation for deep learning. Recent developments in AI, such as ChatGPT and self-driving automobiles, are supported by deep learning's usage of artificial neural network topology. AI has a significant role in 6G communications. Additionally, it is anticipated to enable augmented reality (AR) and extended reality (XR). Researchers are already focussing on the 6G communication architecture even prior to the popularity of 5G. Researchers also have begun incorporating AI in 6G to offer ubiquity and dependability in communication. By combining different technologies and applications, 6G communication technology creates a link with everything. It also makes holographic, submersible, and haptic technologies possible. Because of this, cutting-edge AI can assist 6G communication technology in delivering on its promises. It is necessary to create new technical paths for spectrum efficiency with the lowest possible cost for addressing the limitations of 5G as well as 5G beyond technologies. Multiple access (MA) systems can be quite helpful in this regard to partially address these drawbacks. NOMA will improve the efficiency of spectrum in the upcoming 6G networks. NOMA enables simultaneous use of a single frequency channel by several users inside the same cell. Short transmission delays, flexible channel feedback, and higher spectrum efficiency are all advantages of NOMA [12–14]. NOMA performs better than standard orthogonal multiple access (OMA) since it can serve several users concurrently and with the similar frequencies. Additionally, it uses successive interference cancellation (SIC) to reduce interference and achieve good spectral efficiency. Large-scale connection is offered since it can accommodate the enormous number of users. Current work is an initiative to review so far involvement of NOMA approach in combination with AI for benefit of upcoming 6G Network [15].

9.1.1 Highlights

- A review on employment of NOMA based on AI in 6G networks,
- Overview of 6G and IoT,
- Discussions on AI enablement and 6G, and
- Challenges and Opportunities of NOMA.

9.1.2 Organisation of the Chapter

The chapter has been organised as follows: First, the significance of AI-based NOMA for 6G has been introduced. Then, an insight into 6G AI enablement has been discussed followed by 6G AI enablement and 6G NOMA. Further, the limitations of NOMA have been addressed and also the employment of NOMA in 5G along with the future opportunities and challenges is highlighted.

9.2 INSIGHT TO 6G

In 6G communication, sixth sense communication is everything. With this technology, time, space along with frequency will all be three dimensional. True AI-driven communications will be made possible by 6G. The specifications for 6G communication technology include higher data transfer rate (up to 1 Tbps), higher frequency (1 THz), short start-to-end latency (1 ms), and high portability (1,000 km/h). Further aiding intelligent network communication systems will be holographic communication and enhanced augmented simulation. The 3D sorts of endorsement will be offered by 6G with the help of new technologies including edge innovation, AI, distributed computing along with blockchain. IoT, ubiquitous instant messaging, and other applications are expected to be supported by 6G networks, making them even more diverse than their predecessors. The definition of 6G technology is still being developed as of 2023, and there is no widely acknowledged government or non-government standard. Millimetre-wave technology will be heavily utilised by current 5G and upcoming 6G networks to fulfil their commitments in terms of network performance and connectivity (Figure 9.1).

The demand for more applications means that 6G will have to put out new technological specifications and performance benchmarks. While 6G networks can achieve one to ten TB/s, 5G networks can only achieve a maximum data capacity of 20 GB/s. Utilising AI to enhance network management may result in spectrum efficiency increases of three–five times and an increase in energy efficacy of ten times when compared to 5G. Increases in connection density (by a factor of ten to one hundred) will be brought on by variables, such as expanding high-frequency band bandwidths, a variety of communication scenarios, and the adoption of very "Heterogeneous Networks (HetNets)". To properly examine 6G networks, additional crucial performance metrics must be created, such as those for cost-effectiveness, protection, coverage, as well as the level of intelligence. Significant paradigm modifications will need to be implemented in the 6G communication networks. The most recent paradigm developments make use of full spectrum utilisation (Figure 9.2), worldwide coverage, and strong endogenous security. In order to provide worldwide coverage, integrated space–air–ground–sea networks will replace terrestrial networks in 6G wireless

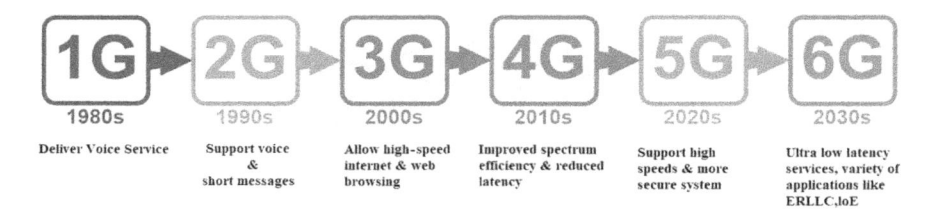

Figure 9.1 Evolution of wireless communication.

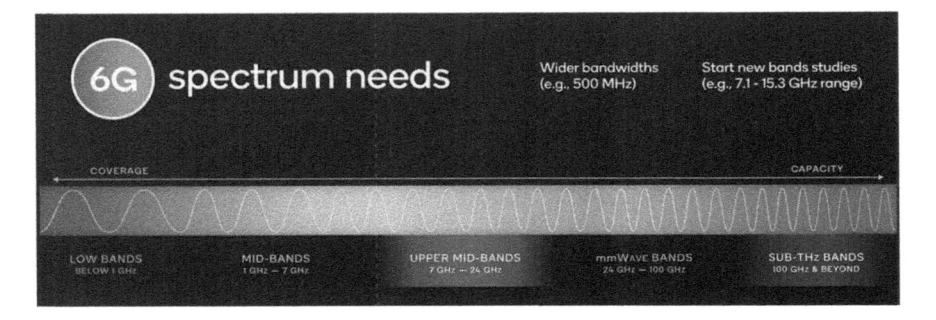

Figure 9.2 Spectrum requirement of 6G.

communication systems. Next-generation apps and significant technologies will be tightly integrated with AI and data science approaches to enable comprehensive applications. AI exhibits potential to suggestively enhance network performance by enabling the efficient and flexible utilisation of networking and computational resources. Physical layer as well as network layer security, as well as comprehensive or built-in network security, will appear more frequently as 6G networks are constructed

When the standards are implemented, they will result in a significant paradigm shift, as seen in the design elements of the 6G system in Figure (Figure 9.3). The primary element that dramatically improves wireless

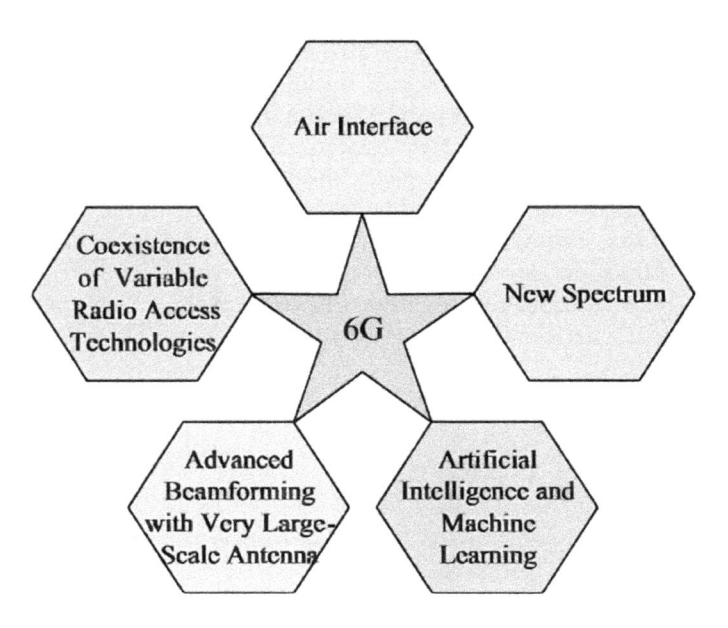

Figure 9.3 Major elements of 6G wireless network.

Table 1. Key Requirement Comparison of 5G and 6G

	5G	6G
Application Services	✓ eMBB ✓ URLLC ✓ mMTC	✓ eUMBB ✓ eURLLC ✓ UmMTC ✓ LDHMC ✓ ELPC
Communication Network Architecture	✓ 5G-NR cellular ✓ mmWave network	✓ AI-empowered network ✓ Aerial network ✓ Terrestrial network ✓ Underwater network ✓ mmWave/THz network
Transmission Spectrum Usage	✓ sub-6 GHz (2.4/3.5/5 GHz) ✓ mmWave (28/39/60 GHz)	✓ sub-6GHz (2.4/3.5/5 GHz) ✓ mmWave (28/39/60 GHz) ✓ THz (above 100 GHz) ✓ Laser ✓ VLC ✓ Non-RF
Peak Data Rate	20 Gbps	1 Tbps
Latency Requirement	1 ms	0.01–0.1 ms
Reliability Demands	99.999 %	99.99999999 %
Connectivity Density	10^6 devs/km^2	10^7 devs/km^2
Mobility Support	500 km/hour	≥1,000 km/hour
Area Spectral Efficiency Compared to 5G	1×	10×
Energy Efficiency Compared to 5G	1×	100×

Figure 9.4 Comparison of key necessities between 5G and 6G.

generations is the air interface. CDMA, which was the primary technology of 3G, was substituted by OFDM, that was crucial in the creation of 4G.

In a similar vein, the deployment of the novel air interface might significantly affect the 6G system's architecture. The 6G was obliged to use a new protocol to communicate because of spectrum congestion. The following Figure 9.4 shows a comparison of key necessities between 5G and 6G.

9.3 AI ENABLEMENT AND 6G

Although there are many alternative methods to classify AI, the two most common ones are based on the technology's capabilities and functionalities. The various divisions of AI are explained in the flowchart that follows (Figure 9.5).

The most public and widely accessible type of AI is narrow AI, which is incompetent at carrying out tasks beyond its domain or subset of constraints as it is only qualified for a particular goal. It is also termed as weak AI. Apple Siri is a good example of narrow AI. Any intellectual task can be completed by general AI just as efficiently as by a person. According

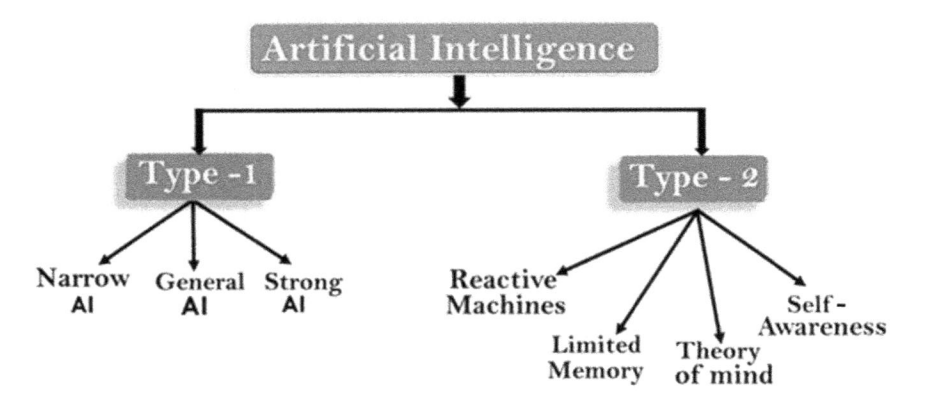

Figure 9.5 Types of AI.

to super AI, computers might surpass humans at performing any specific task due to cognitive competencies. Purely reactive machines are among the simplest types of AI. These AI systems don't keep track of memories or past encounters to draw on in the present. These machines basically keep an eye on the news and react as finest as they can. Machines with little memory can temporarily store past experiences or specific data. Theory of Mind AI need to be capable to cooperate socially with people and recognise their feelings. The future of AI is self-awareness. They will be extremely clever and possess consciousness, feelings, and self-awareness. In order to bring the server nearer to the consumers from the cloud, 6G communication technology would combine edge computing along with AI. Information and communications technology, as well as operational technology, will alter greatly in the next generation. In order to optimise networks and develop new waveforms for 6G technology, AI will be crucial. By using data that is locally kept on the 6G sensor, 6G technology will also enable future AI breakthroughs. Following are a few crucial elements that will make the usage of AI in 6G communication technology stronger [16,17]. In 6G, the significance of physical as well as networking layer will be the same. The boundaries between these layers will thereafter become less defined. Furthermore, this technical advancement is probably going to be greatly influenced by AI. Throughput and network capacities may go up as a result. The complex interactions between physical systems and behaviours may be missed by model- and algorithm-based approaches to network configuration and optimisation. As a result, there can be less-than-ideal options that don't completely satisfy the needs of the network. Due to its size, density, and heterogeneity, 6G networks are too dynamic and complicated to be described using the same methods. AI will be crucial in optimising the upcoming 6G networks because it can handle problems that closed-form models find challenging. The development of a distributed learning architecture and the refinement of a parameter communication mechanism will play a vital role in determining how well AI applications operate on 6G networks. A neural network may learn and train rules autonomously and realise self-evolution by using data-based model structure search, which is another way that auto-machine learning can automatically determine the ideal neural network structure. Reinforcement learning may sense spectrum, energy, cache, computing, and other resources through the interaction between the agent and environment, fostering a greater integration of communication, computation, and storage. To ensure the flexibility of service, arrangement, and management to increase the autonomy of 6G networks, the unpredictability of the environmental model and the long-term aim should be taken into consideration while choosing behaviour solutions. An overall scope of AL in 6G in presented in Figure (Figure 9.6).

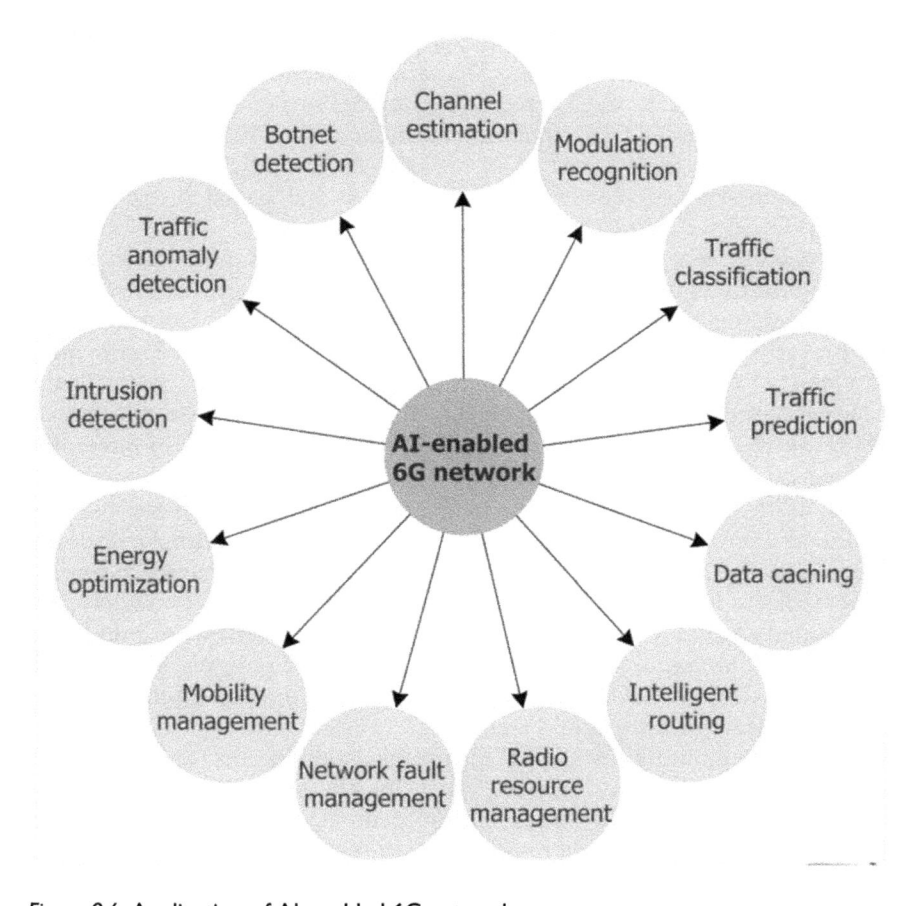

Figure 9.6 Application of AI-enabled 6G network.

9.4 NOMA & 6G

The primary goal of conventional transmission methods is to increase the frequency, temporal, and spatial resource efficiency of the signals. With the aid of several strategies, many beams can be produced in order to meet various quality of service (QoS) criteria while providing service to a big user base in numerous directions. But sophisticated radio resource management (RRM), interference management, and power control are needed. Furthermore, by superimposing all signals in the spatial resource slots, frequency and time, we may be able to enhance the trade-off between spectrum efficiency and energy efficiency. The same methods are used for full-duplex, coordinated multi-point, NOMA, three-dimensional beamforming, and Rate-splitting multiple access (RSMA).

In NOMA, transmitter multiplexes a number of acceptable signals to a solo resource slot of specific resource domain in order to reduce the interference, while the receiver performs SIC. In the most prevalent power-domain NOMA, when more than two signals overlap at a certain strength, the strongest signal is identified using SIC, and the feebler signals are handled as interference. After that, remodulated signal is being removed by the clean, reduced signal. To combine the multiple mmWave/THz resource domains, remove interference in ultra dense network (UDNs), or increase the unmanned aerial vehicle (UAV) coverage as well as satellite networks, better NOMA techniques are necessary.

3D beamforming: Sectorised antennas, which use sophisticated beamforming techniques and antenna arrangements to support users who roam at different angles, are the inspiration for 3D beamforming. Due to the substantial pathloss experienced by mmWave carriers, high-gain beamforming is a must. Even more crucial is high-gain beamforming for pencil beam THz carrier. Furthermore, it's critical to effectively manage the constrained 3D resources in order to satisfy a range of QoS requirements in integrated ground-air-space (IGAS)networks.

FD: With this technique, data can be transmitted simultaneously in an uplink and downlink fashion over the same frequency and time period. But there is a significant issue that calls for complex self-interference mitigation methods: high and low power interference. FD solutions now have the opportunity to increase spectrum efficiency in both terrestrial and aerial networks by leveraging mmWave/THz ultra-massive multiple-input, multiple-output (UM-MIMO) beamforming, thanks to the development of reliable new 6G architectures and technologies [18].

Coordinated multipoint (CoMP): Multiple base stations (BSs) can transmit simultaneously to a single receiver using this technique. Additionally, as an advantage of mmWave/THz beamforming, UM-MIMO CoMP can increase the network's throughput. Due to the fact that the receiver can only be fully synced with one BS, a large-scale CoMP roll-out is prevented by serious synchronisation issues.

IRS: Intelligent reflecting surface (IRS) has been highlighted as a potentially groundbreaking breakthrough in 6G because of its enhanced spectrum, energy, and cost-efficiency [19]. It is a low-cost antenna array having a huge number of configurable reflecting components. When the loss of signal (LOS) connection between BS and the users is impeded by barriers, IRS might be utilised to develop alternative paths to ensure reliable connections and QoS. The authors of [20–22] mix NOMA alongside the IRS as they enjoy its aids, including its higher spectrum efficacy [23,24]. UAV, a prospective 6G technique which can be used to boost IRS flexibility [25].

MIMO: Massive multiple-input multiple-output (m-MIMO) is being intensively investigated as a possible way for increasing the volume of a transmission network by simply accumulating additional antennas. Usage of MIMO over NOMA is highly fascinating. MIMO-NOMA is a viable alternative for assisting communication networks in achieving even better spectral efficacy as well as lesser latency.

9.5 LIMITATIONS OF NOMA

NOMA has some limitations that must be overcome in order to fully realise its advantages, such as: When compared to NOMA, the following factors increase receiver complication and energy consumption: (1) Every user should decode the information of every other user in the same cluster with worse channel gains before decoding its own information; and (2) Whenever an error in SIC occurs at a particular user, the consequent decoding of the information of rest of all users would likely be impacted by the error. This proposes maintaining the user count in every cluster at a tolerable level to lessen the impact of error circulation; (3) In order to reap the purported rewards of power-domain multiplexing, there must be a sizable channel gain difference amidst the strong as well as weak users. Code-domain NOMA is a further division of a number of multiple access algorithms that depend on sparse code multiple access as well as low-density spreading (Figure 9.7).

Because of interference in such structures, the NOMA concept allows several users to be superimposed on an identical resource. Because this novel approach creates additional interference, persisting resource handling as well as interference mitigation solutions, predominantly for ultra-dense networks, must be evaluated. For a similar reason, beamforming in m-MIMO systems & the accompanying concerns (such as precoding [26]) provide additional challenges that must be overcome before these technologies can be effectively utilised. Physical layer modifications are also required to address current channel coding, modulation, and estimation problems. Cognitive, cooperative, and visible light communications paradigms can all be used to compare NOMA systems to conventional systems. The difficulties brought on by embracing this new technology must first be overcome in

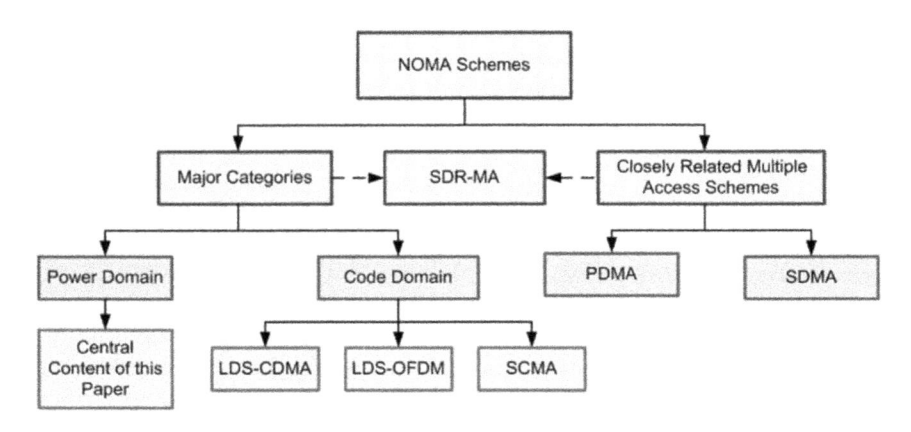

Figure 9.7 Simple classification of NOMA techniques.

order to benefit from these paradigms. The NOMA technique has numerous advantages, but it also increases the risk to security and privacy because it enables more people to more efficiently sense information.

9.6 EMPLOYMENT OF NOMA IN 5G SO FAR

Islam et al.'s research [27] mainly focusses on power-domain NOMA, which uses superposition coding at the transmitter as well as successive interference dissolution at the receiver. Islam [28] also suggested that NOMA's potential is not just restricted to SISO systems; it may also be used in MIMO systems to boost capacity. Other communication technologies, such as mmWave and visible light systems, are also susceptible to the use of NOMA. NOMA's restrictions, such as error propagation and Intercarrier Interference (ICI) in multi-cell networks, should be removed in order to make it more useful. Massive-access scenes provide a very high level of complexity to the traditional NOMA system. In order to allow massive-access situations, Chen et al. [29] presented a user-pairing-based suboptimal method for uplink multiple access. Additionally, an O(nlogn) complicated user-pairing method is also devised. The Monte Carlo simulation indicates that the suggested strategy exhibits just a minor throughput loss when compared to the information—theoretically ideal transmission scheme. To address the issue of high computational complexity and boost system performance in 5G using OFDMA-based NOMA, Saraereh et al. [30] recommended an enhanced lower complexity radio resource allocation algorithm for either user for power allocation optimisation. Belkacem et al. [31] provided an assessment of downlink NOMA systems over Nakagami-m fading channels. Non-SIC NOMA system is then suggested by Chung [32].

Combining NOMA/cooperative NOMA with m-MIMO, single-carrier frequency domain equalization (SC-FDE) signals, and the low complexity memory reference code (MRC) receiver is a good proposal to achieve future 5G evolutions [33]. Song et al. [34] studied the joint resource allocation in hybrid multi-carrier non-orthogonal multiple access (MC-NOMA) systems to achieve Spectral Efficiency (SE)-Energy Efficiency(EE) trade-off. The minimal rate requirement served as a QoS guarantee for each user when the SE-EE trade-off was established as a MUD, object-oriented (MOO) problem. Timotheou et al. [35] studied NOMA from the viewpoint of fairness and investigated Power Allocation (PA) practices that guarantee fairness for downlink users under both (i) instantaneous channel state information at the transmitter (CSIT) and (ii) average channel state information(CSI). NOMA as well as orthogonal frequency division multiplexing (OFDM) methods are recommended as workable solutions to meet the needs of 5G. In view of the likelihood that the polynomial model with memory might be used to reproduce the nonlinear distortions caused by the host protected area (HPA), new theoretical expressions are constructed to calculate the system's bit error rate (BER) [36]. Although the customers' decisions on the intended data rates and power distribution have a significant impact on NOMA's outage performance, NOMA [37–39] is still capable of achieving outstanding performance in terms of ergodic sum rates. NOMA, which may suggestively boost spectral efficacy while servicing a large number of users, has recently come to be seen as a very promising approach for 5G and beyond communications. However, NOMA has a number of limitations, including the need for accurate CSIT and a high level of computing complexity in the receiver. Deep learning (DL) techniques are a fantastic way to overcome the difficulties. The employment of DL techniques in NOMA for 5G and future communications are covered in research by Hasan et al. [40].

9.7 OPPORTUNITIES ALONG WITH CHALLENGES

When it changes from OFDM to NOMA, receiver's end is extended with SIC, which adds considerable technical complexity for signal decoding when compared to OMA schemes. As a result, it reduces receiver performance. The other major problem unique to the NOMA method is that if an error happens in SIC, all user information is encoded incorrectly. Because the mathematical comprehension of the impact of inadequate SIC has become a research topic and it is required to be addressed for NOMA. Another difficult part of NOMA is the creation of the air interface. When the number of user equipment (UEs) rises, the orthogonal signals for reference will not be enough to meet the demand. As a result, it is suggested that semi-orthogonal reference signals be considered. Semi-blind estimation of channels along with semi-blind multiuser detection can be used to accomplish this. Furthermore, how to arrange reference signals in NOMA to limit collision likelihood requires additional investigation. The ICI problem in multi-cell networks could get more severe as BS and consumer device density increase. Diverse networks may face far greater challenges in mitigating interference. Therefore, it is crucial to create interference cancellation and management solutions that are compatible with NOMA. The data rates for the specified end users may be raised even further with carrier aggregation (CA) integrated into the NOMA-based system. For NOMA solutions, undecided is the perfect CA type. Low-complexity resource allocation methods must be developed because they are essential for enhancing NOMA performance. Further research is needed to determine the best NOMA solutions for mmWave networks and large MIMO. It's essential to combine NOMA with other 6G enablers like mmWave and m-MIMOs.

9.8 CONCLUSION

5G will be replaced by the next generation of wireless technology, 6G. It is expected to give 5G better coverage, faster speeds, and lower latency even though it is still in the research and development stage. Following are some potential benefits of 6G over 5G: More than 20 times faster than the 10 gigabits per second that 5G can deliver, 6G has the potential to reach rates of up to 206.25 gigabits per second. The frequency bands used by 6G, which are higher than those of 5G and potentially offer more bandwidth and data speed, may vary from 30 GHz to 3 THz. By utilising a wider range of spectrum and technological platforms, 6G might provide better dependability and coverage. With 6G, new services and applications are made available like holographic communications, AI, high-precision manufacturing, smart societies, etc. Although 6G has many benefits, it also has a number of disadvantages, including high costs, a lack of spectrum, technological complexity, privacy and security issues, etc. As a result, implementing 6G will take time and effort. The planned start date for 6G deployment is 2030. The most recent research offers a thorough description of NOMA together with its coexistence with the promise of 6G technology. Recent developments in AI are being used in current efforts to produce the enormous NOMA approach, which is appropriate for both 5G and 6G networks.

REFERENCES

1. del Peral-Rosado, José A., Ronald Raulefs, José A. López-Salcedo, and Gonzalo Seco-Granados. "Survey of cellular mobile radio localization methods: From 1G to 5G." *IEEE Communications Surveys & Tutorials* 20, no. 2 (2017): 1124–1148.

2. Gawas, Anju Uttam. "An overview on evolution of mobile wireless communication networks: 1G-6G." *International Journal on Recent and Innovation Trends in Computing and Communication* 3, no. 5 (2015): 3130–3133.
3. Sharma, Pankaj. "Evolution of mobile wireless communication networks-1G to 5G as well as future prospective of next generation communication network." *International Journal of Computer Science and Mobile Computing* 2, no. 8 (2013): 47–53.
4. Hui, Suk Yu, and Kai Hau Yeung. "Challenges in the migration to 4G mobile systems." *IEEE Communications Magazine* 41, no. 12 (2003): 54–59.
5. Shafi, Mansoor, Andreas F. Molisch, Peter J. Smith, Thomas Haustein, Peiying Zhu, Prasan De Silva, Fredrik Tufvesson, Anass Benjebbour, and Gerhard Wunder. "5G: A tutorial overview of standards, trials, challenges, deployment, and practice." *IEEE Journal on Selected Areas in Communications* 35, no. 6 (2017): 1201–1221.
6. Nayak, Sabuzima, and Ripon Patgiri, "6G communication: Envisioning the key issues and challenges." *EAI Endorsed Transactions on Internet of Things* 6, no. 24 (2020): e1. doi:10.4108/eai.11-11-2020.166959.
7. Nayak, Sabuzima, and Ripon Patgiri, "6G communications: A vision on the potential applications." *Edge Analytics, Lectutre notes in Electric Engineering* (2020). arXiv:2005.07531.
8. Shen, Li-Hsiang, Kai-Ten Feng, and Lajos Hanzo. "Five facets of 6G: Research challenges and opportunities." *ACM Computing Surveys 55*, no. 11 (2023): 1–39.
9. Whittaker, Meredith, Kate Crawford, Roel Dobbe, Genevieve Fried, Elizabeth Kaziunas, Varoon Mathur, Sarah Mysers West, Rashida Richardson, Jason Schultz, and Oscar Schwartz. *AI Now Report 2018.* New York: AI Now Institute at New York University, 2018.
10. Siriwardhana, Yushan, Pawani Porambage, Madhusanka Liyanage, and Mika Ylianttila. "AI and 6G security: Opportunities and challenges." In: *2021 Joint European Conference on Networks and Communications & 6G Summit (EuCNC/6G Summit)*, pp. 616–621. IEEE, Porto, 2021.
11. Jordan, Michael I., and Tom M. Mitchell. "Machine learning: Trends, perspectives, and prospects." *Science* 349, no. 6245 (2015): 255–260.
12. Saito, Yuya, Yoshihisa Kishiyama, Anass Benjebbour, Takehiro Nakamura, Anxin Li, and Kenichi Higuchi. "Non-orthogonal multiple access (NOMA) for cellular future radio access." In: *2013 IEEE 77th Vehicular Technology Conference (VTC Spring)*, pp. 1–5. IEEE, Dresden, 2013.
13. Makki, Behrooz, Krishna Chitti, Ali Behravan, and Mohamed-Slim Alouini. "A survey of NOMA: Current status and open research challenges." *IEEE Open Journal of the Communications Society* 1 (2020): 179–189.
14. Ogundokun, Roseline Oluwaseun, Joseph Bamidele Awotunde, Agbotiname Lucky Imoize, Chun-Ta Li, AbdulRahman Tosho Abdulahi, Abdulwasiu Bolakale Adelodun, Samarendra Nath Sur, and Cheng-Chi Lee. "Non-orthogonal multiple access enabled mobile edge computing in 6G communications: A systematic literature review." *Sustainability* 15, no. 9 (2023): 7315.
15. Liu, Yuanwei, Wenqiang Yi, Zhiguo Ding, Xiao Liu, Octavia Dobre, and Naofal Al-Dhahir. "Application of NOMA in 6G networks: Future vision and research opportunities for next generation multiple access." (2021). arXiv preprint arXiv:2103.02334.

16. Wu, Yuan, Yuxiao Song, Tianshun Wang, Liping Qian, and Tony QS Quek. "Non-orthogonal multiple access assisted federated learning via wireless power transfer: A cost-efficient approach." *IEEE Transactions on Communications* 70, no. 4 (2022): 2853–2869.
17. Ye, Neng, Xiangming Li, Jianxiong Pan, Wenjia Liu, and Xiaolin Hou. "Beam aggregation-based mmWave MIMO-NOMA: An AI-enhanced approach." *IEEE Transactions on Vehicular Technology* 70, no. 3 (2021): 2337–2348.
18. Imoize, Agbotiname Lucky, Hope Ikoghene Obakhena, Francis Ifeanyi Anyasi, and Samarendra Nath Sur. "A review of energy efficiency and power control schemes in ultra-dense cell-free massive MIMO systems for sustainable 6G wireless communication." *Sustainability* 14, no. 17 (2022): 11100.
19. Sur, Samarendra Nath, and Rabindranath Bera. "Intelligent reflecting surface assisted MIMO communication system: A review." *Physical Communication* 47 (2021): 101386.
20. Zhong, Ruikang, Yuanwei Liu, Xidong Mu, Yue Chen, and Lingyang Song. "AI empowered RIS-assisted NOMA networks: Deep learning or reinforcement learning?" *IEEE Journal on Selected Areas in Communications* 40, no. 1 (2021): 182–196.
21. Nguyen, Thai-Anh, Hoang-Viet Nguyen, Dinh-Thuan Do, and Samarendra Nath Sur. "Performance analysis of two IRS-NOMA users in downlink." In: *International Conference on Communication, Devices and Networking*, pp. 661–674. Singapore: Springer Nature Singapore, 2022.
22. Wu, Qingqing, and Rui Zhang. "Intelligent reflecting surface enhanced wireless network via joint active and passive beamforming." *IEEE Transactions on Wireless Communications* 18, no. 11 (2019): 5394–5409.
23. Ding, Zhiguo, Robert Schober, and H. Vincent Poor. "On the impact of phase shifting designs on IRS-NOMA." *IEEE Wireless Communications Letters* 9, no. 10 (2020): 1596–1600.
24. Jiao, Shiyu, Fang, Xiaotian Zhou, and Haixia Zhang. "Joint beamforming and phase shift design in downlink UAV networks with IRS-assisted NOMA." *Journal of Communications and Information Networks* 5, no. 2 (2020): 138–149.
25. Sur, Samarendra Nath, Debdatta Kandar, Agbotiname Lucky Imoize, and Rabindranath Bera. "An overview of intelligent reflecting surface assisted UAV communication systems." In: Agbotiname Lucky Imoize, Sardar M. N. Islam, T. Poongodi, Lakshmana Kumar Ramasamy, B.V.V. Siva Prasad (eds.), *Unmanned Aerial Vehicle Cellular Communications*, pp. 67–94. Cham: Springer, 2022.
26. Sur, Samarendra Nath, Debdatta Kandar, Adão Silva, Nhan Duc Nguyen, Sukumar Nandi, and Dinh-Thuan Do. "Hybrid precoding algorithm for millimeter-wave massive MIMO-NOMA systems." *Electronics* 11, no. 14 (2022): 2198.
27. Islam, S.M. Riazul, Nurilla Avazov, Octavia A. Dobre, and Kyung-Sup Kwak. "Power-domain non-orthogonal multiple access (NOMA) in 5G systems: Potentials and challenges." *IEEE Communications Surveys & Tutorials* 19, no. 2 (2016): 721–742.
28. Islam, S.M. Riazul, Ming Zeng, and Octavia A. Dobre. "NOMA in 5G systems: Exciting possibilities for enhancing spectral efficiency." (2017). arXiv preprint arXiv:1706.08215.

29. Chen, Shuang, Kewu Peng and Huangping Jin, "A suboptimal scheme for uplink NOMA in 5G systems," In: *2015 International Wireless Communications and Mobile Computing Conference (IWCMC)*, Dubrovnik, Croatia, pp. 1429–1434, 2015. doi:10.1109/IWCMC.2015.7289292.
30. Saraereh, Omar A., Amer Alsaraira, Imran Khan, and Peerapong Uthansakul. "An efficient resource allocation algorithm for OFDM-based NOMA in 5G systems." *Electronics* 8, no. 12 (2019): 1399.
31. Belkacem, Oussama Ben Haj, Mohamed Lassaad Ammari, and Rui Dinis, "Performance analysis of NOMA in 5G systems with HPA non-linearities." *IEEE Access*, 8, (2020): 158327–158334. doi:10.1109/ACCESS.2020.3020372.
32. Chung, Kyuhyuk. "NOMA for correlated information sources in 5G Systems." *IEEE Communications Letters* 25, no. 2 (2021): 422–426. doi:10.1109/LCOMM.2020.3027726.
33. Marques da Silva, Mário, and Rui Dinis. "Power-ordered NOMA with massive MIMO for 5G systems." *Applied Sciences* 11, no. 8 (2021): 3541.
34. Song, Zhengyu, Qiang Ni, and Xin Sun. "Spectrum and energy efficient resource allocation with QoS requirements for hybrid MC-NOMA 5G systems." *IEEE Access* 6 (2018): 37055–37069.
35. Timotheou, Stelios, and Ioannis Krikidis, "Fairness for non-orthogonal multiple access in 5G systems." *IEEE Signal Processing Letters* 22, no. 10 (2015): 1647–1651. doi:10.1109/LSP.2015.2417119.
36. Hilario-Tacuri, Alexander, Jesus Maldonado, Mario Revollo, and Hernan Chambi, "Bit error rate analysis of NOMA-OFDM in 5G systems with non-linear HPA with memory." *IEEE Access* 9 (2021): 83709–83717. doi:10.1109/ACCESS.2021.3087536.
37. Ding, Zhiguo, Zheng Yang, Pingzhi Fan, and H. Vincent Poor. "On the performance of non-orthogonal multiple access in 5G systems with randomly deployed users." *IEEE Signal Processing Letters* 21, no. 12 (2014): 1501–1505.
38. Lv, Lu, Jian Chen, Qiang Ni, and Zhiguo Ding, "Design of cooperative non-orthogonal multicast cognitive multiple access for 5G systems: User scheduling and performance analysis." *IEEE Transactions on Communications* 65, no. 6 (2017): pp. 2641–2656. doi:10.1109/TCOMM.2017.2677942.
39. Zewde, Tewodros A., and M. Cenk Gursoy, "NOMA-based energy-efficient wireless powered communications in 5G systems," In: *2017 IEEE 86th Vehicular Technology Conference (VTC-Fall)*, Toronto, ON, Canada, pp. 1–5, 2017. doi:10.1109/VTCFall.2017.8288114.
40. Hasan, Moh Khalid, Md Shahjalal, Md Mainul Islam, Md Morshed Alam, Md Faisal Ahmed, and Yeong Min Jang, "The role of deep learning in NOMA for 5G and beyond communications," In: *2020 International Conference on Artificial Intelligence in Information and Communication (ICAIIC)*, Fukuoka, Japan, pp. 303–307, 2020. doi:10.1109/ICAIIC48513.2020.9065219.

Chapter 10

Utilization of an AI-assisted lens antenna array in 5G networks

Krishanu Kundu, Ankan Bhattacharya,
Samarendra Nath Sur, and Bappadittya Roy

10.1 INTRODUCTION

High-gain antennas are becoming more and more necessary as wireless communication technology advances. Several strategies to increase antenna gains have been put forth in recent years. The fifth generation of mobile communications and beyond aims to provide extremely high data speeds, very low latency, and enhanced capacity for a range of applications. The availability of spectrum, signal transmission, and interference are among the issues that 5G must also contend with. Multiple-input multiple-output (MIMO) technology and several types of antennas that can function in various frequency bands are needed for 5G/6G in order to overcome these obstacles. Any wireless system that sends and receives electromagnetic (EM) waves needs antennas as a fundamental component. Antennas for 5G must be versatile, effective, small, and interoperable with a range of network and device designs. Since 5G employs many frequency bands for various purposes, it is clear that it needs antennas in order to maximize the performance of each band. MIMO technology is furthermore used in 5G to boost the data throughput, channel capacity, and transmission range of wireless networks. As a result, antennas are essential for 5G to fulfill its promises of a speedier and more connected future. The following are a few of the popular 5G antenna types:

A. **Planar inverted-F antenna (PIFA):** Due to its straightforward, small size, and omnidirectional properties, this quarter-wavelength resonant antenna is commonly employed in mobile devices. With strong performance for both indoor and wider coverage regions, PIFA [1,2] can function in the coverage range below 2 GHz.
B. **Monopole antenna:** This half-wavelength resonant antenna needs a ground plane to act as the other half of the antenna. A monopole antenna [3] can function in the C-band within 2 and 6 GHz and offer a balance among coverage and capacity.
C. **Massive MIMO antenna:** An array of several antennas known as a massive MIMO antenna can use beamforming techniques to

    DOI: 10.1201/9781003517689-10

concurrently transmit and receive various data streams. For applications that need direct line of sight, massive MIMO antennas [4] may operate at frequencies above 6 GHz and offer high bandwidth, low latency, and high capacity.

Antennas, known as lens antennas, concentrate or collimate the EM waves coming from a feed source using a dielectric lens. For millimeter-wave-based 5G communication systems, they can offer high gain, low loss, large bandwidth, and beam directing capabilities. To lower the cost and complexity of production, lens antennas may also be created utilizing 3D printing or meta surfaces. For 5G communication networks, lens antennas have several uses, including: For point-to-point as well as point-to-multipoint communications, lens antennas offer exceptional directivity and efficiency at high frequencies (over 30 GHz). By splitting up a single feed into numerous beams or by employing a variety of feeds behind the lens, lens antennas can enable MIMO systems. The spectral efficiency as well as the capacity of 5G networks may be improved as a result. Utilizing additive manufacturing methods like fused deposition modeling (FDM), lens antennas may be created, which lowers the antenna design's cost, weight, and complexity. This may allow for quick lens antenna development and adaptation for various 5G situations. Meta surfaces are artificial creations that may be merged with lens antennas to alter EM waves in desirable ways. By lowering aberrations, boosting bandwidth, facilitating dual-band or multiband functioning, and achieving reconfigurability, meta surfaces can enhance the efficiency of lens antennas.

10.1.1 Highlights

- A review on Lens Antenna.
- Overview on 5G
- Discussions about the application of Artificial Intelligence (AI) on 5G-enabled Lens Antenna and Lens Antenna array
- Challenges and Opportunities of Lens Antenna

10.1.2 Organization of the chapter

The chapter has been organized as follows: First, types of antennas applicable for 5G have been introduced. Then, an insight into lens antennas has been discussed, followed by their types. Further, the basics of lens antenna arrays have been addressed. The application of AI/DL in the design and analysis of lens antennas for 5G wireless communications, along with future opportunities and challenges, have been portrayed.

10.2 LENS ANTENNA BASICS

Lens antennas are a particular kind of antenna that use the correct lens material to collimate the incident divergent energy and transform it into plane waves. In order to re-radiate the received energy in the desired direction, it is a microwave antenna that employs the same principle of refraction as an optical lens. The lens antenna's operation is based on the idea of wavefront shaping. Similar to how ripples spread out from a pebble tossed into a pond, radio waves travel outward when they are sent or received by an antenna in a spherical pattern. However, because of the omnidirectional design, the signal intensity decreases with increasing distance from the antenna. Lens antennas concentrate the radio waves into a precise, directed beam using a lens constructed of a dielectric substance, such as plastic or glass. The lens features a curved surface that modifies the incoming waves' phase and amplitude, forcing them to converge at a particular location in space. This concentrates light similarly to how a magnifying glass does. Lens antennas are able to perform better than conventional antennas in terms of gain, range, and interference resistance because they concentrate the radio waves into a tight beam. By altering the lens's location or form, the beam's direction may be changed. The lens antenna is based on the idea that when a source is focused (at a distance equal to the focal length along the lens axis), parallel rays (or a plane wavefront) are formed on the other side of the lens following refraction. Figure 10.1 below displays this.

A lens built of dielectric material also worked for radio frequency (RF) or EM sources. In the case of EM waves, we utilize a dielectric lens constructed of polystyrene rather than a glass lens. On the source side of the

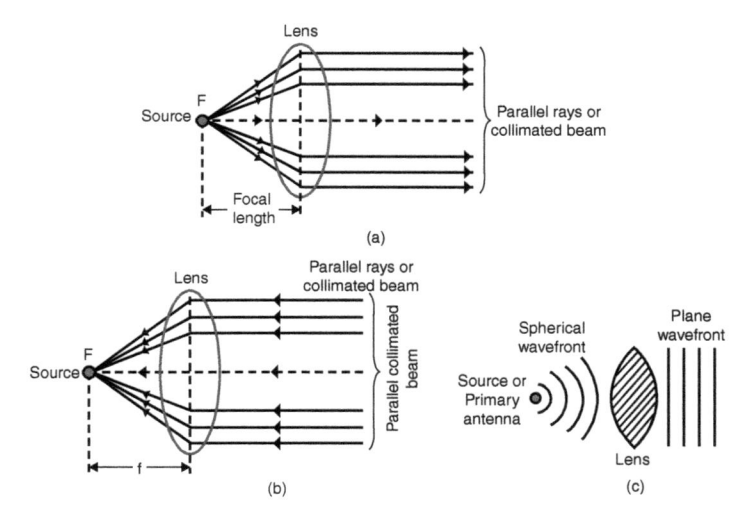

Figure 10.1 Working principle of lens antenna.

lens, a spherical wavefront may be viewed using an EM wave technique. A plane wavefront is necessary to provide the proper phase connection.

This spherical wavefront is straightened out by the lens. By slowing the central component of the wave, this may be accomplished through the lens. Since the dielectric lens at the edge is thinner, portions of the wavefront close to the lens's edges are somewhat retarded.

10.3 LENS ANTENNA TYPES

Delay lenses and fast lenses are two primary types of lens antennas.

Delay lens: In this kind, the lens medium's index of refraction is larger than one, which causes radio waves to move more slowly through it than they would in empty space. This causes the route length to lengthen. This resembles the way a regular optical lens bends light. Convex lenses are converging lenses that focus radio waves, whereas concave lenses are diverging lenses that scatter radio waves. This is because thicker regions of the lens lengthen the path length. Dielectric materials and H-plane plate structures are used to build delay lenses [5] (Figure 10.2).

Fast lens: In this kind, the lens medium's index of refraction is smaller than one because radio waves move through it more quickly than they do in free space. As a result, when light passes through the lens media, the optical path length shortens. When radio waves travel through waveguides at a phase velocity that may exceed the speed of light, it is said to be of this type. A convex lens is a diverging lens because larger portions of the lens diminish the path length. Whereas Concave lenses are converging lenses that concentrate radio waves. E-plane plate structures and negative-index metamaterials are used to make fast lenses [6] (Figure 10.3).

The following are some of the benefits of using a lens antenna: It features a narrow beam width, a low noise temperature, a high gain, and a small number of side lobes. These antennas have a more compact construction.

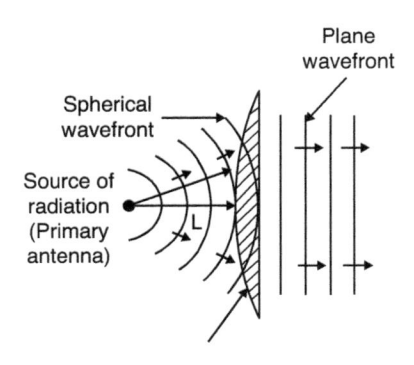

Figure 10.2 H-plane metal delay lens.

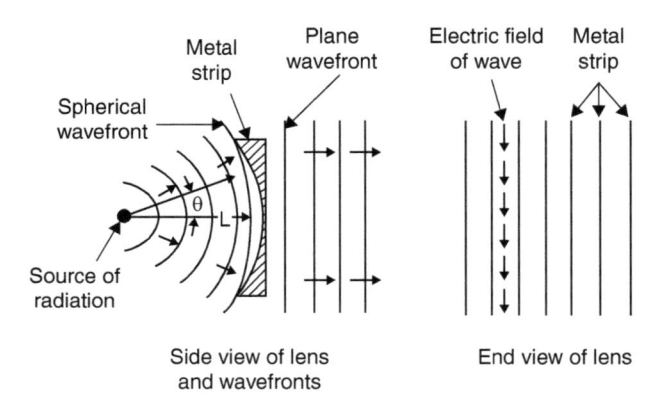

Figure 10.3 E-plane metal lens.

These are lighter than parabolic reflectors and horn antennas. It has increased design tolerance. This antenna's feed and feed support do not impede the aperture. The beam can be moved angularly with regard to the axis. It allows additional freedom within design tolerance, allowing for twisting inside this antenna. It is used in very high-frequency applications. The following are some of the drawbacks of lens antennas. Lenses are particularly thick at lower frequencies. When compared to reflectors, they are more costly for the same qualities. The following are some of the uses for lens antennas. These are appropriate for frequencies exceeding 3 GHz. Used in the same way as the wideband antenna. These are mostly utilized in microwave frequency applications. The converging qualities of this antenna may be exploited to create a wide variety of antennas known as parabolic reflector antennas, which are widely employed in satellite communications. These are used as collimators in high-gain microwave devices such as radio telescopes, millimeter-wave radar, and satellite antennas.

10.4 LENS ANTENNA ARRAY

The lens antenna array [7,8] is typically made up of an EM lens having the capacity to concentrate energy along with a matching antenna array where components are situated in the lens' focal region. In mmWave MIMO systems, the lens antenna array may provide higher gain as well as directivity while lowering the quantity of RF links required. A lens antenna array provides a number of benefits over traditional antenna arrays. By employing a passive lens rather than a dense network of active phase shifters, it is possible to simplify and lower the cost of the system. By making use of route variety and multiplexing gain in the beamspace domain, it can also enhance the system's performance. Third, it can make it possible for novel transmission techniques like path division multiplexing (PDM) and path division multiple access (PDMA), which can enhance user fairness as well as spectral efficiency. Future wireless infrastructures hold great promise for lens antenna array technology. It can offer high data rates, low latency, and high dependability for 5G, automotive networks, and Internet of Things (IoT).

10.5 BENEFITS OF 5G

5G [9] has the potential to deliver faster, more reliable, and more efficient internet connectivity for various devices and applications. Some of the benefits and future of 5G are:

A. **Increased connectivity**: 5G enables more devices to connect to the internet, such as smartphones, tablets, laptops, wearables, smart home appliances, vehicles, drones, and industrial machines. This will create a network of connected things that can communicate and exchange data with each other. 5G also supports new technologies such as IoT, which will allow devices to sense, monitor, and control their environment.

B. **Improved communication**: 5G will improve the quality and speed of communication, both between people and between machines. 5G will allow for faster data transfer, higher-resolution video streaming, lower latency, and better coverage. 5G will also enable new forms of communication, such as virtual and augmented reality (VR/AR), holograms, and telepresence2.

C. **Enhanced experiences**: 5G will enhance the user experience in various domains, such as entertainment, education, healthcare, gaming, and social media. 5G will enable immersive and interactive experiences that can provide realistic sensations and feedback. For example, 5G will allow users to enjoy VR/AR content with high-quality graphics and sound, play online games with minimal lag and high responsiveness, and access personalized content and services based on their preferences and context.

D. **Increased productivity**: 5G will increase the productivity and efficiency of various sectors and industries, such as manufacturing, agriculture, transportation, energy, and public safety. 5G will enable automation, optimization, and innovation in these fields by providing real-time data analysis, remote monitoring and control, predictive maintenance, smart logistics, and intelligent decision-making. For example, 5G will allow factories to use robots that can coordinate and collaborate with each other, farms to use drones that can monitor and optimize crop growth, and vehicles to use sensors that can prevent accidents and optimize traffic flow.

10.6 APPLICATIONS OF ARTIFICIAL INTELLIGENCE/ DL IN 5G-ENABLED ANTENNA TECHNOLOGY

With the introduction of 5G networks, new opportunities for harnessing artificial intelligence (AI) to improve wireless communication systems have emerged. Antenna technology for 5G networks is one critical area where AI may have a huge influence. Antennas play a significant role in signal transmission and reception, thus optimizing their performance is important for 5G success.

10.6.1 AI for antenna design and optimization

Even before actual prototyping, AI can forecast and analyze antenna performance using machine learning. This enables the discovery and correction of flaws during the design process, saving time and money. AI can improve antenna performance by learning from prior designs and performance data. AI enables the development of adaptive 'smart antennas' that can modify settings based on their surroundings. AI simulations can put antennas through a variety of real-world tests. Overall, AI simplifies and accelerates the antenna design process.

10.6.2 AI for beamforming in 5G

Finding appropriate beams is critical for beamforming improvements in 5G. AI can help by reducing the number of candidate beams that are likely to include the best beam. This saves resources while still locating the best beam. AI techniques include multi-target regression as well as multi-class classification.

10.6.3 AI for antenna tilting optimization in 5G

AI utilizing reinforcement learning can optimize antenna tilt angles in 5G networks. The AI agent trains to modify angles according to network KPIs as well as performance. This enhances coverage, capacity, quality, and radio link efficiency. In experiments conducted by Ericsson, AI lowered offline training time from days to hours.

10.6.4 AI for massive MIMO in 5G

In massive MIMO, AI can detect and anticipate user distribution trends. This enables dynamic antenna element weight optimization. AI allows for adaptive optimization that is adapted to specific application cases. AI boosts massive MIMO efficiency in multi-cell installations overall.

10.6.5 AI for network slicing in 5G

It allows for the dynamic provisioning of slices in accordance with user requirements. Data is analyzed by AI to create a panoramic perspective of each slice. This enables intelligent resource scalability and optimization.

10.6.6 AI algorithms for 5G antennas

Deep learning (DL) methods are ideal for antenna optimization. CNNs may be utilized in antenna design for pattern recognition. RNNs are well suited to temporal analysis, such as beam tracking. Antennas may be optimized online using reinforcement learning. Overall, DL opens up new avenues for antenna design. Early research indicates that AI is capable of creating antennas 10 times quicker than traditional approaches. In simulations, AI improved patch antenna configurations by ~2.5×. On prototypes, AI-designed antennas demonstrate great efficiency and gain. The research is still in its early phases, but it is yielding encouraging findings.

10.6.7 Challenges of AI in 5G antennas

The complexity of designing antennas makes AI implementation tough. The vast volume of data involved causes difficulties. The limited coherence time of quantum computers limits the scale of the challenge. To properly leverage quantum AI, more qubits and reduced noise are required.

Finally, AI has a wide range of applications in 5G antenna technology optimization, including design, beamforming, tilting, massive MIMO, and network slicing. Early study yield encouraging outcomes, but problems remain. As AI and quantum computing advance, AI is set to revolutionize antenna engineering for 5G and beyond.

10.7 EMPLOYMENT OF LENS ANTENNA, LENS ANTENNA ARRAY FOR 5G, AND ARTIFICIAL INTELLIGENCE

The rise in data consumption has given rise to the 5G wireless paradigm. Only in higher frequency bands, where the spectra are still being developed, can one collect as much data. In order to use directed antennas, the operational frequency must be increased owing to free space attenuation. Only by reducing the circuitry required for the hardware can such antennas become economically viable. Phase shifters are cost-prohibitive for large-scale manufacture. Therefore, the employment of conventional phases along with active arrays might not be considered an optimum solution. In such cases, lenses as well as leaky wave antennas prove to be best for highly directive antennas with low circuit complexity suitable for 5G communications.

For the construction of a wide-bandwidth met-surface lens suitable for 5G mmWave antennas, a double-layered Huygens' unit cell is suggested [10]. A thin dielectric substrate with two antisymmetric conducting semi-circle arc components on each surface makes up the Huygens' unit cell.

Quevedo-Teruel et al. [11] provided a summary of the recently identified potential uses of geodesic lenses for creating effective antennas for 5G and 6G communications. The relationship between rotationally symmetric graded-index lenses and geodesic lenses has been discussed.

An innovative wideband Luneburg lens antenna with a flared open edge is presented by Wang et al. [12] for use in multibeam scanning at 5G millimeter-wave band. Two parallel PCB groups make up the lens. Each group consists of 12 layers of inexpensive FR4 PCB, held in place by metal screws. There is no dielectric present inside the lens, creating an air-filled structure.

For base station (BS) applications, lens antennas with minimal thickness and curvature are preferred due to their light weight and simplicity of installation. In the study by Ansarudin et al. [13], a novel lens shaping technique for tiny and delicate lens curvature is suggested. Comparisons of antenna constructions using traditional aperture distribution lenses and Abbe's sine lenses are done in order to build the thin-lens antenna. Additionally, the multi-beam radiation patterns of the three different types of lenses are contrasted. This ensures the suggested lens's thin and tiny curvature as well as an effective multi-beam radiation pattern.

For use with 5G, a high-gain lens antenna is created by He et al. [14]. The substrate of the patch antenna, whose primary frequency is 17.5 GHz, was created by RT/Duroid 5880.

For low-priced multiple-beam antennas employing the 26-GHz band for 5G communications, a planar dielectric lens was created [15]. A microstrip antenna with four input ports makes up the lens feeder, allowing it to support four separate, high-directivity radiation beams. With a simulated gain of more than 18 dB for all scanning angles, a maximum steering of 25° was achieved.

Wu et al. [16] offered brand-new discrete lens antenna designs that address Orbital Angular Momentum (OAM) multiplexing as well as the diffractive OAM wave problem.

It is suggested to use a triple-layer multi-beam Luneburg lens antenna in the mmWave spectrum [17]. The suggested lens has a cylindrical, triple-layer all-dielectric construction. The identical arrangement is present on every layer, thanks to hole drilling technology.

Basavarajappa et al. [18] proposed a dual-band, 1 GHz bandwidth millimeter-wave multi-beam antenna that operates between the 28 and 31 GHz frequencies. In order to increase gain, the antenna is constructed using the technique of cylindrical wavefront transformation to planar wavefront. The antenna's dual-band functionality makes it easier to broadcast and receive at two different frequencies. Due to the utilization of the shared aperture across the lens, there are significantly fewer RF chains, which is advantageous.

Poyanco et al. [19] outline the construction of a wideband, 3D-printed planar graded index (GRIN) lens that operates on the Ka band between 25 and 40 GHz. Using a U-shaped thin lens and a low-profile 1×4 antenna array as a feed source, Kim et al. [20] offered a 28-GHz compact antenna array.

Beamspace MIMO, a new millimeter-wave huge MIMO technique that uses lens antenna arrays [21], can significantly minimize the amount of power-hungry RF chains that are needed. As a result, it has been viewed as a viable approach for future communications technologies, including 5G.

By utilizing the angle-dependent energy focusing property of lens arrays [22–25] and the angular sparsity of mmWave channels, authors demonstrated that mmWave lens antenna systems can operate at capacity-optimal performance even for wideband frequency-selective channels with a minimal number of RF chains and single-carrier transmission. According to numerical findings, the lens-based system performs better than the benchmark mmWave communications methods in terms of both spectrum efficiency as well as energy efficiency.

The study by Zeng et al. [26] addresses a multi-user, single-sided, mmWave MIMO system in which the BS is furnished with a full-dimensional lens antenna array, whereas every mobile station supports conventional antenna arrays.

Huang et al. [27] investigated a low-complexity single-carrier broadcast method with route delay pre-compensation performed at the BS by leveraging the lens antenna array's angle-dependent energy focusing capacity [28,29] as well as the angular sparsity concerning mmWave channels.

Liao et al. [30] presented a reconfigurable terahertz holographic discrete dielectric lens (DDL) antenna with DL capabilities. The antenna is made of two separate dielectric lenses that are cascaded and supplied by a static horn.

For a lens antenna array, Hoang et al. [31] look at a data-driven method to count the number of incoming signals. In order to enumerate both multipath and independent signals, an input spectrum is first created utilizing the energy-focusing feature of an EM lens. Next, the authors introduce the power spectrum-based convolutional neural network (PSCNet), a technique for sharp peak detection that is helped by DL.

10.8 FUTURE SCOPE AND RESEARCH CHALLENGES

Lens antennas [32–35] offer several advantages over traditional antennas: Improved beam focusing, reduced size, reduced sidelobes, broadband performance, reduced EM interference, and increased efficiency. Despite their advantages, lens antennas also have some disadvantages that should be considered when choosing an antenna for a particular application: Complexity, limited bandwidth, fragility, cost, sensitivity to environmental conditions, limited radiation pattern. Current antenna optimization methods include manual procedures based on design knowledge and rules of thumb, as well as computer strategies such as evolutionary algorithms as well as surrogate model-based optimization. However, these have constraints in terms of optimality, efficiency, as well as scalability.

DL and AI have the potential to dramatically improve antenna optimization; however, there are several important technical obstacles to overcome:

a. The difficulty of designing antennas with wide search spaces and various competing purposes. AI models require enough training data to span the whole design space.
b. EM simulations have a high computing cost for correct antenna assessment. Surrogate modeling is useful, but balancing model accuracy and efficiency is challenging.
c. Difficulties in incorporating trained AI models into real-world antenna designs. Model adaptation and transfer learning strategies are necessary.

Data availability, interaction with current tools, and acceptance by antenna engineers are all practical issues. Solutions include the development of more complex AI techniques such as neural architecture search and meta-learning, parallel computing, and cooperation between AI specialists and antenna engineers.

10.9 CONCLUSION

Lens antennas [36–42] have applications in various fields, including: Satellite communication, radar systems, wireless communication, medical imaging, astronomy, and automotive radar. Lens antennas are effective in 5G and 6G because they can focus the radiation pattern of the antenna source and increase the antenna gain. This is important for high-frequency communication systems such as 6G, which require low-loss and high-speed transmission. Lens antennas can also be used for beam steering, which is the ability to change the direction of the radiation beam without physically moving the antenna. This can improve the coverage and reliability of wireless communication. A lens antenna array is a combination of a dielectric lens and an array of radiating elements, such as microstrip patches or annular slots. The lens can focus the radiation pattern of the array and increase the antenna gain, which is important for long-distance transmission and high data rates. Current work is basically an effort to justify the importance of lens antenna and lens antenna array in the fields of AI and 5G.

REFERENCES

1. Chattha, Hassan T., Yi Huang, Muhammad K. Ishfaq, and Stephen J. Boyes. "A comprehensive parametric study of planar inverted-F antenna." *Wireless Engineering and Technology* 3, no. 1 (2012): 1–12.
2. Kundu, Krishanu, Avneesh Dubey, Ankur Dhama, and Narendra Nath Pathak. "Planar inverted F antenna, PIFA array in 5G applications." *Journal of Physics: Conference Series* 2062, no. 1 (2021): 012002.
3. Agrawall, Narayan P., Girish Kumar, and Kamla Prasan Ray. "Wide-band planar monopole antennas." *IEEE Transactions on Antennas and Propagation* 46, no. 2 (1998): 294–295.
4. Li, Yixin, Yong Luo, and Guangli Yang. "12-port 5G massive MIMO antenna array in sub-6GHz mobile handset for LTE bands 42/43/46 applications." *IEEE Access* 6 (2017): 344–354.
5. Kock, Winston E. "Metallic delay lenses." *Bell System Technical Journal* 27, no. 1 (1948): 58–82.
6. Sauleau, Ronan, C. A. Fernandes, and J. R. Costa. "Review of lens antenna design and technologies for mm-wave shaped-beam applications." In: *11th International Symposium on Antenna Technology and Applied Electromagnetics [ANTEM 2005]*, pp. 1–5. IEEE, Saint Malo, 2005.
7. Imbert, Marc, Jordi Romeu, Mariano Baquero-Escudero, Maria-Teresa Martinez-Ingles, Jose-Maria Molina-Garcia-Pardo, and Lluis Jofre. "Assessment of LTCC-based dielectric flat lens antennas and switched-beam arrays for future 5G millimeter-wave communication systems." *IEEE Transactions on Antennas and Propagation* 65, no. 12 (2017): 6453–6473.
8. Qu, Zhishu, Shi-Wei Qu, Zhe Zhang, Shiwen Yang, and Chi Hou Chan. "Wide-angle scanning lens fed by small-scale antenna array for 5G in millimeter-wave band." *IEEE Transactions on Antennas and Propagation* 68, no. 5 (2020): 3635–3643.
9. Gupta, Akhil, and Rakesh Kumar Jha. "A survey of 5G network: Architecture and emerging technologies." *IEEE Access* 3 (2015): 1206–1232.
10. Xue, Chunhua, Qun Lou, and Zhi Ning Chen, "Broadband double-layered Huygens' metasurface lens antenna for 5G millimeter-wave systems." *IEEE Transactions on Antennas and Propagation* 68, no. 3 (2020): 1468–1476. doi:10.1109/TAP.2019.2943440.
11. Quevedo-Teruel, Oscar, Qingbi Liao, Qiao Chen, Pilar Castillo-Tapia, Francisco Mesa, Kun Zhao, and Nelson J.G. Fonseca, "Geodesic lens antennas for 5G and beyond." *IEEE Communications Magazine* 60, no. 1 (2022): 40–45. doi:10.1109/MCOM.001.2100545.
12. Wang, Xi, Yang Cheng, and Yuandan Dong, "A wideband PCB-stacked air-filled luneburg lens antenna for 5G millimeter-wave applications." *IEEE Antennas and Wireless Propagation Letters* 20, no. 3 (2021): 327–331. doi:10.1109/LAWP.2021.3049432.
13. Ansarudin, Farizah, Tharek Abd Rahman, Yoshihide Yamada, Nurul Huda Abd Rahman, and Kamilia Kamardin. "Multi beam dielectric lens antenna for 5G base station." *Sensors* 20, no. 20 (2020): 5849.

14. He, Bo, Yong-Chang Jiao, and Huan He, "Design of high-gain lens antenna for 5G application." In: *2017 Sixth Asia-Pacific Conference on Antennas and Propagation (APCAP)*, Xi'an, China, pp. 1–3, 2017. doi:10.1109/APCAP.2017.8420863.

15. Garcia-Marin, Eduardo, Dejan S. Filipovic, Jose Luis Masa-Campos, and Pablo Sanchez-Olivares. "Low-cost lens antenna for 5G multi-beam communication." *Microwave and Optical Technology Letters* 62, no. 11 (2020): 3611–3622.

16. Wu, Geng-Bo, Ka Fai Chan, Kam Man Shum, and Chi Hou Chan, "Millimeter-wave and terahertz OAM discrete-lens antennas for 5G and beyond." *IEEE Communications Magazine* 60, no. 1 (2022): 34–39. doi:10.1109/MCOM.001.2100523.

17. Wang, Xi, Yongsheng Pan, and Yuandan Dong, "An E-plane-focused triple-layer multibeam luneburg lens antenna for 5G millimeter-wave applications." *IEEE Antennas and Wireless Propagation Letters* 21, no. 2 (2022): 227–231. doi:10.1109/LAWP.2021.3124129.

18. Basavarajappa, Vedaprabhu, Alberto Pellon, Ignacio Montesinos-Ortego, Beatriz Bedia Exposito, Lorena Cabria, and Jose Basterrechea, "Millimeter-wave multi-beam waveguide lens antenna." *IEEE Transactions on Antennas and Propagation* 67, no. 8 (2019): 5646–5651. doi:10.1109/TAP.2019.2916388.

19. Poyanco, Jose-Manuel, Francisco Pizarro, and Eva Rajo-Iglesias, "3D-printed dielectric GRIN planar wideband lens antenna for 5G applications." In: *2021 15th European Conference on Antennas and Propagation (EuCAP)*, Dusseldorf, Germany, pp. 1–4, 2021. doi:10.23919/EuCAP51087.2021.9411342.

20. Kim, Eugean, Seung-Tae Ko, Young Ju Lee, and Jungsuek Oh, "Millimeter-wave tiny lens antenna employing U-shaped filter arrays for 5G." *IEEE Antennas and Wireless Propagation Letters* 17, no. 5 (2018): 845–848. doi:10.1109/LAWP.2018.2819022.

21. Xie, Tian, Linglong Dai, Derrick Wing Kwan Ng, and Chan-Byoung Chae, "On the power leakage problem in millimeter-wave massive MIMO with lens antenna arrays." *IEEE Transactions on Signal Processing* 67, no. 18 (2019): 4730–4744. doi:10.1109/TSP.2019.2926019.

22. Zeng, Yong, and Rui Zhang, "Cost-effective millimeter-wave communications with lens antenna array." *IEEE Wireless Communications* 24, no. 4 (2017): 81–87. doi:10.1109/MWC.2017.1600336.

23. Imbert, Marc, Jordi Romeu, Mariano Baquero-Escudero, Maria-Teresa Martinez-Ingles, Jose-Maria Molina-Garcia-Pardo, and Lluis Jofre. "Assessment of LTCC-based dielectric flat lens antennas and switched-beam arrays for future 5G millimeter-wave communication systems." *IEEE Transactions on Antennas and Propagation* 65, no. 12 (2017): 6453–6473.

24. Xue, Chunhua, Qun Lou, and Zhi Ning Chen, "Broadband double-layered Huygens' metasurface lens antenna for 5G millimeter-wave systems." *IEEE Transactions on Antennas and Propagation* 68, no. 3 (2020): 1468–1476. doi:10.1109/TAP.2019.2943440.

25. Cho, Yae Jee, Gee-Yong Suk, Byoungnam Kim, Dong Ku Kim, and Chan-Byoung Chae. "RF lens-embedded antenna array for mmWave MIMO: Design and performance." *IEEE Communications Magazine 56*, no. 7 (2018): 42–48.

26. Zeng, Yong, Lu Yang, and Rui Zhang, "Multi-user millimeter wave MIMO with full-dimensional lens antenna array." *IEEE Transactions on Wireless Communications* 17, no. 4 (2018): 2800–2814. doi:10.1109/TWC.2018.2803180.

27. Huang, Wei, Yongming Huang, Yong Zeng, and Luxi Yang, "Wideband millimeter wave communication with lens antenna array: Joint beamforming and antenna selection with group sparse optimization." *IEEE Transactions on Wireless Communications* 17, no. 10 (2018): 6575–6589. doi:10.1109/TWC.2018.2860963.

28. Gao, Xinyu, Linglong Dai, Shidong Zhou, Akbar M. Sayeed, and Lajos Hanzo. "Beamspace channel estimation for wideband millimeter-wave MIMO with lens antenna array." In: *2018 IEEE International Conference on Communications (ICC)*, pp. 1–6. IEEE, Kansas City, MO, 2018.

29. Yang, Jie, Yong Zeng, Shi Jin, Chao-Kai Wen, and Pingping Xu, "Communication and localization with extremely large lens antenna array." *IEEE Transactions on Wireless Communications* 20, no. 5 (2021): 3031–3048. doi:10.1109/TWC.2020.3046766.

30. Liao, Dashuang, Manting Wang, Ka Fai Chan, Chi Hou Chan, and Haogang Wang, "A deep-learning enabled discrete dielectric lens antenna for terahertz reconfigurable holographic imaging." *IEEE Antennas and Wireless Propagation Letters* 21, no. 4 (2022): 823–827. doi:10.1109/LAWP.2022.3149861.

31. Hoang, Dai Trong, and Kyungchun Lee, "Deep learning-aided signal enumeration for lens antenna array." *IEEE Access* 10 (2022): 123835–123846. doi:10.1109/ACCESS.2022.3224608.

32. Chen, Zhi Ning, Teng Li and Wei El Liu, "Microwave metasurface-based lens antennas for 5G and beyond." In: *2020 14th European Conference on Antennas and Propagation (EuCAP)*, Copenhagen, Denmark, pp. 1–4, 2020. doi:10.23919/EuCAP48036.2020.9135285.

33. Ala-Laurinaho, Juha, Jouko Aurinsalo, Aki Karttunen, Mikko Kaunisto, Antti Lamminen, Juha Nurmiharju, Antti V. Räisänen, Jussi Säily, and Pekka Wainio, "2-D beam-steerable integrated lens antenna system for 5G E-band access and backhaul." *IEEE Transactions on Microwave Theory and Techniques* 64, no. 7 (2016): 2244–2255. doi:10.1109/TMTT.2016.2574317.

34. Ballesteros, Christian, Marcos Maestre, Maria C. Santos, Jordi Romeu, and Luis Jofre. "A 3D printed lens antenna for 5G applications." In: *2019 IEEE International Symposium on Antennas and Propagation and USNC-URSI Radio Science Meeting*, pp. 1985–1986. IEEE, Atlanta, GA, 2019.

35. Poyanco, José-Manuel, Francisco Pizarro, and Eva Rajo-Iglesias. "Cost-effective wideband dielectric planar lens antenna for millimeter wave applications." *Scientific Reports* 12, no. 1 (2022): 4204.

36. Liu, Kunning, Shiwen Yang, Shi-Wei Qu, Yikai Chen, Ming Huang, and Jun Hu. "A low-profile wide-scanning fully metallic lens antenna for 5G communication." *International Journal of RF and Microwave Computer-Aided Engineering* 31, no. 5 (2021): e22584.

37. Aljaloud, Khaled, Yosef T. Aladadi, Majeed AS Alkanhal, Wazie M. Abdulkawi, and Rifaqat Hussain. "A wideband GRIN dielectric lens antenna for 5G applications." *Micromachines* 14, no. 5 (2023): 997.

38. Yang, Xujun, Yuan Ji, Jun Hu, Lei Ge, Yujian Li, and Kwai Man Luk. "Wideband quasi-spherical lens antenna module with two-dimensional switched beams for 5G millimeter-wave IoT applications." *IEEE Internet of Things Journal* 11, no. 1 (2023): 1217–1227.

39. Moreno, Resti Montoya, Juha Ala-Laurinaho, and Ville Viikari. "Plastic-filled dual-polarized lens antenna for beam-switching in the Ka-band." *IEEE Antennas and Wireless Propagation Letters* 18, no. 12 (2019): 2458–2462.

40. Kodnoeih, Mohammad Reza Dehghani, Yoann Letestu, Ronan Sauleau, Eduardo Motta Cruz, and André Doll. "Compact folded fresnel zone plate lens antenna for mm-wave communications." *IEEE Antennas and Wireless Propagation Letters* 17, no. 5 (2018): 873–876.

41. Garcia-Marin, Eduardo, Dejan S. Filipovic, José Luis Masa-Campos, and Pablo Sanchez-Olivares. "Ka-band multi-beam planar lens antenna for 5G applications." In: *2020 14th European Conference on Antennas and Propagation (EuCAP)*, Copenhagen, Denmark, pp. 1–5. IEEE, 2020. https://ieeexplore.ieee.org/document/9135364

42. Shalini, Goutha Reddy, D. Rama Krishna, Gs Karthikeya, and Shiban K. Koul. "Design of 3D printed lens antenna for 5G applications." In: *2021 IEEE MTT-S International Microwave and RF Conference (IMARC)*, Kanpur, India, pp. 1–4. IEEE, 2021. https://ieeexplore.ieee.org/document/9714615

Chapter 11

AI, ML, and IoMT integrated approach to DNA sequence analysis and prediction

*Manoj Kumar Dey, Sayantika Das,
Madhushree Ghosh, Koustubh Majumdar,
Manab Kumar Saha, and Ankan Bhattacharya*

11.1 INTRODUCTION

The genetic material known as DNA is where a cell's creation and maintenance instructions are stored. The four nucleotides that make up DNA sequences are adenine (A), guanine (G), cytosine (C), and thymine (T). The configuration of these nucleotides determines the genetic make-up of the organism. The function of DNA sequence analysis is information extraction from DNA sequences. The DNA sequence data has to be preprocessed in order for machine learning (ML) algorithms to handle it. This involves encoding the sequences numerically and managing errors. For the purpose of building accurate prediction models, we will make use of a number of ML methods. These methods will be shown using labeled DNA sequence data, where the labels denote certain biological properties or consequences. Once the models have been trained, they may be used to predict unknown DNA sequences and infer their most likely functional implications. To improve our models' usability and reliability, we will employ ensemble learning. By integrating the forecasts of several models, we may increase the accuracy and reliability of our predictions. Utilizing cross-validation techniques, we'll also assess how effectively our models generalize to new, untested data. The outcomes of this study have a wide range of applications in genetic research and medical treatment. Our ML technique can help with the identification of genetic variations related to illnesses, the finding of regulatory networks, the annotation of functional regions within genomes, and the prediction of gene expression levels. Researchers and medical professionals now have powerful tools for DNA sequence analysis and prediction thanks to the developed models, which may also be included into user-friendly apps. Ultimately, the goal of our work is to forecast and assess DNA sequences using ML. By spotting patterns and making predictions, we can improve our understanding of genetic data and enhance genomics research and personalized medicine.

DOI: 10.1201/9781003517689-11

11.2 LITERATURE SURVEY

An important component of current research may include the sequencing DNA. We often use it in conjunction with genetics, meta-genetics, and phylogenetics to further a variety of different fields. Within a specific domain, where the sample size is a polynomial, courageous learning is examined in this study. Not much attention has been given to these concepts since they are trivially polynomial-learnable. In the past, academics have concentrated on the capacity for learning mental categories, whose example regions supposing that (else the subject would be easy), super polynomial [1]. Finding the nucleotide sequence and the frameability of a DNA segment as a maximum-likelihood sequence detection (MLSD) issue in sequencing-by-synthesis systems. Even for relatively small DNA sequences, it is prohibitive to solve them using exhaustive searches. While theoretically viable, partial-MLSD approaches and symbol-by-symbol approaches fall short in terms of performance. In order to effectively handle sequencing-by-synthesis systems, we discuss the MLSD issue and alter the so-called sphere decoding technique in this study. We examine the algorithm's anticipated complexity and use simulations to show that it performs noticeably better than heuristic methods [2]. Modeling of the DNA sequencing-by-synthesis process and associated non-idealities uses a noisy switching linear system with the unidentified DNA pattern as a parameter. It is thus possible to frame the base-calling trouble with parameter detection. The computational cost of doing accurate maximum-likelihood decoding on this device, which has a potentially large memory, is too high. Using experimental Pyrosequencing data, an approximative ML approach shows that valid read lengths may surpass 200 bases, which is far longer than what is currently possible using other techniques [3]. The DNA electrophoresis time series underlying statistics are crucially responsible for accurately obtaining the basic sequence from automated DNA sequencing. The DNA sequencing techniques used today are heuristic in nature and sparing with statistical data. The best ML processor is built in this study using the DNA time series in a formal statistical model. The peak parameter estimate, whitened waveform comparison, and multiple hypothesis processing are all aspects of the Kalman prediction of peak locations method. Utilizing the M-algorithm, these are all characteristics of the DNA-ML method. With the use of both simulated and actual data, the algorithm's properties are investigated. Critical model parameters are identified, along with how they affect various error processes, including insertions and deletions. Statistical analysis of the foundation for further research and improvement of DNA sequencing methods is provided by DNA time-series and the architecture of the DNA-ML method [4]. The study of DNA sequences requires appropriate parallel computer resources and techniques since it is a data and computationally expensive job. In this article, we outline a technique that is optimized for a heterogeneous platform for rapid DNA sequence analysis by an

Intel Xeon Phi processor. These platforms typically include a single or two general-purpose CPUs for the host plus one or more Xeon Phi devices. In order to shorten the total analysis time, we provide a parallel approach that divides the job between the host CPUs and the Xeon Phi device of DNA sequence analysis. For automated work-sharing, we employ supervised ML, which forecasts the performance of DNA sequence analysis on the host and device and, in response, maps portions of the DNA sequence to the host and the gadget. On a heterogeneous setup made up of two Intel Xeon E5 CPUs with 12 cores each and a 61-core Intel Xeon Phi 7120P device, we empirically test our methodology using real-world DNA segments for human and different animal species [5]. *Mycobacterium tuberculosis (MTB)* is the causative agent of *tuberculosis*, a severe infectious illness that predominantly affects the lungs. It is well known that a number of MTB strains are resistant to treatment-related medications. This condition highlights the significance of identifying and halting new medication resistance, which will lower the death rate. The traditional molecular diagnostic test costs a lot of money, takes a lot of time, and is not very predictive. This study intends to investigate the use of ML to reliably anticipate medication resistance, which provides a much quicker and less expensive solution than the traditional technique. Several ML algorithms, including C4.5, Random forest, and LogitBoost, were used in experiments on 3393 MTB isolates. Rifampicin (RIF) and isoniazid (INH) are two of the several medications examined in this case along with pyrazinamide (PZA), ethambutol (EMB), and INH. The model's ability to predict drug resistance with an accuracy of 99% and an Area Under the Curve (AUC) that approaches (near) 1 was shown by the results of 10-fold cross-validation. This finding implies that the ML technique has a potential outcome in predicting treatment resistance for tuberculosis [6]. Due to the genetic variety of cancers and the scarcity of molecules produced from tumors, using cell-free DNA (cfDNA) as a biomarker in cancer is difficult. In this article, we provide and illustrate a unique ML-driven panel design technique for enhancing the identification of tumor mutations in cfDNA. This method was used to create a model to categorize and assess candidate variations for inclusion on a targeted sequencing panel for prostate cancer. Then, in both in silico and hybrid capture environments, we employed this panel to screen tumor variations from prostate cancer patients with localized illnesses [7]. By using RNA sequencing, allele-specific expression (ASE), which concerns the differential expression levels of alternative alleles, is quantified. According to several research, ASE influences the intensity or penetrance of phenotypes, which in turn affects how heritable disorders manifest themselves. However, because genome diagnostics is focused on DNA sequencing, it ignores mechanisms that control gene expression, such as ASE. It must be predicted using only DNA variation if ASE is to be used without RNA sequencing. We've built ASE models that utilize DNA properties to forecast ASE using data from BIOS ($n=3432$) and GITEX ($n=369$). These models demonstrate the complicated control that underpins ASE and are very repeatable as they contain a wide variety of feature types. Three genes in which ASE has a clinically significant function were subjected to the BIOS-trained model for population variant analysis, BRCA2, RET, and NF1 are examples. For 27 variations, of which 10 were known pathogenic variants, the ASE effects were thus projected. Through the use of ML, we have shown that ASE can be predicted from DNA characteristics. Future work may enhance sensitivity and convert these models into a novel kind of genome diagnostic tool that prioritizes potential pathogenic variants or their regulators for RNA sequencing validation. On GitHub and Zenodo, you can find all the utilized code and ML models [8]. How to quantitatively analyze the structures and functions of the deluge of biological sequences created in the post-genomic age is one of the most challenging issues. In this discipline, ML techniques have a major impact. Feature extraction, predictor generation, and performance evaluation are typically the three primary processes in the creation of predictors using ML techniques. Although a number of standalone programs and Web servers have been created to aid in the study of biological sequences, many only concentrate on a single phase. An effective Web server with the name BioSeq-Analysis was used in this investigation. The three essential processes for building a predictor may be carried out automatically using BioSeq-Analysis. The benchmark dataset must only be uploaded by the user [9]. As the first web-based server to do sequence-level analysis of a wide variety of biological sequences using ML approaches, the BioSeq-Analysis has played a pivotal role in the creation of various powerful predictors in the field of computational biology. Yet, the BioSeq-Analysis can only be used to activities requiring analysis at the sequence level, limiting its usefulness. This highlights the need for a smart instrument that can create several predictors for biological sequence analysis, both at the residue and the sequence levels. Because of this, we've decided to develop BioSeq-Analysis2.0, an improved server that covers 90 features at the sequence level and 26 features at the residue level and simply requires users to submit the benchmark dataset [10]. Many cyberattacks on the Internet, such as cybercrime, fraud, scams, and nation-state cyberwar, are carried out using malicious software, or malware. Viruses, Trojans, worms, spyware, botnet malware, ransomware, Rootkits, and many more types of malicious software fall into this broad category. Ransomware is a type of malicious software that encrypts a user's files, rendering them unavailable until the user pays a ransom. Different ransomware variations use one or more approaches, such as ML algorithms, in their assault cycle to prevent detection. Therefore, it is important to comprehend the processes involved in developing and deploying ransomware [11]. Computers may learn through experience, by doing, and by analogy, which is called ML. Therefore, it is a field of techniques that offers, in one way or another, intelligent information processing capacities for dealing with real-life situations. The use of ML in bioinformatics is one of them. By using computer science and information technology, bioinformatics is an interdisciplinary field of study that interprets biological data. ML is focused on obtaining knowledge automatically from data collection. ML takes into account the data's ability to be gradually learned, the learning pace, and the convergence guarantee. Artificial neural networks (ANNs), genetic algorithms (GAs), fuzzy systems, and hybrid approaches, which combine some of these methods, are the types of techniques we most frequently discuss. A significant issue is distinguishing between the genuine genes that are contaminated by various disorders and the normal genes. The activities of novel proteins are discovered in genomic research by categorizing DNA sequences into pre-existing groups. As a result, it's critical to classify and identify those genes. Here, we employ ML approaches in order to categorize the genes into the normal genes and the genes that are infected. The mechanics of gene sequence categorization using ML approaches are reviewed in this work, along with a quick overview of bioinformatics, a literature review, and the major problems with ML-assisted DNA sequencing [12]. For genetic sequencing and identification, a developing alternative to conventional polymerase chain reaction (PCR) procedures is the all-electronic single molecule break junction (SMBJ) approach. The present spectra obtained from SMBJ experiments have distinctive characteristics that can be used to distinguish known sequences from a dataset, according to previous research. The substrate, sample, environment, and measurement device interact in unpredictable and complicated ways, which makes the spectra highly noisy in most cases. As a result, hundreds or thousands of experiments are required to provide trustworthy and accurate findings [12]. Four typical uses of ML in DNA alignment data—DNA progression organization, pattern, clustering, and design mining—were deconstructed and analyzed in this research. The basis and significance of their natural-world applications were dissected as well. The document goes on to discuss a few significant difficulties in the area of DNA batching DM as well as a few potential research topics and models. In order to deliver mining results that are simpler to utilize, a forthcoming survey has accepted that more organic spaces and artificial intelligence (AI) will be merged [13]. For DNA collection requests, Label and K-mer coding were used in this study together with CNN, CNNLSTM, and CNN Bi-direction LSTM structures. Different data collection methods are utilized to evaluate the model. According to the results of the exploratory analyses, the accuracy of the test data for CNN and CNN bidirectional LSTMs with K-mer 20 encoding is 93.16 and 93.13%, respectively. Three powerful learning techniques were examined in this article: bidirectional CNN-LSTM using name coding and K-mer coding, CNN, and LSTM. They discovered, to their surprise, that CNN, which encodes marks, outperforms other models. The test's accuracy is poor. The best technique to get exceptional test and approval precision is via K-mer encoding. Accurate measurements are not possible for this dataset's evaluation. There should be several indicators, such as fit, review awareness, and particularity [14].

The categorization of DNA sequences is a significant problem in a wide computational setting for biological data processing. This goal has been successfully completed in recent years using a variety of ML approaches. In order to prevent an epidemic like COVID-19, viral identification and categorization are crucial. The process of feature selection is still the most difficult part of the problem, though. Sequences lack clear characteristics and the most widely used representations make the problem of high dimensionality worse. It also aids in the detection of the impact of viruses and medicine formulation. Deep learning (DL) models have recently improved their ability to spontaneously extract features from input. Herein, we used label and K-mer encoding to classify DNA arrays using CNN, CNN-LSTM, and CNN-Bidirectional LSTM architectures. Various metrics are used to assess the models [15].

11.3 INSPIRATIONS AND OBJECTIVES

We can anticipate a person's future by utilizing the vast amount of information found in the medical field. Although "models of ML and DL" are being used to rapidly sequence DNA and make predictions based on the sequence, medical practitioners still require state-of-the-art tools to apply the data to the model and make predictions. The ML model's implementation will rely on the kind of data we use and the details of the columns or features we're utilizing. The mechanism's starting condition is to have the practitioner approach, which takes time; therefore, we cannot rely on its accurate. Another critical component of the "ML and DL models" is time, which will be used in DNA research. The goal is to improve the prediction model's accuracy and even find insights in order to achieve the most correct values for CNN, transform learning, Naive Bayes, Random forest, decision trees, and Random forest algorithms. Comparison of the outcomes of the ML and DL algorithms, followed by the selection of the top-performing algorithm among the five different algorithms.

11.4 A SURVEY OF AI, ML, AND IOMT

The term "AI" describes a method for creating robots that are intended to fail on both sides. Robots are unable to precisely respond to questions from humans in symbolic AI studies, whereas cybernetic AI uses ANNs that are unable to do so. As a result, both sides' efforts cannot be successful when interpreted literally. Specialized AI exercises have developed as a result of failures in symbolic and cybernetic AI research on opposing sides. These actions will carry on with a single aim in mind rather than a variety of branches and minds. There have been a number of difficulties as a result of AI research, despite the fact that the idea has inspired

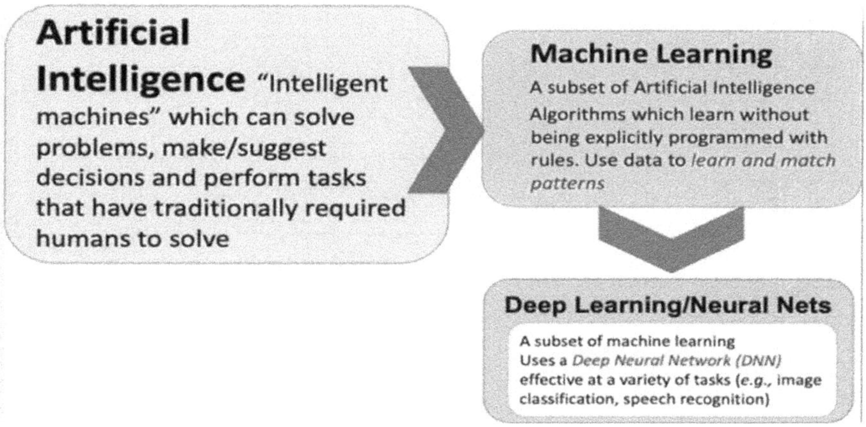

Figure 11.1 The relationship between AI, ML, and DL.

studies in the field from opposite sides, due to failures in the studies of symbolic and cybernetic AI. These actions will carry on with a single aim in mind rather than a variety of branches and minds. Despite the fact that the idea of AI has stimulated research in the field, there have been a number of difficulties since current AI products don't fully comprehend the technology they are using. A billion-dollar budget is needed to complete successful projects, as the AI industry that developed over the coming years demonstrated. However, the commercial level of AI has been reached by inventors of AI who have offered logical answers to new problems. Research on AI has recently shown that language is essential. In Figure 11.1, we quickly discuss how to merge AI, ML, and DL to prioritize language. The notion underpinning AI is the construction of intelligent machines. ML, a type of AI, may be utilized to create AI-powered apps. A model is trained in DL, a branch of ML, utilizing massive amounts of data and advanced algorithms.

11.5 THE PROPOSED SYSTEM'S GENERAL CONCEPT

The dataset is difficult to grasp, despite the proposed architecture appearing to be straightforward to apply. The DNA sequence is the only part of the dataset that we are unable to comprehend. We need to apply the pre-processing methods in a few instances when undesirable symbols are present. The proposed technique will be architecturally explained in Figure 11.2.

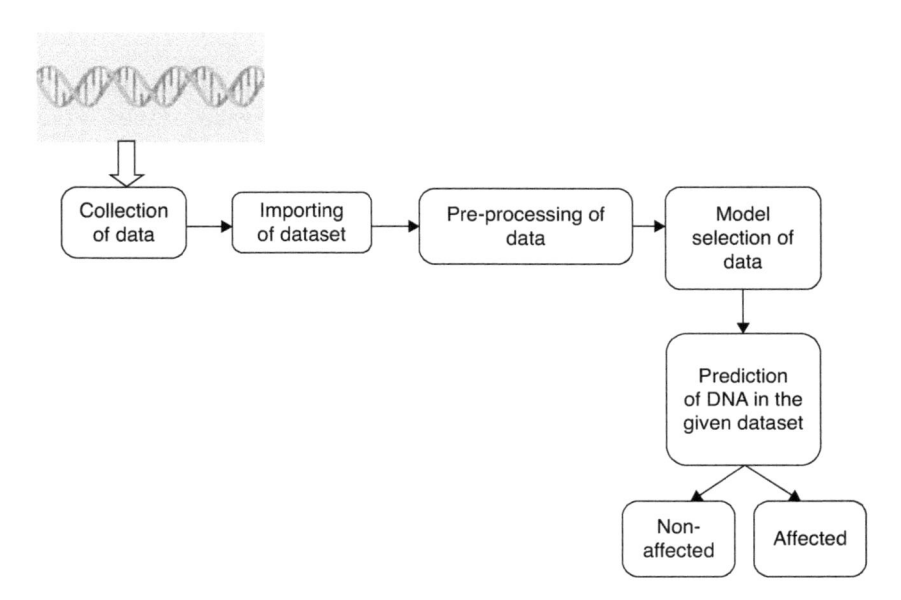

Figure 11.2 Methodology process.

11.5.1 Data collection and preprocessing

1. Locate and collect from reputable sources the publicly accessible DNA sequence data for humans, chimpanzees, and dogs.
2. Preprocess the data by deleting unneeded details, addressing missing values and, if necessary, normalizing the sequences. Figure 11.3 provides a quick explanation of the data processing portion.

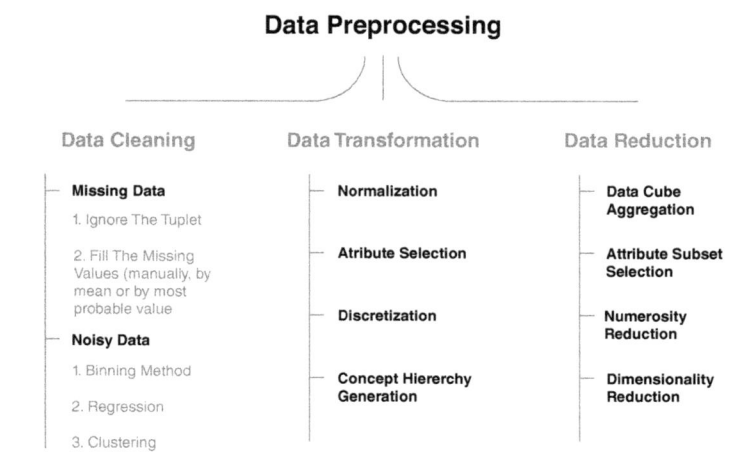

Figure 11.3 Data preprocessing.

11.5.2 Splitting data

When a dataset is split into two or more subsets for the purposes of developing, testing, and assessing a ML model, this is known as data splitting. The most popular method of data division is into a training set and a testing set. The training set is used to train the model, while the testing set is used to evaluate its performance on data that has not yet been seen. Use an 80–20 split, where 80% of the data is used for training, whereas 20% is used for testing. Data splitting helps to avoid overfitting and guarantees that the model can generalize to new data, making it an essential phase in the ML process. As seen in Figure 11.4, ML algorithms splitting data into two groups.

11.5.3 Model selection and training

1. Choose relevant ML methods such as Random forests and support vector machines (SVMs), decision tree, and NNs that are suited for DNA sequence analysis.
2. Separate training and testing datasets in order to appropriately assess model performance.
3. Use the labeled DNA sequences to train the chosen models.

11.5.4 Model evaluation and prediction

1. Use appropriate performance criteria, like as accuracy, precision, recall, or F1 score, to evaluate trained models.
2. Compare and contrast the DNA sequences of humans, chimpanzees, and dogs using the trained models to anticipate genetic variants.
3. Examine and analyze the findings to learn more about evolutionary connections and potential genetic differences between the species.

11.5.5 Result visualization and reporting

1. Develop visualizations that show the genetic differences and connections among humans, chimpanzees, and dogs, such as heatmaps, sequence alignments, or phylogenetic trees.
2. Create a clear and succinct report that summarizes the discoveries and learnings from the DNA sequence analysis.

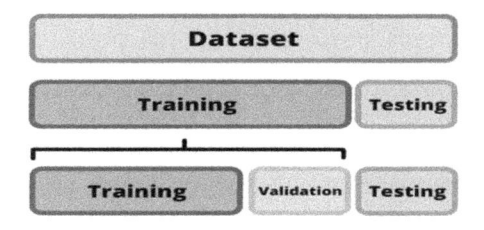

Figure 11.4 Splitting data process.

11.5.6 Algorithms in the proposed system are implemented

In order to comprehend how the models run on the dataset and the purpose of the issue statement solution is to find a solution technique, let's explore each ML model that the suggested system states independently.

11.5.6.1 Decision Tree

Decision trees are a type of supervised learning, which may be used for classification as well as regression issues; however, the former is where they shine. It's a classifier organized like a tree, with internal nodes serving as substitutes for aspects of the dataset, branching for the reasoning process, as well as leaf nodes for the final verdict. A decision tree has two kinds of nodes: decision nodes and leaf nodes. Leaf nodes are the outcomes of decisions and don't branch off into anything else, whereas decision nodes are where the decisions are actually made and have multiple offshoots. Decision trees are so named because their structures resemble trees, growing from a central node outward along branches. A decision tree just asks a question and then splits the tree into subtrees based on the answer (Yes/No).

STEPS:

Step 1: Begin with the tree at the base node, which includes the complete dataset, according to S.

Step 2: Make use of the attribute selection measure (ASM) to determine the top attribute in the dataset.

Step 3: Divide the S into subgroups with possible benefits for the finest attributes.

Step 4: Build the decision tree node with the best characteristic.

Step 5: Create new decision trees iteratively using the dataset subsets created in step 3.

A decision tree is illustrated in the diagram showing the distance between a river, a lake, and a beach. Figure 11.5, which also demonstrates that the distance between the river and the lake is larger than 8 km.

11.5.6.2 Support Vector Machine (SVM)

SVM is a popular supervised learning method that is applied to address classification and regression concerns. Nonetheless, it is mostly used in ML classification problems. The goal of the SVM method is to find the best line or decision boundary that can be split n-dimensional space into classes, allowing us to categorize new information points fast in the future. This ideal decision border, referred to as SVM, selects the most extreme vectors and points that will help create the hyperplane. The SVM method derives

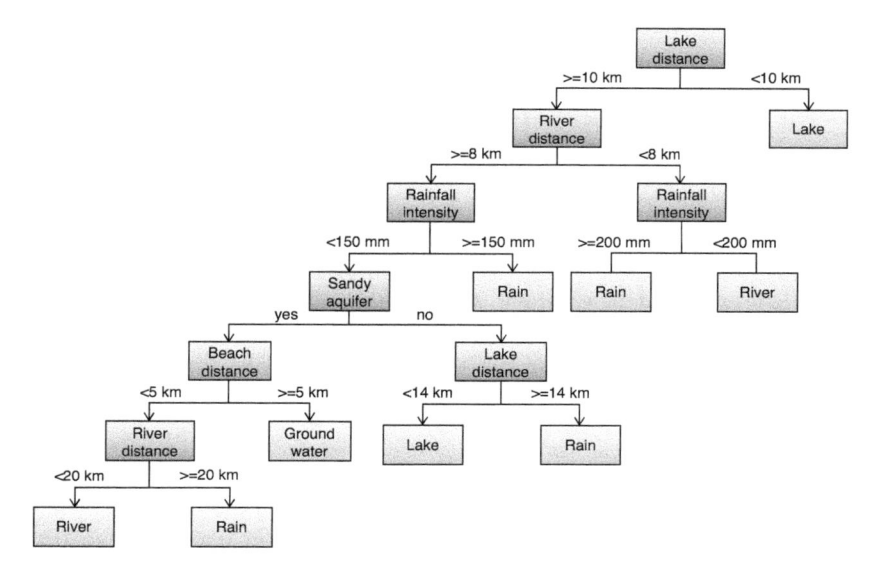

Figure 11.5 Example of decision tree.

its name from the support vectors used to represent the extreme cases. The diagram of the margin between the decision boundary and hyperplanes in Figure 11.6 shows that the margin between the two classes is maximized by the support vectors.

As we search for the largest separating hyperplane between the many classes present in the target feature, SVM methods are particularly efficient.

11.5.6.3 Naive Bayes

To handle classification challenges, the supervised learning approach known as the Naive Bayes algorithm, which is predicated on the Bayes theorem, is utilized. The Naive Bayes Classifier is a straightforward and efficient way

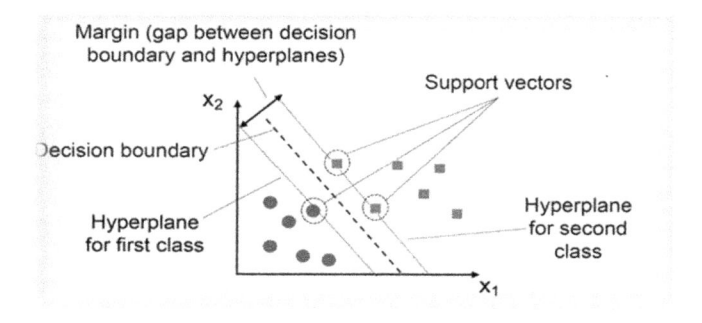

Figure 11.6 Example of SVM.

of categorizing that assists in the building of rapid ML models capable of generating correct predictions. As a probabilistic classifier, it provides predictions based on the likelihood that an object will occur.

What is the significance of the name "Naive Bayes"?

The words the Naive and Bayes algorithms that comprise the Naive Bayes algorithm are as follows:

Naive: It holds that the presence of one characteristic is unconnected to the presence of other features, which is why it is called naïve. If a red, spherical, sweet Fruit is recognized depending on its color, shape, and flavor, it is recognized as an apple. So, without relying on one another, each aspect adds to recognizing it as an apple.

Bayes: It gets its name from the assumption of Bayes' Theorem.

STEPS:

Step 1. Pre-processing of data
Step 2. Applying Naive Bayes on the training dataset
Step 3. Predicting the exam outcome
Step 4. Evaluate the outcome's precision (build a confusion matrix).
Step 5. Displaying the test set results.

The formula of Bayes theorem:

$$P(A/B) = \frac{P(B/A) \cdot P(A)}{P(B)} \tag{11.1}$$

Where in Equation 11.1,

$P(A|B)$ represents the Posterior Probability: The likelihood of hypothesis A on observed data occurrence B.

$P(B|A)$ represents the Likelihood Probability: The likelihood of evidence supplied that a theory is likely to be true.

$P(A)$ represents the Prior Probability: Before looking at the data, consider the hypothesis's probability.

$P(B)$ represents the Marginal Probability: The probability of evidence.

Random Forest: Random forest is a classifier that, as the name indicates, employs numerous decision trees based on certain groups of the input dataset and averages them to boost the dataset's prediction accuracy. Instead of relying on a single decision tree, the Random forest utilizes forecasts from each tree and predicts the outcome based on the majority of votes. The growing number of trees in the forest inhibits increased accuracy and overfitting.

STEPS:

Step 1: At random, select K data points from the training set.
Step 2: Build the decision trees that are related to the selected data subsets.

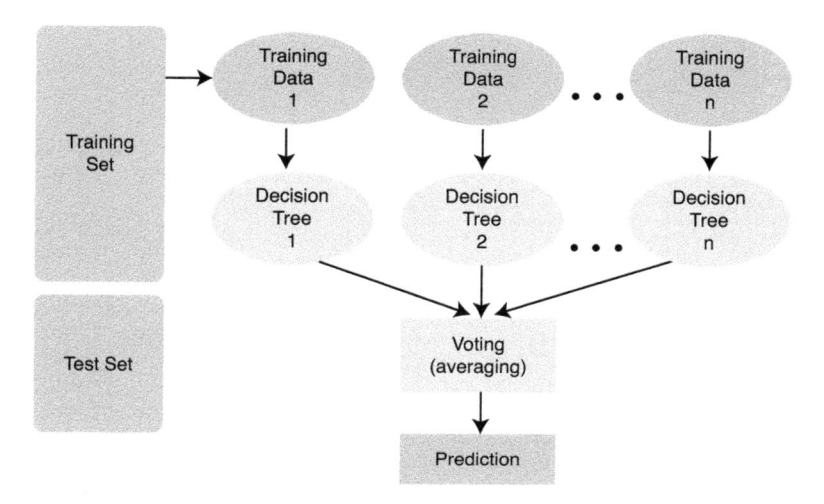

Figure 11.7 **Example of random forest process.**

Step 3: Enter N as the size of the decision trees you want to build.
Step 4: Repetition of Steps 1 and 2.
Step 5: Assign new data points to the category with the highest votes by locating the predictions for the new data points in each decision tree.

An illustration of a decision tree model in Figure 11.7 demonstrates how predictions are made using the model after it has been trained on a training set.

11.5.6.4 XGBoost

XGBOOST (Extreme Gradient Boosting) is a strong and extensively used ML algorithm noted for its efficiency and good performance in a broad range of data science applications, particularly those using structured/tabular data, such as regression, classification, and ranking. It is a type of ensemble learning that combines the predictions of numerous weak learners (usually decision trees) to generate a strong learner.

The mathematical equation for XGBOOST is as follows:

$$L\big(y,\,F(x)\big) = \sum_{\{i=1\}}^{n} L\big(yi,\,F(xi)\big) + \Omega(F) \tag{11.2}$$

Where in Equation 11.2,

$L\,(y, F(x))$ is the loss function, which measures the difference between the predicted value $F(x)$ and the actual value y.

n is the number of data points.

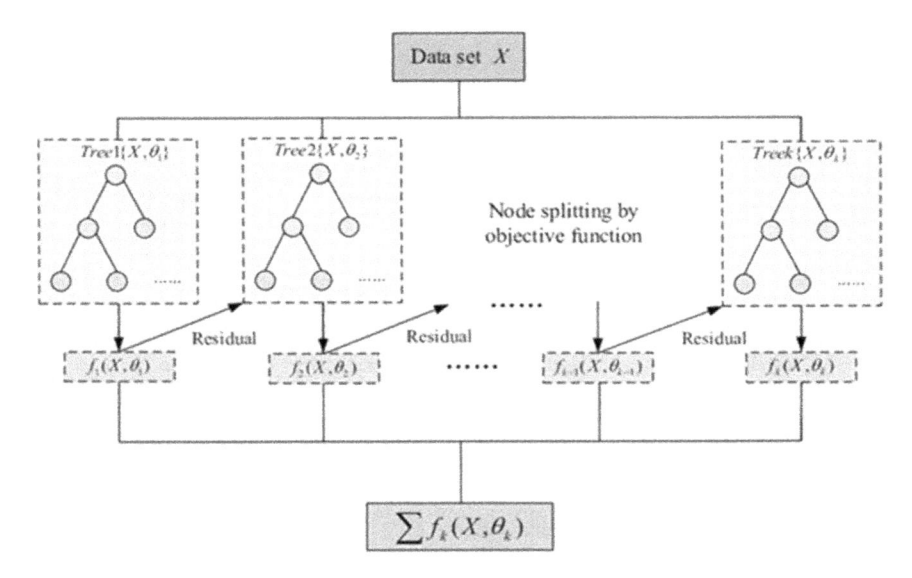

Figure 11.8 Example of XGBoost.

$\Omega(F)$ is the regularization term, which penalizes the complexity of the model.
STEPS:

Step 1: This involves setting the hyperparameters, such as the number of trees, the learning rate, and the regularization strength.

Step 2: This can be a simple prediction, such as the mean of the target variable.

Step 3: The residuals are the difference between the actual target values and the initial prediction.

Step 4: The decision tree is built to minimize the sum of the squared residuals.

Step 5: The predictions are updated to include the contribution of the decision tree.

Step 6: Repeat steps 3–5 until the model converges. The model converges when the predictions no longer improve. The XGBoost method in Figure 11.8 is depicted in the schematic of a node splitting by goal function.

11.5.6.5 Neural Network

NNs are a type of DL algorithm that is inspired by the structure and function of the human brain. They are sophisticated models that can learn complicated patterns from data and have demonstrated outstanding performance

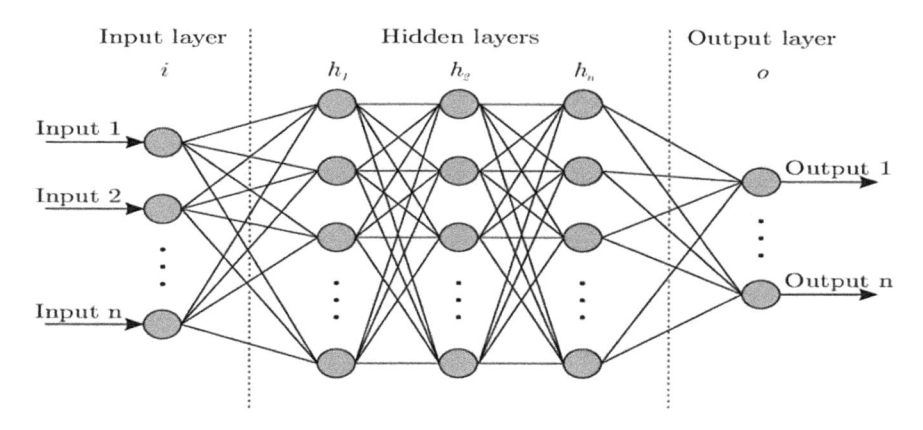

Figure 11.9 Process of NN.

in a variety of ML applications such as image recognition, natural language processing, and others. The artificial neuron, commonly known as a perceptron, is the fundamental building unit of a NN.

STEPS:

Step 1: This involves specifying the number of layers, the number of neurons in each layer, and the activation function for each layer.

Step 2: The weights and bias vectors are randomly initialized.

Step 3: The training data is a set of input-output pairs.

Step 4: The NN is trained using gradient descent, which is one example of an optimization method. In the training phase, a loss function is minimized, which minimizes the discrepancy between the actual and expected output and the actual output.

Step 5: The NN is evaluated on a test dataset. The test dataset is a set of input-output pairs that the NN has not seen before.

Step 6: The NN can be deployed to production to make predictions on new data. Figure 11.9 shows that the input layer, hidden layers, and output layer are all depicted in the diagram of a NN with hidden layers.

A NN is made up of neurons that are connected to one another, and each connection in our NN has a weight that, when multiplied by the input value, determines how important the relationship is to the neuron.

11.5.6.6 AdaBoost

AdaBoost, using an ensemble learning approach called (Adaptive Boosting), several weak learners—usually decision trees—are combined to produce a strong learner. It is designed to improve the performance of weak learners by giving more weight to misclassified instances, allowing them to focus on difficult-to-classify examples. AdaBoost has been widely used in various classification tasks and is known for its ability to handle complex datasets and avoid overfitting.

STEPS:

Step 1: Each and every weight is initially set to equal $1/N$, where N is the number of data points.

Step 2: A weak learner is educated using the data and its predictions are made.

Step 3: The mistake rate represents the percentage of data points that the weak learner misclassified.

Step 4: The weights of the weak learners are updated based on their error rate. The weak learners that make the fewest mistakes are given higher weights, and the weak learners that make the most mistakes are given lower weights.

Step 5: Steps 2–4 should be repeated until the halting requirements are met. The stopping criteria can be the proportion of subpar students, the error rate, or the time limit. A diagram of the stages of a trained classifier in Figure 11.10 shows how a boosting algorithm works.

11.6 IMPLEMENTATION

Our research primarily focuses on figuring out how we'll accomplish DNA sequencing utilizing ML, DL, and other approaches and algorithms. We are aware that the DNA in humans is made up of sequences of the type ATGC or another sort. Basically, we found a dataset on Kaggle and began working on this particular project because of it. We used DNA sequencing to create a classification system that can determine the gene class to which a given sequence belongs, for example, in humans. NumPy, Pandas, and Matplotlib are only a few of the imported libraries. The dataset was acquired from Kaggle and is

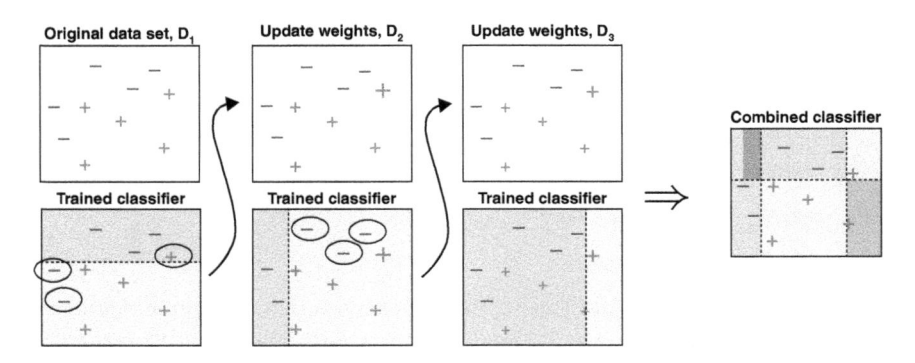

Figure 11.10 Example of AdaBoost.

Table 11.1 Gene family table

Gene family	Number	Class label
G protein-coupled receptors	531	0
Tyrosine kinase	534	1
Tyrosine phosphatase	349	2
Synthetase	672	3
Synthase	711	4
Ion channel	240	5
Transcription	1343	6

known as human_data.txt. After processing this dataset, in addition to their class, we have the human DNA sequence. In essence, sequence and class are now available. This specific dataset should be able to identify the class to which a sequence belongs when read based on the sequence. These sequences are often divided into many types and may be the DNA or a particular human's gene sequence. In addition to this specific dataset, we also have other datasets from Kaggle that we obtained, such as chimpanzee and dog data. These other datasets have the same thing—a sequence and a class—as this particular dataset. The table of gene family numbers and class labels for genes in Table 11.1 shows the number of genes in each gene family and their class labels.

What we've done here is simply find the sequence and identify its category. For example, if the pattern has the label "class 0," it indicates that it falls to the g protein-coupled receptor gene family; the count "531" indicates that it is designated "class 0," and so on. K-mer counting is used. When utilizing DNA sequences, we often resort to a method called K-mer measurement, which practically converts DNA sequences into a language. Hexamers, commonly known as six-letter words, are what we employed in this instance. Four hexamer words are created from this sequence as a final result.

NLP will also be used to transform our sequence into a vector. The following step will be to divide this specific collection of information into 6 groups, each with a length of 6. This is what we did so that we could apply the word count or word bag to this specific set of data. Chimpanzee and dog research has been conducted. As a further step, we must turn the stringization of the list of K-mers for each gene phrases. Due to the ease with which it may be turned into a bag of words, we must now merge all the sequences. For dog and chimpanzee data up to this point, the same processes will be performed. We'll take those actions at a later time. We shall then attempt to convert the string phrases created from the list of K-mers for every gene. Therefore, because it is simple to turn all the sequences into a bag of words, we now need to mix them together. For chimpanzee and dog data up until this point, the identical processes will be repeated. These actions will be performed later. Next, we'll attempt to apply a bag of words using a count vectorizer to the strings to transform them. What we did since our distinctive characteristic was by way of strings, and as NLP prohibits the direct use of data key strings in models, we had to turn the ties into many words using the vectorizing counts. I'm only attempting to do that since we now check to see if the data collection is balanced. Figure 11.11a shows the class balance of human DNA; Figure 11.11b and c shows class balance of chimpanzee and dog DNA.

Here, we can see that all of the classes are roughly balanced, and this is extremely clear. We can practically utilize this directly, and with this, we can also handle low datasets because other classes are roughly balanced when some datasets were low. Oversampling can be used to address an issue with an unbalanced dataset. According to the test size, the train test is divided and will take 20%. After that, we tested the accuracy on our three datasets using all of our classification methods. The estimation of these classification methods uses many classification measures. These indicators were all calculated using the confusion matrix as a basis. When used with human data, the model performs effectively. Chimpanzees also experience it. Given that we are already aware of this individual, it is hardly shocking, and chimpanzees have the same ancestor. However, the model could not produce satisfactory results when applied to dog data since both man and dog are not substantially. In Figure 11.12a, the researcher found the best accuracy at this model to be 97%, whereas in Figure 11.12b, we discover the highest accuracy to be 99%.

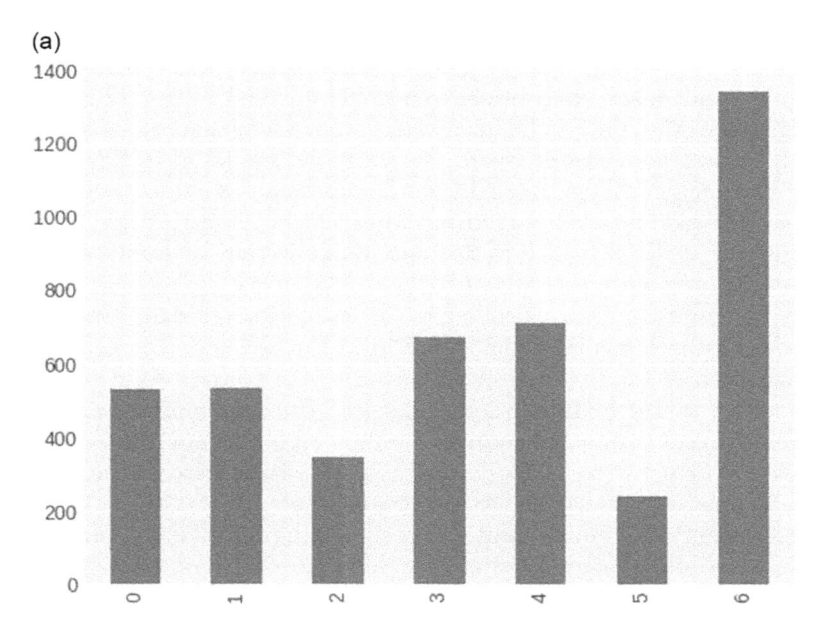

Figure 11.11 (a) Class balance of human. (b) Class balance of chimpanzee. (c) Class balance of dog.

(Continued)

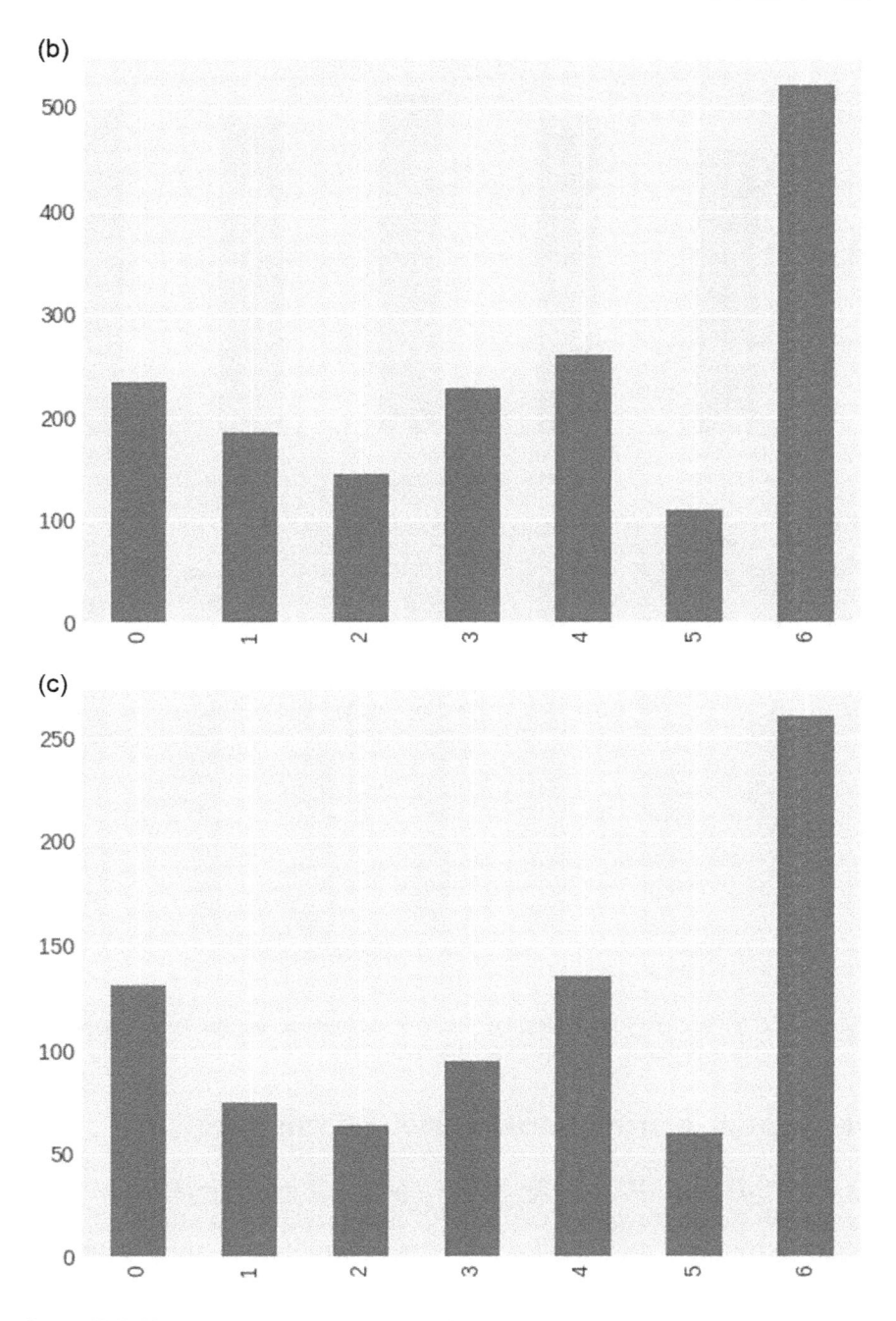

Figure 11.11 (Continued) (a) Class balance of human. (b) Class balance of chimpanzee. (c) Class balance of dog.

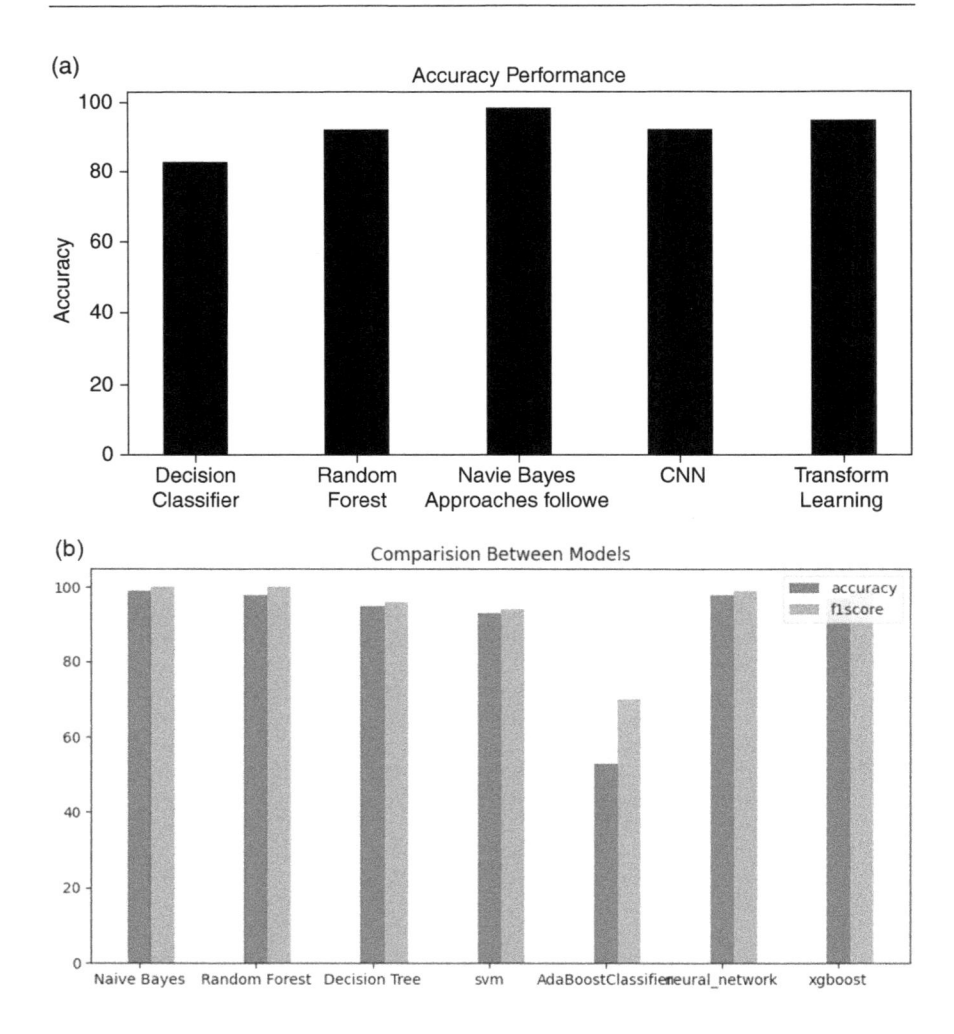

Figure 11.12 (a) Accuracy performance result of related works. (b) Accuracy performance result of our proposed work.

11.7 MODEL COMPARISON AND ANALYSIS

Several "ML and DL techniques" are compared and contrasted with the proposed system to prove the model's accuracy in this study. Decision trees, Naive Bayes, Random forests, NNs, and transform learning are only some of the proposed models. For example, the transform learning algorithm generated higher accuracy for DL at 94.57%, while the Naive Bayes approach produced better accuracy for ML at 99.00%. The seven different classes may be classified accurately using this method. For the decision

Table 11.2 Accuracy performance of all algorithms

Algorithm	Accuracy
Decision Tree	95.00
XGBoost	98.00
Naive Bayes	99.00
NN	98.00
Random Forest	98.00
SVM	93.00

tree, Random forest, and NN, respectively, we achieved accuracy values of 95.00, 98.00, and 98.00. It is abundantly obvious from the experimental results shown above that the recommended model is successful in classifying DNA sequences. Nevertheless, Naïve Bayes performed the best, scoring 99.00. The results of our algorithms' accuracy are shown in Table 11.2.

11.8 RESULTS AND DISCUSSION

The fundamental goal of this work is to better understand DL, ML, and algorithms may be used to do DNA sequencing. Three datasets from Kaggle, including those for humans, chimpanzees, and dogs, were used in this study. Each of our datasets, which are made up of sequences and labels, is split into training and testing portions at a ratio of 80–20% for ML algorithms and 80–20% for DL algorithms. Additionally, for these datasets, we performed DNA sequencing. For classification, we used five different DL and ML algorithms, including decision trees, Random forests, Naive Bayes, CNN, and transform learning. Sequence encoding has been done using Label, K-mer, and one-hot encoding. For both DL and ML, we used K-mer encoding. ML was used to translate DNA sequences into human languages for each dataset. The K-mer counting method was employed, and the K-mer size was set to 6. The resulting string sentences were then converted to a bag of words format using a count vectorizer. For all of our DL datasets, we have encoded labels and used one-hot encoding. Classification methods are estimated using a variety of criteria, including the F1 score, accuracy, recall, and precision. The confusion matrix was used to determine all the specified metrics. DNA sequencing uses both DL methods (labeled encoding) and ML techniques (K-mer counting), and the confusion matrix for these two types of calculations is shown below. As we can see from the confusion matrix, the number of trials is shown on the y-axis, while the number of steps is shown on the x-axis. The various colors stand for various trials.

90 times were spent doing the first trial (0 steps), 100 times were spent performing the second trial (1 step), and so on. Three times were given to the final trial's five stages, which is 205 times. The predicted number of steps is displayed in a light grey shade, while the actual number of steps is displayed

in darker shades. A mathematical model is used to compute the anticipated number of steps. (See Figures 11.13a-c for the matrices and their levels.)

The proportion of various G protein-coupled receptor (GPCR) types in the human body is shown in a pie chart. Transcription factor-activating GPCRs make up the largest category (30.7%), followed by ion channel-activating GPCRs (5.5%), tyrosine kinase-activating GPCRs (12.2%), tyrosine phosphatase-activating GPCRs (8.0%), synthetase-activating GPCRs (15.3%), synthase-activating GPCRs (16.2%), and GPCRs with unknown functions (12.1%). In many facets of human biology, including vision, olfaction, taste, metabolism, and the immune system, GPCRs, a huge and varied family of proteins, play a crucial role. Hormones, neurotransmitters, and other signaling molecules are only a few of the many ligands that can activate them. When GPCRs are activated, a series of subsequent events are set off, eventually altering how cells operate. Whereas GPCR types in the chimpanzee body are shown in a pie chart. Transcription factor-activating GPCRs make up the largest category (31.09%), followed by ion channel-activating GPCRs (6.5%), tyrosine kinase-activating GPCRs (11.09%), tyrosine phosphatase-activating GPCRs (8.59%), synthase-activating GPCRs (15.5%), synthetase-activating GPCRs (13.69%),

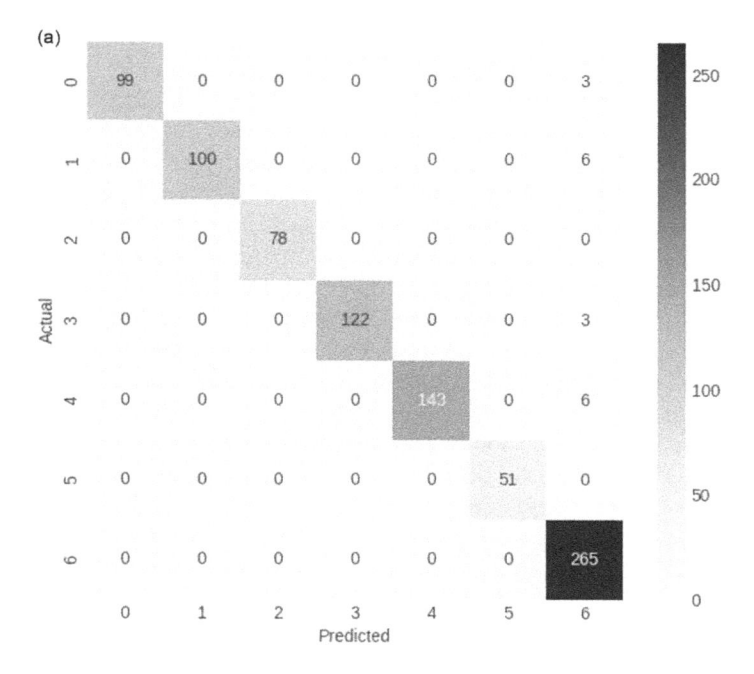

Figure 11.13 (a) Confusion matrix of K-mer. (b) Confusion matrix of K-mer encoding for human dataset encoding for chimpanzee dataset. (c) Confusion matrix of K-mer encoding for dog dataset.

(Continued)

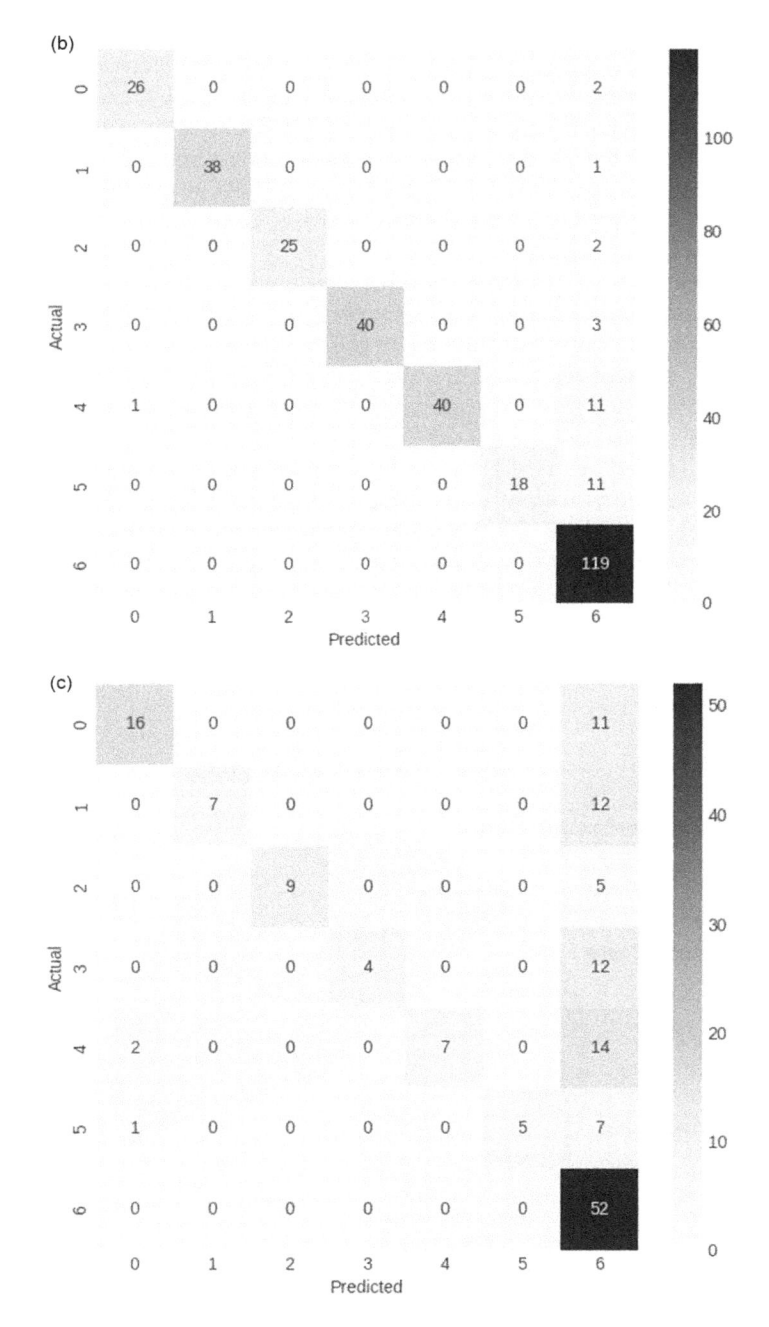

Figure 11.13 (Continued) (a) Confusion matrix of K-mer. (b) Confusion matrix of K-mer encoding for human dataset encoding for chimpanzee dataset. (c) Confusion matrix of K-mer encoding for dog dataset.

and GPCRs with unknown functions (13.9%). Also, GPCR types in the dog body are shown in a pie chart. Transcription factor-activating GPCRs make up the largest category (31.7%), followed by ion channel-activating GPCRs (7.3%), tyrosine kinase-activating GPCRs (9.1%), tyrosine phosphatase-activating GPCRs (7.8%), synthase-activating GPCRs (16.5%), synthetase-activating GPCRs (11.6%), and GPCRs with unknown functions (16.0%). Figure 11.14a–c shows the all-pie chart of GPCR in humans, chimpanzees, and dogs.

Here are a few disadvantages of our project:

The genes included in our analysis were few. In order to draw more conclusive results, we must examine more genes.

Our research was restricted to chimpanzees, humans, and dogs. To comprehend how genes link to features in various species, we must examine other species.

We only looked at one ML algorithm in this study. To see if additional algorithms can enhance our outcomes, we must put them to the test.

A dataset of 40,000 DNA sequences from each species was used to train the model. Several sources, including openly accessible databases and private collections, were used to compile the dataset. In this instance, the model was applied to identify the DNA sequence patterns unique to each species.

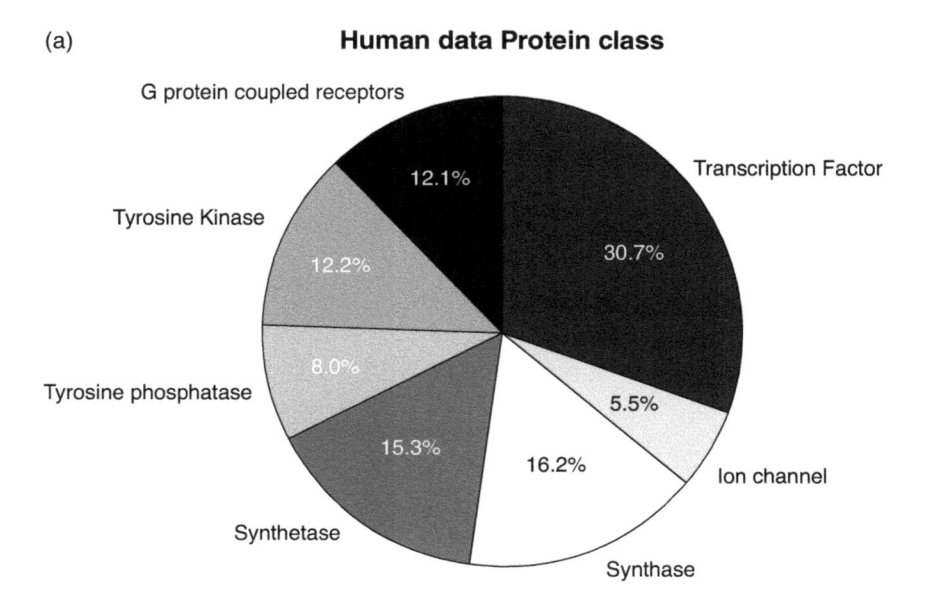

Figure 11.14 (a) Human protein class. (b) Chimpanzee protein class. (c) Dog protein class.

(*Continued*)

(b)   **Chimpanzee data Protein class**

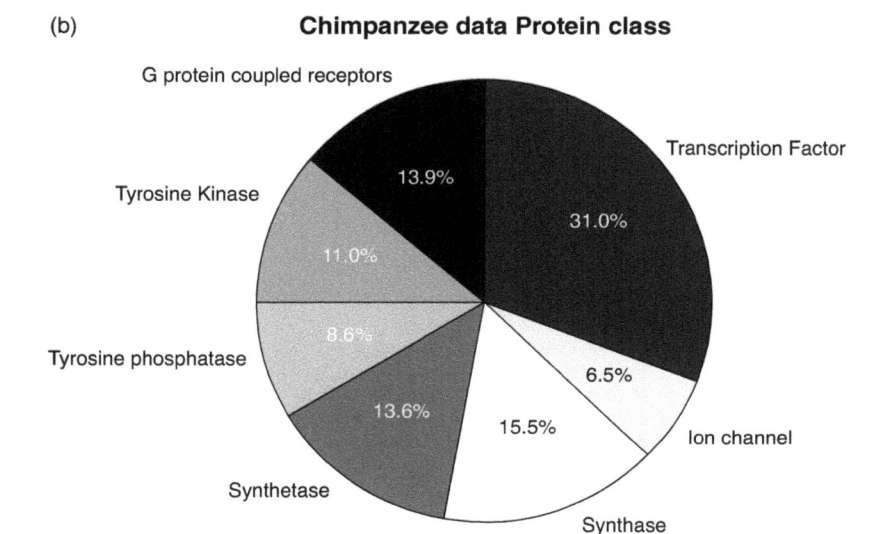

(c)   **Dog data Protein class**

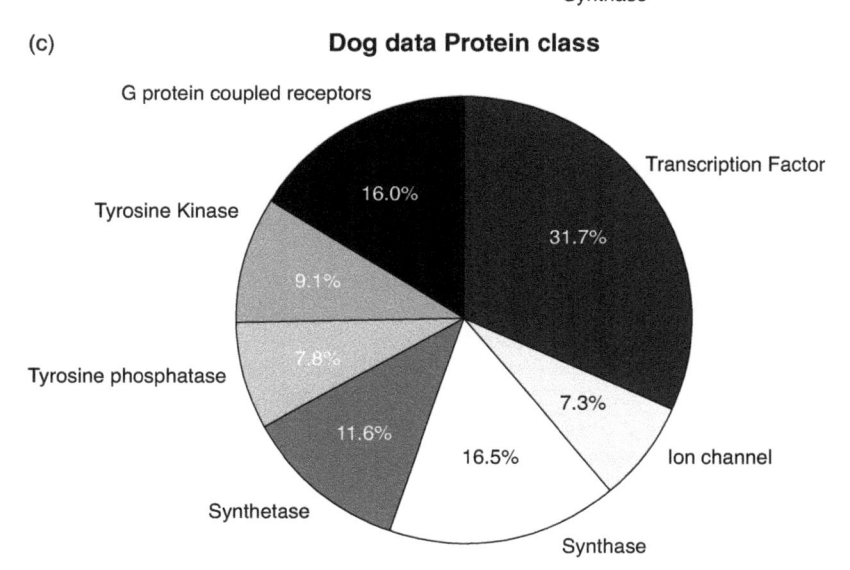

Figure 11.14 (Continued) (a) Human protein class. (b) Chimpanzee protein class. (c) Dog protein class.

11.9 CONCLUSION

Seven different classification strategies from ML and DL (decision tree, Naive Bayes, Random forest, NNs, transform learning with K-mer, and label encoding) were evaluated and contrasted in this chapter. We presented, compared, and analyzed the K-mer counting, one-hot encoding, and label encoding strategies for encoding DNA sequence strings. The NLP bag of words is an approach to analyze DNA sequence strings that was developed using a count vectorizer. We found that Naive Bayes with K-mer encoding, a ML technique, is more accurate than the alternative algorithms. Transform learning with the encoding of labels outperforms CNN with label encoding in terms of accuracy. However, the suggested approach, transform learning, has done better than all other algorithms, with Naive Bayes having the greatest accuracy of 99.00%. In light of this, when should we transform learning? As a result, K-mer counting outperformed label encoding for our datasets. Precision, recall, and accuracy are some of the many measures used to analyze this dataset.

REFERENCES

1. Li, M. Towards a DNA sequencing theory (learning a string). In: *Proceedings [1990] 31st Annual Symposium on Foundations of Computer Science*, St. Louis, MO, 1990, vol.1, pp.125–134. doi:10.1109/FSCS.1990.89531.
2. Wu, T. and Vikalo, H. Maximum likelihood DNA sequence detection via sphere decoding. In: *2010 IEEE International Conference on Acoustics, Speech and Signal Processing*, Dallas, TX, 2010, pp. 586–589. doi:10.1109/ICASSP.2010.5495564.
3. Eltoukhy, H. and Gamal, A. El Modeling and base-calling for DNA sequencing-by-synthesis. In: *2006 IEEE International Conference on Acoustics Speech and Signal Processing Proceedings*, Toulouse, France, 2006, pp. II–II. doi:10.1109/ICASSP.2006.1660522.
4. Davies, S. W., Eizenman, M., and Pasupathy, S. Optimal structure for automatic processing of DNA sequences. *IEEE Transactions on Biomedical Engineering*, vol. 46, no. 9, pp. 1044–1056, 1999. doi:10.1109/10.784135.
5. Memeti, S. and Pllana, S. A machine learning approach for accelerating DNA sequence analysis. *The International Journal of High-Performance Computing Applications*, vol. 32, no. 3, pp. 363–379, 2018. doi:10.1177/1094342016654214.
6. Hadikurniawati, W., Anwar, M. T., Marlina, D., and Kusumo, H. Predicting tuberculosis drug resistance using machine learning based on DNA sequencing data. *Journal of Physics: Conference Series*, vol. 1869, no. 1, p. 012093, 2021.
7. Cario, C. L., Chen, E., Leong, L., Emami, N. C., Lopez, K., Tenggara, I., and Witte, J. S. A machine learning approach to optimizing cell-free DNA sequencing panels: with an application to prostate cancer. *BMC Cancer*, vol. 20, no. 1, pp. 1–9, 2020.
8. Zhang, Z., van Dijk, F., de Klein, N., van Gijn, M.E., Franke, L.H., Sinke, R.J., Swertz, M.A., and van der Velde, K.J. Feasibility of predicting allele specific expression from DNA sequencing using machine learning. *Scientific Reports*, vol. 11, no. 1, p. 10606, 2021. doi:10.1038/s41598-021-89904-y.
9. Liu, B. BioSeq-Analysis: A platform for DNA, RNA and protein sequence analysis based on machine learning approaches. *Briefings in Bioinformatics*, vol. 20, no. 4, pp. 1280–1294, 2019. doi:10.1093/bib/bbx165.

10. Liu, B., Gao, X., and Zhang, H. BioSeq-Analysis2.0: An updated platform for analyzing DNA, RNA and protein sequences at sequence level and residue level based on machine learning approaches. *Nucleic Acids Research*, vol. 47, no. 20, p. e127, 2019. doi:10.1093/nar/gkz740.

11. Khan, F., Ncube, C., Ramasamy, L. K., Kadry, S., and Nam, Y. A digital DNA sequencing engine for ransomware detection using machine learning. *IEEE Access*, vol. 8, pp. 119710–119719, 2020. doi:10.1109/ACCESS.2020.3003785.

12. Dixit, P. and Prajapati, G. I. "Machine learning in bioinformatics: A novel approach for DNA sequencing." In: *2015 Fifth International Conference on Advanced Computing & Communication Technologies*, Haryana, India, 2015, pp. 41–47. doi:10.1109/ACCT.2015.73.

13. Wang, Y., Alangari, M., Hihath, J., Das, A.K., and Anantram, M.P. A machine learning approach for accurate and real-time DNA sequence identification. *BMC Genomics*, vol. 22, p. 525, 2021.

14. Yang, A., Zhang, W., Wang, J., Yang, K., Han, Y., and Zhang, L. Review on the application of machine learning algorithms in the sequence data mining of DNA. *Frontiers in Bioengineering and Biotechnology*, vol. 8, p. 1032, 2020.

15. Gunasekaran, H., Ramalakshmi, K., Rex Macedo Arokiaraj, A., Deepa Kanmani, S., Venkatesan, C. and Suresh Gnana Dhas, C. Analysis of DNA sequence classification using CNN and hybrid models. *Computational and Mathematical Methods in Medicine*, 2021. doi:10.1155/2021/1835056.

Chapter 12

AI-based wireless channel prediction for 6G

Avishek Bhattacharjee

12.1 INTRODUCTION

Wireless communication[1] has become an inherent aspect of our interconnected lives. The dependability and the efficiency of wireless networking have become essentials as society relentlessly advances toward a period characterized by IoT[2], Industry 4.0, and high data consumption. As of 2023, 5G communication has been established in all tier 1 and tier 2 countries. 5G frequency bands[3] are defined as Sub-1 GHz[3] for wide range coverage; 3.5 GHz dense in urban areas for prime 5G midband; and 26, 28, and 39 GHz for hotspot. The spectrum allocation for 6G is still under consideration, and the spectrum allocation of 6G will be defined by international regulators such as the International Telecommunication Union (ITU), Federal Communications Commission (FCC), and European Conference of Postal and Telecommunications Administrations (CEPT). The frequency band of 6G is estimated to be terahertz, so the channel should be more robust and optimistic for achieving this speed. This chapter embarks on a profound intellectual journey, delving into the synergy between the convergence of artificial intelligence (AI)[4] and wireless communication[1,5], culminating in the captivating discipline of AI-based wireless channel prediction. From the word intelligence, we think about quick learning capabilities, perceiving information from the environment and applying it in the future, reasoning abilities, self-consciousness, and critical thinking, but there is no uniform definition of intelligence. As technology upgrades, we are dependent on AI in every field of discipline. The word artificial intelligence (AI) was first used by John McCarthy[6] in 1955, and he was also the core organizer of the Dartmouth Conference[6] in 1956. During this meeting, eleven scientists, including J. McCarthy[6], Alan Turing[6], Marvin Minsky[6], and Claude Shannon[6], declared AI autonomous research[6] field. There are also so many controversies about AI definition. Still, we can define AI via these keywords: human thinking, human actions, remembering large amounts of data, emulating human neurons, fast action, auto-decision and imagination. AI is also divided into different categories like weak AI, strong AI, artificial narrow intelligence (ANI), artificial super intelligence (ASI), and

DOI: 10.1201/9781003517689-12

artificial general intelligence AGI. AI's prowess in processing vast datasets and discerning intricate patterns equips it to revolutionize wireless channel estimation. Mitigating errors from the channel is the main target of channel estimation, which can be obtained by different channel coding techniques. The core objective of the 6G channel coding technique is to go beyond the Shannon limit with a short length of codeword, whereas the Low-density parity-check code (LDPC) channel coding technique used in 5G is up to the Shannon limit. As a result, for intricate channel situations such as multipath fading, rapid time variation, and nonlinear deep fading circumstances, the performance of standard model-based channel estimation techniques is limited. These conditions make it difficult to create precise mathematical channel estimation models that may not completely account for the channel properties. The relationship between various system variables can be both model-free and model-driven methods of compression using AI-based training algorithms, which can conquer those situations. Combining the forecasting powers of AI with the dynamic nature of wireless channels, we unlock a new realm of possibilities in communication technology. In this chapter, we developed and trained a CNN[7] model based on a large parameter set of data, and the journey from utilizing historical data to anticipating the dynamics of future AI-based wireless channels is captured in the pages that follow, illuminating the complex balance between empirical knowledge and computational prowess. Our exploration navigates through the corridors of data-driven discovery, algorithmic brilliance, and real-world application, all converging toward a unified objective: elevating the very foundation of wireless communication. Last, we compared different parameters of an AI-based wireless channel between our proposed model and all conventional AI models.

The current work makes the following contributions:

1. Different types of wireless channels and their models, like stochastic channel models, millimeter-wave (mmWave) channel models, and THz channel models, have been reviewed briefly in this chapter.
2. Deep learning (DL) and AI-based channel estimations have been reviewed here.
3. A CNN is made using the MATLAB DL toolbox using three layers of 64, 128, and 244 filters, respectively, with a kernel size of 3.
4. A large number of unsupervised datasets are trained with this neural network to predict the wireless channel parameters.
5. L2 regularization is used here to remove the overfitting of the training samples.

Section 12.1 describes the introduction and related works; Section 12.2 describes the overview of channels with stochastic channel models, mmWave channel models, THz channel models, and multi-antenna channel models.

Section 12.3 describes the AI-based channel estimation; Section 12.4 briefs about open research challenges; and Section 12.5 tells the proposed model and its analysis.

12.2 RELATED WORKS

In 1951[6], Marvin Minsky[6] developed the stochastic neural analog reinforcement calculator (SNARC)[6], which is considered to be the first artificial neural network (ANN) with synapses[6] and vacuum tubes[6]. Thereafter, Frank Rosenblatt designed the perceptron model, based on ANN, in 1958 to understand how the human visual system works and processes data. The development of neural networks[7] in the 2000s gave wireless channel prediction fresh life. More precise predictions were made possible by neural networks' capacity to recognize complex patterns in data. Researchers started experimenting with various neural network architectures to model complex channel dynamics. By the 2010s, DL, a subset of machine learning (ML), gained prominence. DNN, particularly recurrent neural networks (RNNs)[7,8] and CNNs[8,9], showed remarkable promise in predicting wireless channel conditions. The Institution of Electrical and Electronics Engineers (IEEE) and Third-Generation Partnership Project (3GPP) have defined the wireless channel for 6G communication. The field of multicarrier system channel estimation has been the subject of numerous surveys and reviews[3,10,11]. Without mentioning AI integration, they primarily explore traditional channel estimate techniques. Additionally, specialized works on channel estimate for orthogonal frequency division multiplexing (OFDM)[12] systems offer an in-depth evaluation of high technology at the time of their publication[13,14]. After a careful examination of OFDM[12] systems, several scholars discussed the channel estimate methods[12]. It offers a comprehensive examination of the waveforms used in new generation networks, like OFDM[12], filter bank multicarrier (FBMC)[13,15], generalized frequency division multiplexing (GFDM)[11], and universal filtered multicarrier (UFMC)[3] schemes[15]. Recently, channel estimate methods for millimeter-wave and 5G communication systems, such like but not restricted to OFDM[14] systems, have been proposed. For intelligent wireless communication, the channel estimation method based on AI was examined. To increase the channel state information (CSI) collection accuracy, the approximation of the channel procedure was provided as a physical layer[4] application using AI algorithms[16]. There are existing overviews of ML techniques for resolving various wireless network difficulties, together with a conversation of the ML classifications and a list of solutions. In the case of nonlinear deep fading and high mobility, a regression-assisted method was proposed for channel estimation. It includes a comprehensive overview of ML in the context of vehicle networks and evaluates and discusses AI-based channel approximation techniques in connection to OFDM systems with high motility[3]. Furthermore, RNN-based channel prediction was investigated in paper[17], and performance study of ML-based estimation of the channel was completed in this paper[18]. Lastly, it offered a succinct overview of DL for channel estimation without concentrating on multicarrier systems.

12.3 OVERVIEW ON CHANNEL

A channel in a wireless network is a frequency band or logical medium to transmit data. Multiple channels are created by sending data concurrently as subcarriers. Each frequency band and the total number of subcarriers in a channel define its maximum data rates.

A wireless channel[19] is the physical medium through which the radio waves[20] of this network propagate. It depends upon a number of elements, such as the environment, the distance of the Tx. and Rx., and the existence of different wireless networks. These factors can cause the wireless channel to vary over time and space, which can affect the quality of the wireless signal. In Figure 12.1, LOS and NLOS channels have been shown.

1. **Free space[19] wireless channel:** It is the simplest form of wireless channel, and it is only affected by the distance between the receiver and the transmitter.
2. **LOS[1] channel:** It is another type of wireless channel where communication occurs between the base station[5] (BS) and the subscriber station[1,5] (SS) without any obstacles to get maximum throughput.
3. **NLOS[1,5] wireless channel:** It is another type of wireless channel. This type of wireless channel is affected by obstacles in the environment, such as buildings, trees, and others.
4. **Multipath wireless channel:** This type of wireless channel is affected by multiple reflections[1,20] of the radio waves from the environment, and communication is achieved through multiple paths. As seen in Figure 12.2, a multipath wireless channel has been shown.

In the case of the NLOS[22] channel, there is a reflecting path between SS and BS. Different signal copies come at various times and with varying amplitudes. ISI/ICI is the outcome of this. Using the cyclic prefix (CP) in conjunction with OFDM and OFDMA approaches helps to prevent this issue.

Examples of LOS[1] communication: P2P microwave communication[23] and P2P joint between SS and BS.

Examples of NLOS[1] communication: Mobile subscriber station[24], BS, and SS.

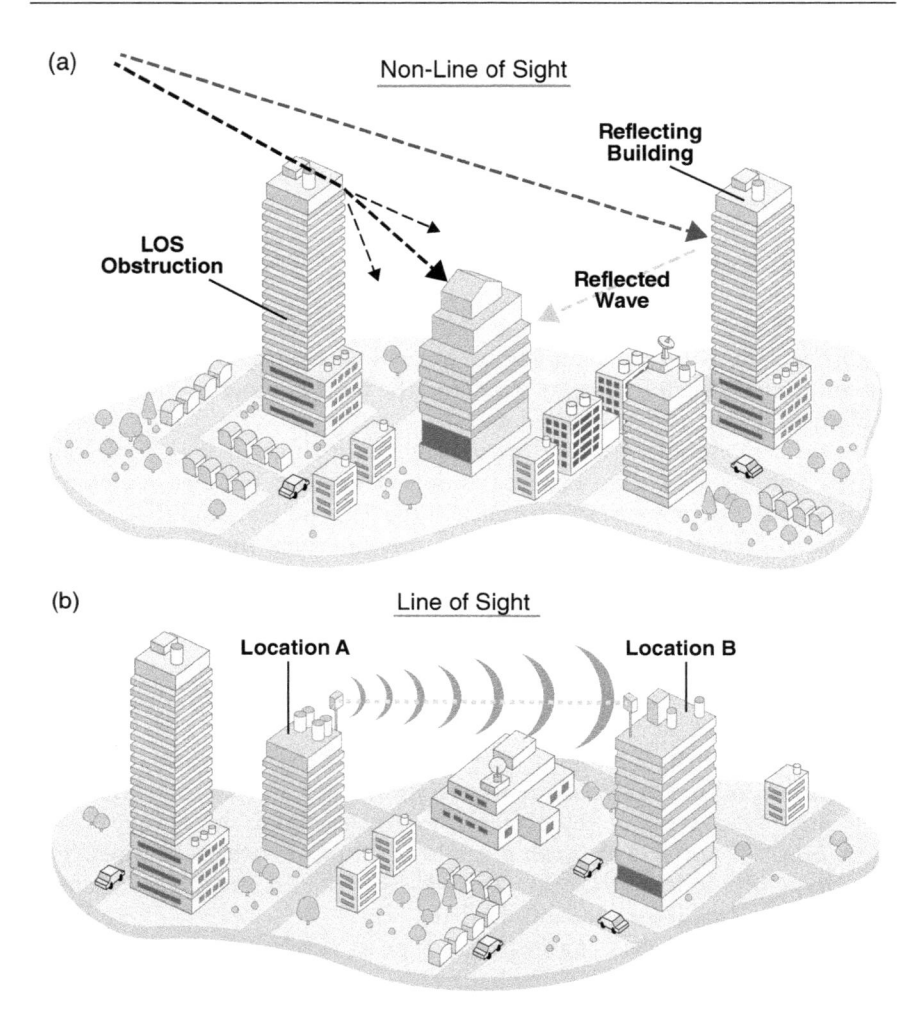

Figure 12.1 (a) A LOS channel prototype and (b) an NLOS channel structure.[20,21]

12.4 DIFFERENT CHANNEL MODELS

This chapter divides different channel models into four types: stochastic channel model, millimeter channel model, THz model, and multi-antenna model. In Sections 12.4.1–12.4.4, detailed information has been followed.

12.4.1 Stochastic channel model

The random variable is referred to as stochastic. A sort of wireless channel model known as a stochastic channel model makes use of statistical techniques to depict the random changes in the wireless channel. Most

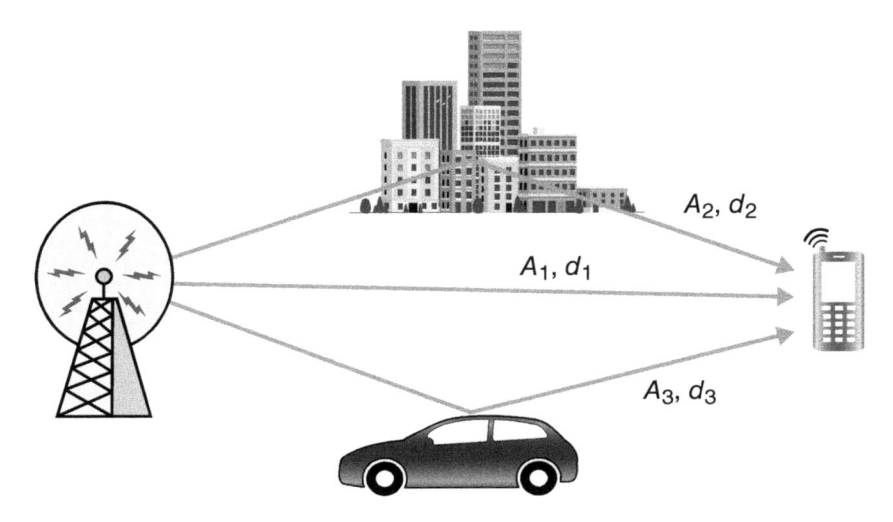

Figure 12.2 Multiple wireless channel in an urban environment.[5]

of the time, these models are predicated on the idea that a set of random characteristics, such as route loss, delay spread, and Doppler spread, may adequately describe a channel.

The tapped delay line (TDL) model is one of the most used stochastic channel models. The channel is represented by the TDL model as a group of taps; each of this represents a different propagation path[19]. Here, amplitude and delay of each tap are considered to be random variables.

Another well-liked stochastic channel model is the Saleh–Valenzuela (SV) model. A more intricate model that takes into consideration the scattering of radio signals by surrounding objects is the SV model. The SV model illustrates the effects of multipath fading and Doppler spread.

12.4.2 Millimeter-wave (mmWave) channel model

mmWave channels are determined by a number of unique features, including high path loss, high directivity, and high frequency selectivity. Since these features hamper the design of mmWave channels, additionally, they open up fresh possibilities for wireless communication networks. A critical component to consider while modeling mmWave channels is the route loss model. The weakening of the signal power during transmission between the sender and the destination is referred to as path loss. Compared to lower frequency channels, path loss in mmWave channels is noticeably larger. This is because mmWaves have a shorter wavelength, which increases their susceptibility to scattering and absorption. The directionality of the signal is a crucial factor in mmWave channel modeling. mmWave signals are highly directional, meaning that they are concentrated in a narrow beam.

This can be used to lower interference and increase the signal to noise ratio (SN ratio) in wireless communication systems. Lastly, there must be a significant degree of frequency selectivity in mmWave channels. This indicates that there are large variations in the channel response over the frequency range. Intersymbol interference (ISI) may result from this, impairing the functionality of communication systems.

The majority of models for mmWave channels have been developed to represent their unique characteristics. Several popular models for mmWave channels are as follows:

- The New York University (NYU) channel model
- The 3GPP mmWave channel model
- The COST 2100 mmWave channel model

These models take into account a variety of factors, such as the path loss, directionality, and the channel's selection for frequencies. It is used to generate synthetic channel realizations that should be applied to evaluate the performance of communication systems under different channel conditions. mmWave channel models are an important tool corresponding to the design and development of mmWave wireless communication systems. By using these models, engineers can gain a better understanding of the behavior of mmWave channels and design systems that are able to operate reliably in a variety of environments.

12.4.3 Multi-antenna channel model

A wireless channel model[1] that considers the existence of many antennas at the Tx. and Rx. is known as a multi-antenna[22] channel model[19]. Although multi-antenna[22] channel models are more intricate than single-antenna[19] channel models[19], their predictions of how wireless channels will behave in actual situations are more accurate.

There are a numerous multi-antenna channel models, each with its own strengths and weaknesses. Here, some of the most common multi-antenna channel models are as follows:

- **Correlation-based stochastic models (CBSMs):** CBSMs are based on the determination of how the channel[19] correlation matrix can be used to characterize the fading statistics of the channel. Although CBSMs are comparatively easy to deploy, in circumstances where estimating the channel correlation matrix is challenging, they may not yield correct results.
- **Geometry-based deterministic models (GBDMs):** GBDMs simulate how radio waves travel over a channel by using geometric data about the transmitter and reception antennas. Although GBDMs are more

accurate than CBSMs, their implementation can be challenging and necessitate in-depth understanding of the propagation environment.

- **Geometry-based stochastic models (GBSMs):** The benefits of GBDMs and CBSMs are combined in GBSMs. GBSMs model the channel's large-scale[19] route loss and fading characteristics using geometric information, and they represent the channel's small-scale[19] fading characteristics using stochastic approaches. Although GBSMs are the most accurate kind of multi-antenna channel model, their implementation might also be the most difficult.

12.4.4 THz channel model

Terahertz (THz) channel models are still under development, as the THz band is a relatively new frontier for wireless communication. However, a number of THz channel models have been proposed, based on measurements and simulations. One common approach to THz channel modeling is to use the ray tracing method. Ray tracing is a technique that models the propagation of electromagnetic waves by tracing the path of individual rays. Ray tracing can be used to model the complex propagation environment in THz bands, including the effects of diffraction, scattering, and reflection. Another approach to THz channel modeling is to use ML. ML algorithms can be trained on measured or simulated THz channel data to learn the statistical properties of the channel. This information can then be used to generate synthetic THz channel realizations that can be used to evaluate the performance of communication systems.

12.5 DL-BASED CHANNEL ESTIMATION

DL[11] has attracted a lot of interest recently as a brand-new paradigm to call the channel estimation inconvenience. It is exceedingly challenging to make a closed-form analytical solution for a complicated system with complex inputs/outputs interactions and a highly nonlinear internal structure. In fact, DL may be useful when an analytical solution is not ideal due to a system limitation (e.g., when the solution is not linear). The goal of DL is to automate the complex, typically highly nonlinear[8] relationship[7] between the input dataset[7,8] and the intended output[8] (DL). To summarize, end-to-end[8] learning and data-driven[8] training of the black box[7] are the two primary ways that DL-based channel estimation differs from classical channel estimation (see Figure 12.2). During the training phase, the model of DL approximates the function of a channel instead of using the analytical approach[7]. During the training time, the DL algorithms[7] are changed in order to determine the end to end[8] (weights and biases). After the training is completed, DL provides the wireless channel[10] for the input signal[11]

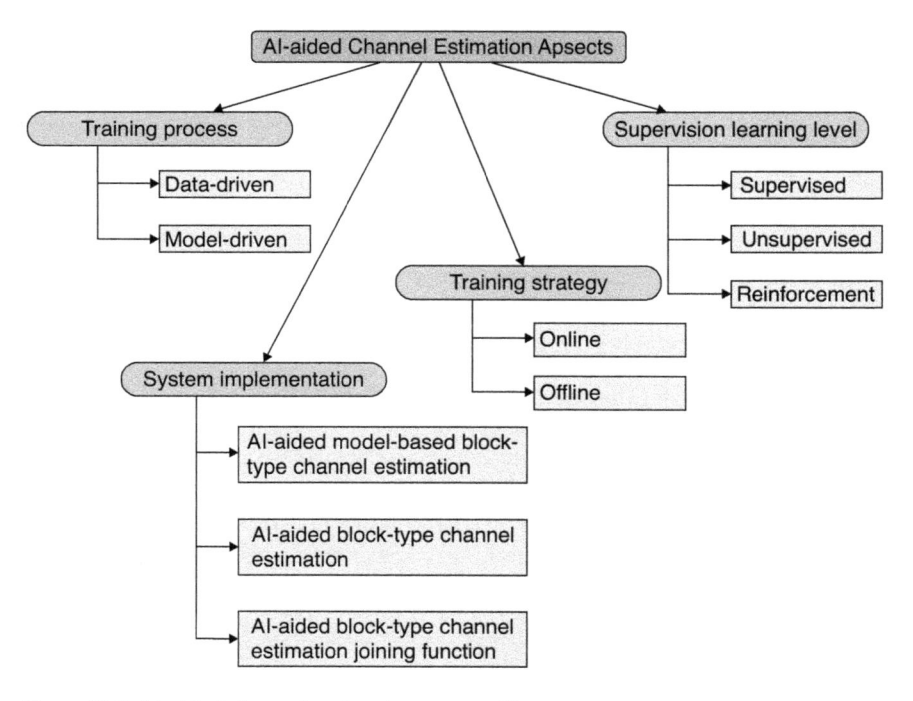

Figure 12.3 AI-aided channel estimation aspects.[24]

during the inference phase[13]. Thus, the main challenge at hand is to feed a well-constructed neural network with a well-prepared training dataset[21]. Even while it sounds easy, getting the best results takes a great deal of practical skill. Figure 12.3 also depicts this.

Supervised learning: The basic objective of supervised learning[8] is to find a label that connects the input dataset to the desired result. In order to determine the degree of design quality of an ANN and communicate it throughout the weight update process, we need a loss function[7] that evaluates the distance between the expected channel using the label. The difference between two, represented as MSE25 or cross-entropy, is used as a loss function. Supervised learning often comes in two flavors: Regression yields a numerical value, while classification determines the input's categorical orientation. The estimation problems, which comprise path delay, path gain, and angle estimates, are well adapted for the regression problem. The classification task is suitable for identifying issues such as time-domain channel tap detection, LOS/NLOS detection, and indoor/outdoor detection. For instance, in the time-domain channel tap detection task, we consider the problem as a multi-label classification problem by identifying a small number, say k, of labels among N classes, and among all potential tapped delay lines, we identify a few nonzero (dominating) taps.

Unsupervised learning: For example, unsupervised learning is used when the problem is nonconvex or nonlinear, making the ground-truth label inaccessible. It is exceedingly challenging to identify the ideal pilot pattern that maximizes the optimum performance parameter in 4G[11] LTE and 5G[3] because the problem of pilot signal allocation is a highly nonlinear mixed-integer programming problem. The loss function might incorporate the goal mean square error (MSE) or the difference between the original and reconstructed channels.

12.6 OVERVIEW OF MULTICARRIER SYSTEM

The multicarrier modulation (MCM)[25] technique, for example, splits a wide channel into contiguous narrowband subcarriers in order to maximize spectral efficiency[5] and throughput[21]. Additionally, the system is resistant in the case of intersymbol interference (ISI), impulsive noise interference, and multipath fading channels. With the advancement of digital signal processing (dsp)[26], the MCM[26] has been included into a variety of wireless communication systems. For instance, the long-term evolution[1] (LTE) system with air interface[1] has been used with the OFDM modulation. The OFDM[5] modulation was also adopted by the 3GPP[26] and 5G[10] network technical specifications for early implementation at the air interface of the new radio (NR). Other multicarrier systems, such as GFDM[18], FBMC[26,16], and UFMC[12], are also being considered for the beyond 5G[4] (B5G) and sixth-generation[16] mobile networks. The OFDM uses the Fourier transform[26] and the inverse fast Fourier transform[26], respectively, to demodulate and modulate the using low complexity signal. In order to reduce ISI, traditional OFDM[15] introduces a CP[13] to its symbol. High peak-to-average power ratio[21], frequency offset[4] sensitivity, and out-of-band[12] leakage characteristics[11] are some drawbacks of the OFDM[12] waveform. However, several methods are added to OFDM[13] systems to lessen these problems, as was previously stated. The multicarrier[11] transmissions are susceptible to carrier frequency[16] offsets introduced by wireless communication systems' high-mobility receivers or by an incompatibility between the local transmitter and reception oscillators. The Doppler effect[24] phenomenon[14], which causes ICI[1] and system performance[8] loss, is amplified by high-mobility[16] receivers. Channel noise sources, long- and short-term fading, and other types of distortion are also present in multicarrier transmissions[19]. As a result, in order to recover data, the channel impairments[26] should be received and acknowledged by the receiver. Through the use of channel estimation[26] and equalization techniques[15], which involve a mathematical model[16] with a channel matrix[5] indicating the relationship between the transmitted[7] and received signals[21], this procedure is carried out. According to signal knowledge at the receiver, regular channel estimate techniques for multicarrier[13] system are broken into two main categories: blind[13]- and non-blind[14]-based techniques. To avoid sending data training sequences while communicating, blind-based[27] channel estimation pulls statistical attributes from the signals that have been received. However, it necessitates a lot of received data, which lowers performance over fast-fading[3] channels. Pilot symbols are the data that non-blind[9] techniques use to estimate the channel at the receiver. Due to the transmission of the pilots' symbols, this outperforms blind approaches at the expense of decreasing spectral efficiency[19]. Semi-blind[15] channel estimation is a procedure that combines blind[27] and non-blind[3] methods. Following the transmission of training data to establish the estimator, blind detection methods are employed. It is challenging to estimate radio channels. These conditions make it difficult to create effective mathematical channel estimation models that may not completely account for the channel properties. The loss of accurate channel estimate caused by this degradation reduces the performance of a multi-carrier system. The link between various system variables can be studied using either a model-driven or model-free method by AI-based learning algorithms, which can, however, overcome those conditions.

12.7 AI-BASED CHANNEL ESTIMATION

AI gives machines the ability to decide for themselves based on the dataset they have trained on and previous learning. Devices reach their best operational state by adapting their parameters to changing circumstances rather than requiring manual adjustment. Additionally, the learning algorithms take advantage of the channel complexity without making irrational assumptions to outperform traditional methods on channels that are similar. Figure 12.3 shows the tree diagram of an AI-based channel.

As a result, AI algorithms do not require precise mathematical models to estimate channels. This enables them to follow parameter fluctuations over complicated settings, which surely includes those channels that have been extensively characterized. As a result, AI-based channel estimation updates existing techniques and develops new ones. As a result, AI-based channel estimate techniques go beyond the restrictions of traditional methodologies, offering a high level of estimation accuracy and enhancing the performance of communication systems. Since AI is regarded as one of the pillars of upcoming 6G networks, investigations on AI-aided channel estimate are pertinent to 5G[11] and 6G[3] communications.

In order to overcome bandwidth constraints and offer better speed, 5G[6] and 6G[6] networks are anticipated for operating at millimeter and terahertz[6] frequencies. As a result of higher attenuation[1] (considering rain attenuation[1]) and high atmospheric[5] absorption rates[1], radio communication systems in the future will encounter channels with a greater degree of complexity. Beyond that, further essential 5G/6G[5] technologies like massive MIMO[5] (mMIMO[5]) and channel bandwidth[5] enhancement will increase the complexity of Tx. architecture and present fresh difficulties for channel estimation.

However, the availability of millimeter and terahertz spectrum globally may encourage the use of frequency division duplex[16], eliminating all reciprocities from the uplink and downlink channels and increasing the necessity for cyclical CSI feedback.

The design of regular wireless communication systems is based on Tx. and Rx. devices that resemble blocks. As seen in Figure 12.4, the various system functions can be thought of as mathematical models and are implemented as independent building blocks. This method offers block-by-block optimization to improve system efficiency as a whole. Additionally, the channel estimation procedure is used as a standalone block feature.

As a result, model-based designs can use AI algorithms to increase channel estimation efficiency through the calculation of channel models, illustrating the AI-based model-type channel prediction approach. Additionally, the model-based channel estimation can be replaced by the AI algorithm, leading to the AI-aided base-type channel prediction approaches. Different combined functionalities at the Tx. or Rx. are produced by the learning capacity of AI, using signal detection and channel estimation together, for instance. This method is known as a base-type channel prediction joining method with the presence of AI. From this idea, both the transmitter and receiver can be modeled with separate AI networks that resemble auto-encoder structures. The model is observed like an end-to-end solution as a unique structure from the perspective of implementation; however, the channel is the part of a hidden layer. Since channel estimate is an intrinsic function, this approach is also taken into account throughout the review process. According to the traditional classification of ML algorithms, AI-based[27] channel estimation techniques are divided into three categories: supervised[7], unsupervised[7], and reinforcement learning. The result is sufficient, but it needs a labeled set of data for training. Unsupervised learning examines a random set of data to obtain patterns. This AI technique is very helpful when the system data are large. Finally, using feedback incentives and penalties, reinforcement learning[7] provides interactions between the system and its interaction performance.

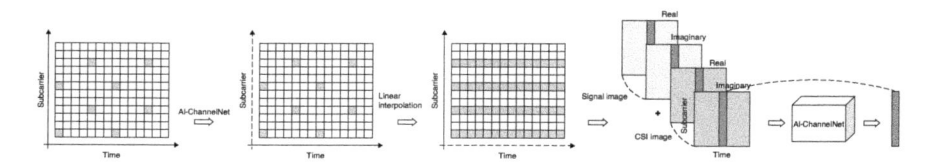

Figure 12.4 The workflow of AI-Channel Net.[24]

For AI-aided channel estimation methodologies, online and online training strategies are taken into consideration. While the latter calls for instruction with a set of static data, the final analysis involves training with a huge amount of data. Due to the nature like static models for training inability to trace the impacts of channel fluctuation, offline training is not practical for complicated situations. Additionally, for systems of communication associated with the training dataset, the static characteristic lowers the training of AI networks. Furthermore, by monitoring the variation in channel effects and expanding the application to various real-world settings, real-time AI network parameter updating is introduced through online training[27].

12.8 AI-CHANNEL NET

The architecture and dataflow, the training techniques, and the workflow were three facets from which the AI-Channel Net was introduced in this part.

12.8.1 Workflow

Due to the rapid time variation[12] of a channel and the linear Gaussian Naive Bayes (GNB)[9,16,21] configuration in high-speed[27] V2I[27] scenarios, the channel between nearby reference signals (RSs) operates temporally regularly in the time domain.

The channel between adjacent subcarriers varies slightly in the frequency domain. As a result, a time-domain prediction AI scheme was devised, and a frequency-domain prediction is a linear interpolation approach. The following four processes make up the AI-Channel Net workflow, which is illustrated in Figure 12.4 below:

1. The AI-Channel Net anticipates all out-of-reference locations from the obtained channel periodicity in the time domain.
2. The channel information[19] dependent on the non-reference[19] position at the frequency domain[19] is predicted using linear interpolation. Remember that excessive mobility has its worst consequences in the time domain[1], and the found channel consistency of AI may not be generalized along the subcarriers.
3. From the CSI, each pixel in these images corresponds to an OFDM symbol and the associated channel data.

12.8.2 Dataflow and its architecture

The AI-Channel Net architecture is depicted in the figure below, which includes the CNN[8] and long short-term memory (LSTM[7]) for recovering received signals and predicting channel information, respectively.

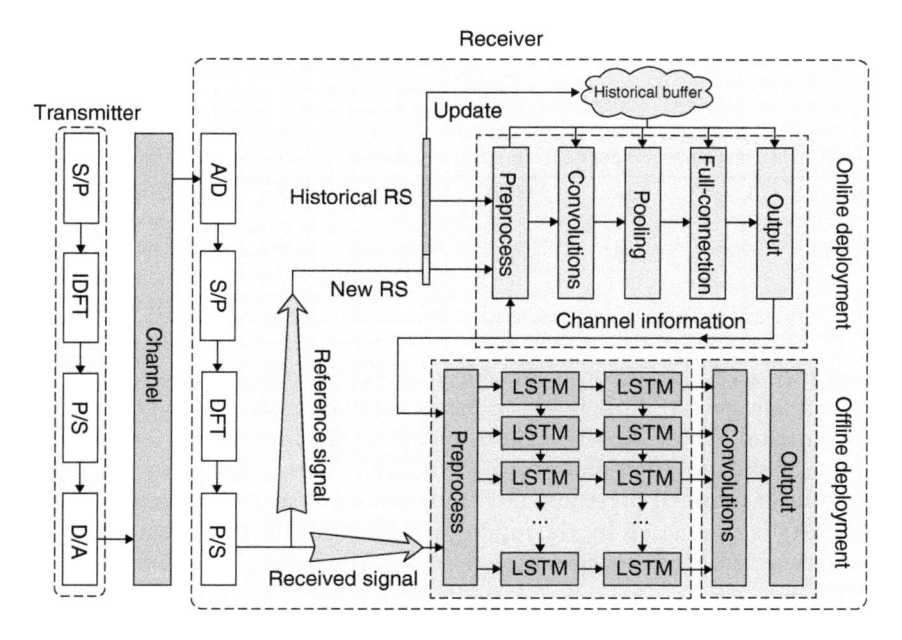

Figure 12.5 AI-Channel Net dataflow architecture.[4]

Convolution, pooling, and fully connected layers in CNN are designed to simplify the architecture and make it more suited for online[27] training. The LSTM[9] layer and convolution layer[9] are generated to understand the connection between the information of the channel[10] and the received signal[1], and broadcast signal with the highest possible degree of complexity.

As seen in Figure 12.5, here, dataflow is divided into two types: RS-based[27] dataflow and received signal-based[27] dataflow. Following that, it is kept for a while in the historical buffer, where it serves as the set of data for the online[27] training algorithm. The RS[27] needs to be prepared beforehand using 2D prior to being fed, digest, and adjust into AI-Channel Net.

The two-dimensional procedure divides the information from the intricately key channel into actual and hypothetical parts. The change is divided the original data by a historical window of fixed length into multiple continuous time series. In Table 12.1, the dissimilarity between online[27] and offline training[27] is given.

In Table 12.1, you will have a clear idea of online and offline deployment. The channel value on the forecast payload's channel and RS[27] orientation symbol position is changing respectively.

12.8.3 Dataflow of received signal

Alternatively, the signal's flow of data, along with projected channel information, is put into AI-Channel Net. In order to produce the sent signal,

Table 12.1 Dissimilarity between online[27] and offline[27] training

Deployment	Temporal channel regularity	Channel value on RS position	Prediction payload symbol position	Function	Network baseline	Training method
Online	Low	Low	High	Tracking	CNN	Online RS-based training
Offline	High	High	Low	Recovery	LSTM+CNN	Offline classification training

the process of restoration is tackled as a classification issue after LSTM[25] extracts the primary link between the signal that is received and the channel information. There are three gates, including the input gate[7], forget gate[25], and output gate[25], that make up a LSTM unit. This forget gate regulates all amount of retained previous state. How much fresh data are connected to the state is calculated by the input gate. How much of the state is output is determined by the output gate shown in the below Equations 12.1 and 12.2.

$$C_t = f_t \times c_{t-1} + i_t \times \hat{c}_t \tag{12.1}$$

$$s_t = 0_t \times \tan h(c_t) \tag{12.2}$$

Here, C_t denotes the cyclic state and i_t and f_t denote input gate and forget gate, respectively. The convolution layer generates the recovery signal using the temporal connection that the LSTM extracted. The modulation order is indicated by the category label, which is {1, 2, 3, and 2^M}. The output layer is intended to be a softmax layer, where K is the value of categories and is represented by the value of nodes. Assume that p_i is the i-th node's output value. The likelihood that i is present in the output layer is then as follows on Equation 12.3:

$$p_i = \frac{e^{p_i}}{\sum_{j=1}^{k} e^{p_j}} \tag{12.3}$$

$$L(\text{Cross} - \text{Entropy}) = -\frac{1}{N} \sum_{n=1}^{N} \sum_{i=1}^{K} I_{i,n} \log p_{i,n} \tag{12.4}$$

From this Equation 12.4, the cross-entropy[24] function is the loss function[7] of training, where the quantity of training samples is N and $i_{i,n}$ is the vector of indicators (1 signifies the nth sample that corresponds to category i and 0 denotes not).

Different weight matrices and bias vectors are implied here. All states take input from previous states and update weights and biases. After passing from different intermediate layers, we have the probabilities for different input vectors. At the output layer, we add all the values to get the final predicted value of the dataflow for the received signal.

12.9 NUMERICAL ANALYSIS

In Figure 12.6, the accuracy and reliability of AI-Channel Net are assessed in comparison with conventional approaches. A channel model used in wireless communication networks is called Wireless Innovation Forum for Europe (WINNER) D2a[27]. It is a simulation of the downlink (Device-to-Device (D2D))[24] direction radio propagation channel, which runs from the Base Station (BS) to the Mobile Station (MS).

The WINNER II channel model, a popular channel model for wireless communication systems, serves as the foundation for the WINNER D2a[27] channel network. In order to more accurately find the properties regarding the downlink channel in high-speed train environments, the WINNER

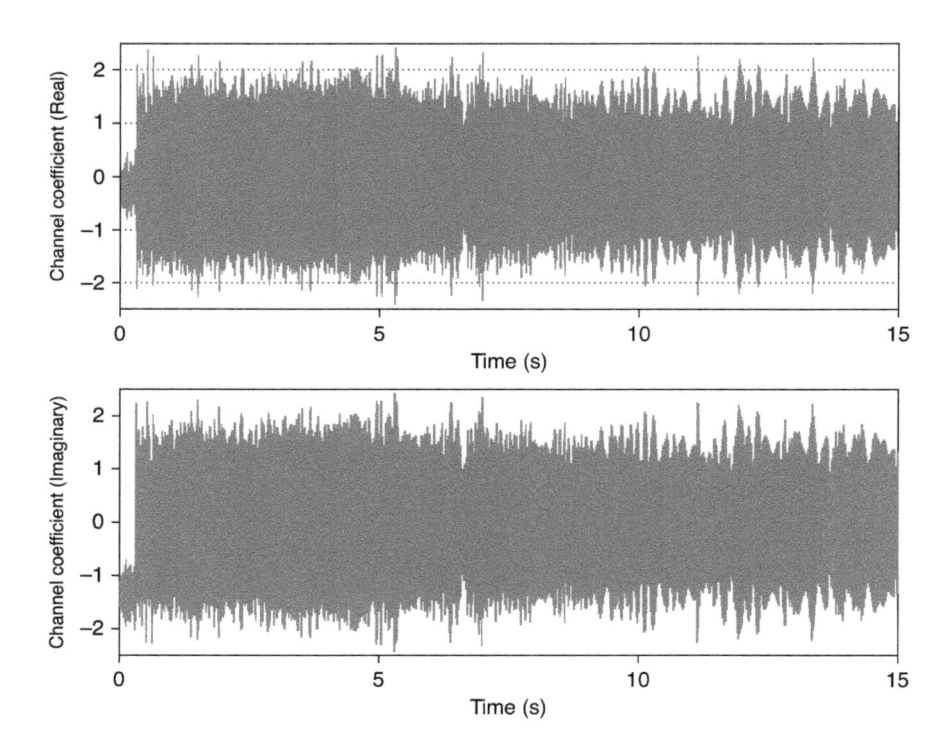

Figure 12.6 The WINNER D2a channel's quick time-varying feature.[24]

D2a13[27] channel model augments the WINNER II channel model with new features. Additionally, the experiment tests the data from the experiment produced by the WINNER D2a[27] channel. Channel features of the WINNER D2a13 model are shown in Figures 12.7 and 12.8. Since there are currently few conventional channel models, particularly under high speed, the experiment in this section used an approximation of the WINNER channel model. Broadband communication systems like LTE and 5G networks can be simulated at the link and system levels using the WINNER II channel model. Radio propagation in situations where the user's equipment is travelling very quickly in a rural area is represented by propagation scenario D2a (Figures 12.9 and 12.10).

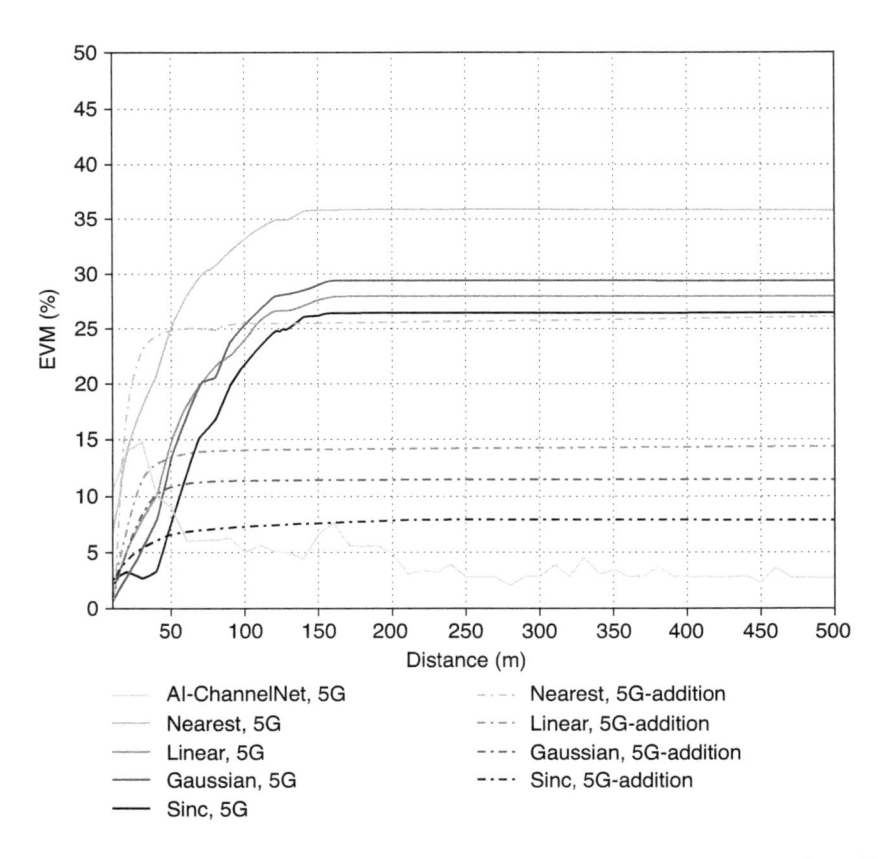

Figure 12.7 The property of the WINNER[27] D2a[27] channel : Changes in Doppler[1] shift[28] with distance of the gNB.[1]

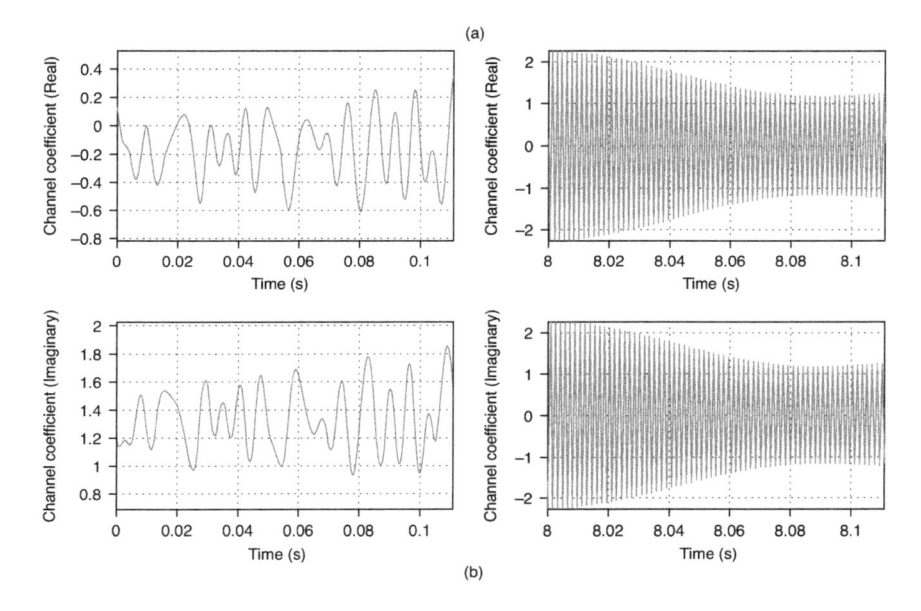

Figure 12.8 The performance of AI-Channel Net's traditional methods' predictions as a function of distance from the graph neural bandits (gNB).[7] (a) Channel Co-efficient (Real). (b) Channel Co-efficient (Imaginary).

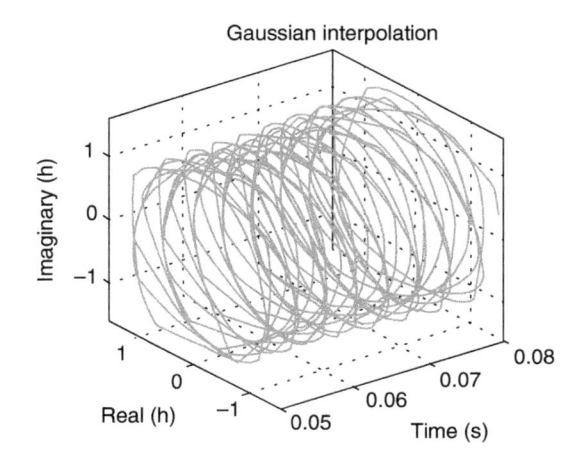

Figure 12.9 Gaussian interpolation.[4,7]

12.10 OPEN RESEARCH CHALLENGES

AI-based channel prediction is a recently emerging field with the potential to revolutionize wireless communication systems. However, there are still a number of open research challenges in this area, including

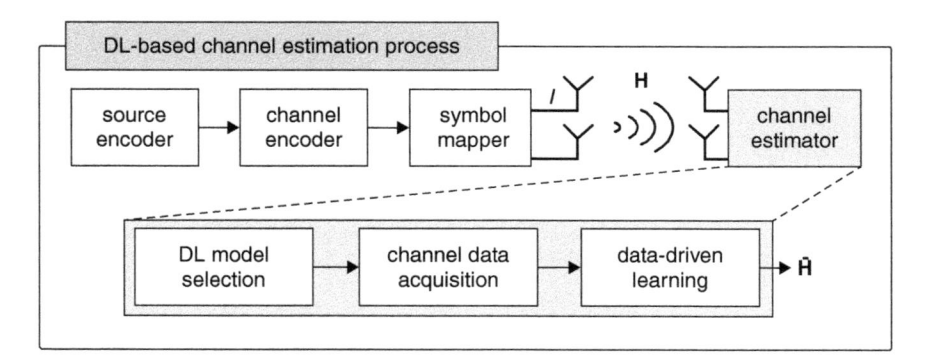

Figure 12.10 DL-based channel estimation.[8]

- **Labeling and data gathering:** The main challenge in developing AI-based channel prediction models is collecting and labeling a large dataset of channel measurements. This is because channel measurements are often expensive and time-consuming to collect, and they can be difficult to label accurately.
- **Model errors in reading:** It might be challenging to comprehend how AI-based models perform predictions because they are frequently intricate and ambiguous. This may be an issue for wireless communication systems, since reliable operation depends on an understanding of the channel prediction process.
- **Robustness to interference and uncertainty:** Wireless channels are innately noisy and uncertain. AI-based channel prediction models need to be robust to these conditions in order to provide error free predictions.
- **Scalability:** AI-based channel prediction models need to be scalable to large networks having a high user count. This can be a challenge, as AI-based models can be computationally expensive to train and deploy.

In addition to these general challenges, there are a number of specific challenges related to different types of AI-based channel prediction models. For example, DL models are often data-intensive and require a huge amount of training data. Reinforcement learning systems can be problematic to train in complex wireless environments.

Despite these challenges, AI-based channel prediction has the capacity to greatly enhance the performance of wireless communication systems. Researchers are actively working on addressing the open research troubles in this field, and it is expected that AI-based channel prediction will play an increasingly essential role in wireless communication systems in the future.

Here are some specific examples of open research challenges in AI-based channel prediction:

- Developing AI-based channel prediction models that are able to generalize to new environments and scenarios.
- Developing AI-based channel prediction models that are able to predict channel dynamics in real time.
- Developing AI-based channel prediction models that are able to operate in low-power and low-latency environments.
- Developing AI-based channel prediction models that are able to be integrated with existing wireless communication system components.

12.11 PROPOSED MODEL

In our proposed model on Figure 12.11, we have developed a convolutional neural network (CNN) using the MATLAB DL toolbox to train our dataset. Here, three two-dimensional (2D) convolution layers of size 3 are taken. In the first convolution layer, we have chosen 64 filters with a kernel of 3×3 grid size; in the next layer, 128, 254 filters are used with the same kernel size; and in the next two layers, using three convolution layers, the input data are passed, processed, and given a unified output by removing all interferences. Here, we used the same padding for fixing special dimension[7] of output data[8] as input data[8]. As seen in Figure 12.3, input data are given on convolution layers and then go to batch normalization to remove overfitting from the training data, and then, it move to the rectified linear unit, which makes our training graph nonlinear to support different types of data. In Figure 12.4, an analysis of the training network has been given. Max-pooling is a down-sampling technique that decreases the dimension of the data by allocating maximum value within a local window with stride 2. This specifies the size of the max-pooling window. In this case, it is set to '2', which means the max-pooling operation will be applied using a 2×2 window. Stride means the rate of moving the pooling window vertically and horizontally. Here, stride 2 signifies that the max-pooling window moves 2 units (Figure 12.12).

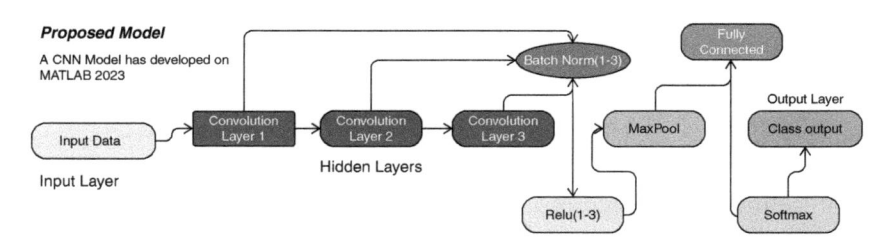

Figure 12.11 Architecture of proposed model of neural network.

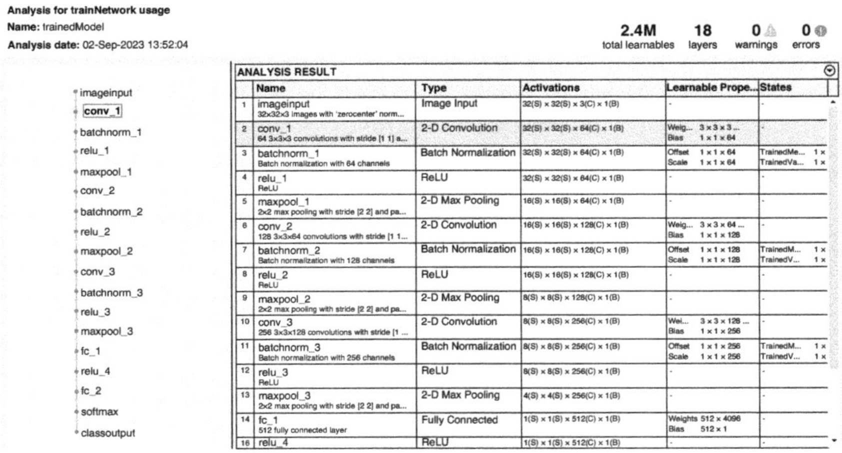

Analysis for trainNetwork usage
Name: trainedModel
Analysis date: 02-Sep-2023 13:52:04

2.4M total learnables   **18** layers   **0** warnings   **0** errors

	Name	Type	Activations	Learnable Prope...	States
1	imageinput
32x32x3 images with 'zerocenter' norm...	Image Input	32(S) × 32(S) × 3(C) × 1(B)	-	-
2	conv_1
64 3x3x3 convolutions with stride [1 1] a...	2-D Convolution	32(S) × 32(S) × 64(C) × 1(B)	Weig... 3 × 3 × 3 ...
Bias 1 × 1 × 64	-
3	batchnorm_1
Batch normalization with 64 channels	Batch Normalization	32(S) × 32(S) × 64(C) × 1(B)	Offset 1 × 1 × 64
Scale 1 × 1 × 64	TrainedMe... 1 ×
TrainedVa... 1 ×
4	relu_1
ReLU	ReLU	32(S) × 32(S) × 64(C) × 1(B)	-	-
5	maxpool_1
2x2 max pooling with stride [2 2] and pa...	2-D Max Pooling	16(S) × 16(S) × 64(C) × 1(B)	-	-
6	conv_2
128 3x3x64 convolutions with stride [1 1...	2-D Convolution	16(S) × 16(S) × 128(C) × 1(B)	Weig... 3 × 3 × 64 ...
Bias 1 × 1 × 128	-
7	batchnorm_2
Batch normalization with 128 channels	Batch Normalization	16(S) × 16(S) × 128(C) × 1(B)	Offset 1 × 1 × 128
Scale 1 × 1 × 128	TrainedM... 1 ×
TrainedV... 1 ×
8	relu_2
ReLU	ReLU	16(S) × 16(S) × 128(C) × 1(B)	-	-
9	maxpool_2
2x2 max pooling with stride [2 2] and pa...	2-D Max Pooling	8(S) × 8(S) × 128(C) × 1(B)	-	-
10	conv_3
256 3x3x128 convolutions with stride [1 ...	2-D Convolution	8(S) × 8(S) × 256(C) × 1(B)	Wei... 3 × 3 × 128 ...
Bias 1 × 1 × 256	-
11	batchnorm_3
Batch normalization with 256 channels	Batch Normalization	8(S) × 8(S) × 256(C) × 1(B)	Offset 1 × 1 × 256
Scale 1 × 1 × 256	TrainedM... 1 ×
TrainedV... 1 ×
12	relu_3
ReLU	ReLU	8(S) × 8(S) × 256(C) × 1(B)	-	-
13	maxpool_3
2x2 max pooling with stride [2 2] and pa...	2-D Max Pooling	4(S) × 4(S) × 256(C) × 1(B)	-	-
14	fc_1
512 fully connected layer	Fully Connected	1(S) × 1(S) × 512(C) × 1(B)	Weights 512 × 4096
Bias 512 × 1	-
15	relu_4
ReLU	ReLU	1(S) × 1(S) × 512(C) × 1(B)	-	-

imageinput
conv_1
batchnorm_1
relu_1
maxpool_1
conv_2
batchnorm_2
relu_2
maxpool_2
conv_3
batchnorm_3
relu_3
maxpool_3
fc_1
relu_4
fc_2
softmax
classoutput

Figure 12.12 Analysis for train network usage.

12.11.1 Proposed algorithm

In Figure 12.13, we have drawn our proposed algorithm flow chart that shows our algorithm steps and gives the proper direction on how to train the datasets based on the channel estimation.

12.11.2 Explanation of the algorithm

Generate a sample dataset/select dataset:

- Set the number of samples (**numSamples**).
- Generate random input data (**inputData**) with dimensions (32, 32, 3, numSamples).
- Generate random binary labels (**labels**) for the samples.

Split the dataset:

- Define the ratios for training, validation, and test sets (**trainRatio, valRatio, testRatio**).
- Call the **splitData** function to split **inputData** and **labels** into training, validation, and test sets.

Define the CNN model architecture:

- Define layers including an input layer, convolutional layer, max-pooling layer, fully connected layer, ReLU activation layer, softmax layer, and classification layer.

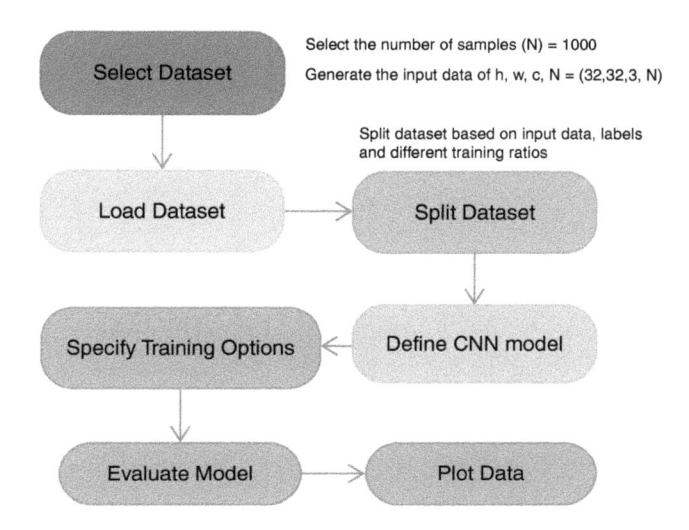

Figure 12.13 Flow chart of the proposed algorithm.

Specify training options:

- Set options by using stochastic gradient descent[3] method with momentum (**sgdm**)[21], maximum epochs (**MaxEpochs**), mini-batch size (**MiniBatchSize**), initial learning rate (**InitialLearnRate**), validation data and labels (**ValidationData**), validation frequency (**ValidationFrequency**), and plot training progress.

Train the CNN:

- Call the **trainNetwork** function with training data (**trainData**), categorical labels (**categorical(trainLabels)**), defined layers, and specified options. It has been shown in Figure 12.13.

Evaluate the model:

- Classify the test data using the trained network and get predicted labels (**predictedLabels**).
- Calculate accuracy by comparing predicted labels with actual test labels (**testLabels**).
- Display the test accuracy.

Function to split data (**splitData**):

- Define a function that takes input data, labels, and ratios as arguments.
- Shuffle the data.
- Split the data into training, test sets, and validation with the provided ratios (Figure 12.14).

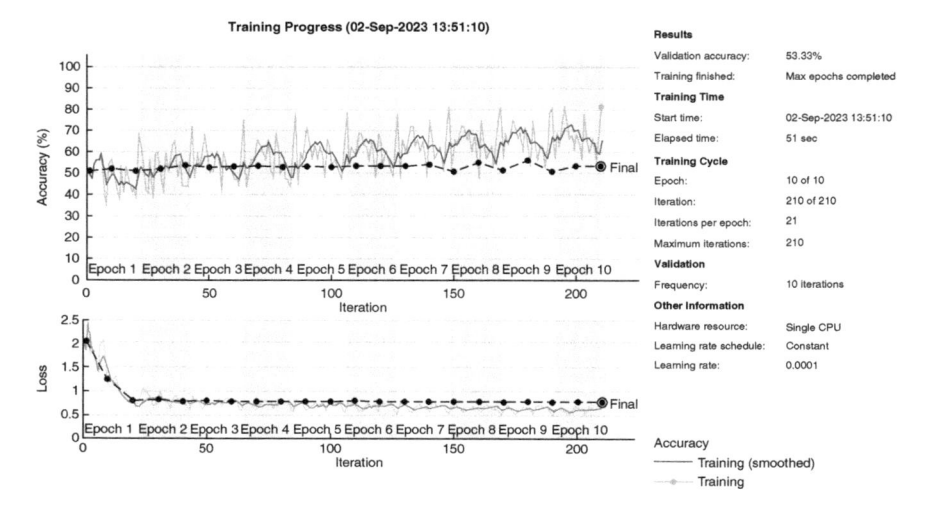

Figure 12.14 Training progress graph.

12.12 CONCLUSION

In conclusion, the utilization of AI-based wireless channel prediction demonstrates enhanced accuracy in estimating maximum data transmission rates and effectively mitigating single-bit errors. The incorporation of L2 (Ridge) regularization in our algorithmic approach proves pivotal in combating overfitting, resulting in a commendable validation accuracy of 53.3%. This shows a higher degree of precision in our estimations. The chapter undoubtedly commences by delving into the historical backdrop of AI-driven wireless channel prediction, subsequently establishing a foundational understanding of DL-based analytical techniques. This pedagogical approach imparts knowledge with remarkable clarity and sophistication. The inclusion of a training algorithm explanation further bolsters the practical applicability of the chapter's content. As we draw this chapter to a close, we turn our gaze toward the potential parameters that may shape the landscape of 6G technology in the future. This forward-looking perspective adds a visionary dimension to our exploration of AI-driven wireless channel prediction, setting the stage for the next era of wireless communication advancements.

ABBREVIATIONS

ITU   International Telecommunication Union[29]
FCC   Federal Communications Commission[29]

CEPT	European Conference of Postal and Telecommunications Administrations[29]
ANI	artificial narrow intelligence[24]
ASI	artificial super intelligence[4]
AGI	artificial general intelligence[24]
CSI	channel state information[11]
DL	deep learning[8]
RS	reference signal
MIMO	multiple-input multiple-output[22]
LSTM	long short-term memory[7]
WINNER	Wireless Innovation Forum for Europe[10]
CNN	convolutional neural network[8]
RNN	recurrent neural network[7]
LTE	long-term evolution[1]
VOLTE	voice over LTE[1]
5G	fifth generation[1]
6G	sixth generation[1]

REFERENCES

1. Raghunandan, K. *Introduction to Wireless Communications and Networks*. Springer International Publishing, 2022. doi:10.1007/978-3-030-92188-0.
2. Kim, H. Low power routing and channel allocation method of wireless video sensor networks for Internet of Things (IoT). In: *2014 IEEE World Forum on Internet of Things (WF-IoT)*, pp. 446–451, IEEE, 2014. doi:10.1109/WF-IoT.2014.6803208.
3. Shaik, N. & Malik, P. K. A comprehensive survey 5G wireless communication systems: Open issues, research challenges, channel estimation, multi carrier modulation and 5G applications. *Multimedia Tools and Applications* 80, 28789–28827 (2021).
4. Vilas Boas, E. C., e Silva, J. D. S., de Figueiredo, F. A. P., Mendes, L. L. & de Souza, R. A. A. Artificial intelligence for channel estimation in multicarrier systems for B5G/6G communications: A survey. *EURASIP Journal on Wireless Communications Networking* 2022, 116 (2022).
5. Djordjevic, I. B. *Advanced Optical and Wireless Communications Systems*. Springer International Publishing, 2022. doi:10.1007/978-3-030-98491-5.
6. Kim, H. Historical sketch of artificial intelligence. In: *Artificial Intelligence for 6G*, pp. 3–14, Springer International Publishing, 2022. doi:10.1007/978-3-030-95041-5_1.
7. Jiang, W., Dieter Schotten, H. & Xiang, J. Neural network-based wireless channel prediction. In: *Machine Learning for Future Wireless Communications*, pp. 303–325, Wiley, 2020. doi:10.1002/9781119562306.ch16.
8. Moon, S., Kim, H. & Hwang, I. Deep learning-based channel estimation and tracking for millimeter-wave vehicular communications. *Journal of Communications and Networks* 22, 177–184 (2020).

9. Li, Z., Wang, C.-X., Huang, J., Zhou, W. & Huang, C. A GAN-LSTM based AI framework for 6G wireless channel prediction. In: *2022 IEEE 95th Vehicular Technology Conference: (VTC2022-Spring)*, pp. 1–5, IEEE, 2022. doi:10.1109/VTC2022-Spring54318.2022.9860457; https://ieeexplore.ieee.org/document/9860457.

10. Ijiga, O. E., Ogundile, O. O., Familua, A. D. & Versfeld, D. J. J. Review of channel estimation for candidate waveforms of next generation networks. *Electronics (Basel)* 8, 956 (2019).

11. Arora, K., Singh, J. & Randhawa, Y. S. A survey on channel coding techniques for 5G wireless networks. *Telecommunication Systems* 73, 637–663 (2020).

12. Hwang, T., Yang, C., Wu, G., Li, S. & Ye Li, G. OFDM and its wireless applications: A survey. *IEEE Transactions on Vehicular Technology* 58, 1673–1694 (2009).

13. Ozdemir, M. & Arslan, H. Channel estimation for wireless ofdm systems. *IEEE Communications Surveys & Tutorials* 9, 18–48 (2007).

14. Savaux, V. & Louët, Y. LMMSE channel estimation in OFDM context: A review. *IET Signal Processing* 11, 123–134 (2017).

15. Kang, S. G., Ha, Y. M. & Joo, E. K. A comparative investigation on channel estimation algorithms for ofdm in mobile communications. *IEEE Transactions on Broadcasting* 49, 142–149 (2003).

16. Vilas Boas, E. C., e Silva, J. D. S., de Figueiredo, F. A. P., Mendes, L. L. & de Souza, R. A. A. Artificial intelligence for channel estimation in multicarrier systems for B5G/6G communications: A survey. *EURASIP J Wirel Commun Netw* 2022, 116 (2022).

17. Mei, K., Liu, J., Zhang, X., Rajatheva, N. & Wei, J. Performance analysis on machine learning-based channel estimation. *IEEE Transactions on Communications* 69, 5183–5193 (2021).

18. Jiang, W. & Schotten, H. D. Neural network-based fading channel prediction: A comprehensive overview. *IEEE Access* 7, 118112–118124 (2019).

19. Mahlobogwane, Z., Owolawi, P. A. & Sokoya, O. Multiple wavelength propagation in free space optical wireless channel. In: *2018 International Conference on Advances in Big Data, Computing and Data Communication Systems (icABCD)*, pp. 1–6, IEEE, 2018. doi:10.1109/ICABCD.2018.8465406.

20. Ellingson, S. W. *Electromagnetics*, vol. 1, Virginia Tech Publishing, 2018. https://vtechworks.lib.vt.edu/items/6f75e22e-ea6a-48a3-b441-65ce5ebf5331

21. Alkhateeb, A., Jiang, S. & Charan, G. Real-time digital twins: Vision and research directions for 6G and beyond. *IEEE Communications Magazine*, 61(11), 128–134 (2023). doi:10.1109/MCOM.001.2200866.

22. Mawatwal, K., Sen, D. & Roy, R. A semi-blind channel estimation algorithm for massive MIMO systems. *IEEE Wireless Communications Letters* 6, 70–73 (2017).

23. Liu, T. & Lu, D. The application and development of IOT. In: *2012 International Symposium on Information Technologies in Medicine and Education*, pp. 991–994, IEEE, 2012. doi:10.1109/ITiME.2012.6291468.

24. Simmons, N., Gomes, S.B.F., Yacoub, M.D., Simeone, O., Cotton, S.L. & Simmons, D.E. AI-based channel prediction in D2D links: An empirical validation. *IEEE Access* 10, 65459–65472 (2022).

25. Mahadevaswamy, U. B. & Swathi, P. Sentiment analysis using bidirectional LSTM network. *Procedia Computer Science* 218, 45–56 (2023).
26. Mucchi, L., Shahabuddin, S., Albreem, M.A., Abdallah, S., Caputo, S., Panayirci, E. & Juntti, M. Signal processing techniques for 6G. *Journal of Signal Processing Systems* 95, 435–457 (2023).
27. Zhang, Z., Xiong, L., Yao, D. & Wang, Y. The AI-based channel prediction scheme for the 5G wireless system in high-speed V2I scenarios. *Wireless Communictions and Mobile Computing* 2022, 1–17 (2022).
28. Basu, P. K. & Dhasmana, H. *Electromagnetic Theory.* Springer International Publishing, 2023. doi:10.1007/978-3-031-12318-4.
29. Palaios, A. Empirical spatio-temporal characterization of radio environment properties. Dissertation, RWTH Publications (2017). https://publications.rwth-aachen.de/record/721548

Chapter 13

Cost effective IoT-enabled wireless video surveillance model for Next-Gen communication systems

*Subham Das, Sumana Hazra, Sayantan Mallik,
Sourav Nandi, and Ankan Bhattacharya*

13.1 INTRODUCTION: BACKGROUND AND DRIVING FORCES

Robots can be used in a wide range of situations, and their mechanical designs determine their capabilities. Robots must be able to operate on a range of surfaces and adjust to diverse cross-sectional shapes. The main challenge here is that the controller is limited distance routing, but we are developing long-distance routing through mobile network routing with high data speeds for faster communication. Thus, we made our own communication application into that controller for better reach. Here, a robot has been developed for video monitoring and is controllable via graphical user interface (GUI). A video transmit option is available for the control mechanism, which can be used for video surveillance and monitoring. The transmit video option is available for the control mechanism. High-speed image transmission is a realistic means of achieving video communication. To begin with, the robot will have a camera that will record scenes and upload the pictures to a server where users may control and view a live feed and send all the data to the user. Robotic cameras serve a crucial role in assisting organizations in maintaining security, safeguarding sensitive regions, and reacting to change ground conditions, from target acquisition to situational awareness, from border surveillance to vehicle tracking. The Raspberry Pi serves as the circuit's brain. This single-board minicomputer has a four-core ARM Cortex-A7 central processing unit running at 900 MHz with 1 GB of RAM. Additionally, because, well, you're not there, it has two infrared obstacle sensors that let it see obstructions that you are blind to. It will detect an object and alter its direction if one is detected. The robot, as previously mentioned, is equipped with a number of sensors. It also contains a camera that can be used to watch people on video. A motion sensor has also been utilized to identify human movement. The system will detect a human and enable the camera. The L239 motor driver is used to generate power for the motors by means of a Raspberry pi general purpose

270   DOI: 10.1201/9781003517689-13

input/output (GPIO). We have set up an Internet of Things (IoT) server that allows robots to be controlled remotely. The system is already available on the market, but we are making a new, cost-effective system for the general peoples as well, who are maintaining their daily works through that robotic system. Similar IoT-based surveillance systems have been discussed in [1–5]. However, the system proposed here has few added features, like an Android app. Based on communication at 5G speed, motor speed (BLDC) 40–60 Kmph, no heating issue, and cost effectiveness.

13.2 SYSTEM DESIGN AND CONNECTION OF MODULES

We have discussed the system design and connection of modules for the aforesaid system below.

Figure 13.1 depicts the block diagram, which is based upon the Raspberry Pi minicomputer, which enables a camera and USB devices. It shows how the various modules of the system are interconnected and how data is interchanged between the IoT server and the Cloud. Low profile microstrip antennas [6–15] can be used for transmission and reception purposes.

13.3 REMOTE CONTROL MECHANISM AND CONNECTIONS

Following is the circuit of the proposed remote controller of the system.

Figure 13.2 describes this diagram, which is based on an IC and connected via four switches with a radio frequency (RF) Tx-link receiver. The block diagram is proposed and verified for the remote controller for the project.

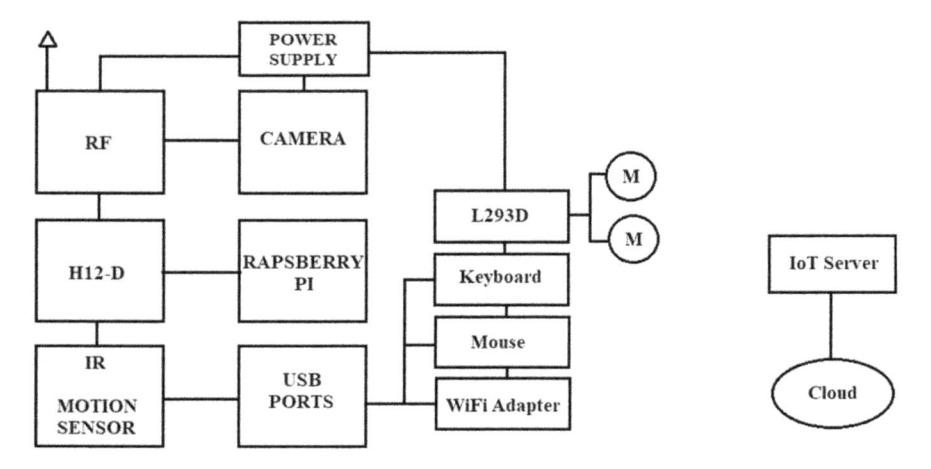

Figure 13.1 System design and connection of modules of the prototype.

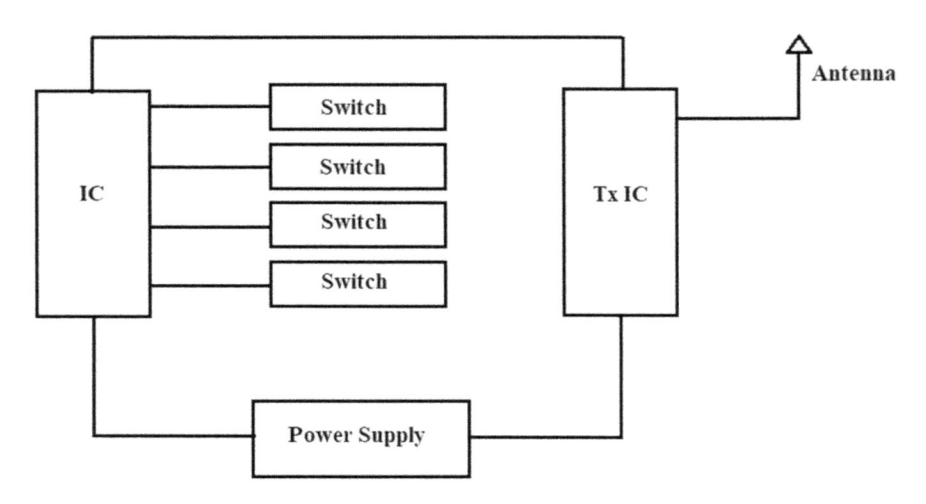

Figure 13.2 Module connections of the remote controller.

13.4 DESCRIPTION OF THE COMPONENTS USED

a. **Raspberry Pi:** - It is a mini computer, with a processor, input/output ports, memory, and expansion pins for hardware interfacing.

b. **Camera:** - The Raspberry Pi camera is a compatible mini video-camera for single-board computers.

c. **Power Supply:** - For mobile or outdoor surveillance systems, a rechargeable battery or a combination of batteries may be used. Consider the capacity (mAh or Wh) to determine how long the system can operate without recharging.

d. **L293D:** - The L293D is designed to control the direction and speed of two DC motors independently. It provides an H-bridge configuration for each motor, allowing the motor to rotate in both forward and reverse directions.

e. **RF Rx:** - RF Receivers typically have an antenna input to capture the incoming RF signals effectively.

f. **HT12-D:** - The HT12-D is specifically designed to decode the serial data received from an RF or IR (infrared) receiver. It is commonly used in remote control systems for applications like home automation, wireless keyless entry systems, and other wireless communication projects.

g. **IR and Motion Sensor:** - IR sensors often use a combination of an IR emitter (LED) and an IR detector (photodiode or phototransistor). The emitter emits IR light, and the detector senses the reflected or emitted IR light.

h. **Wi-Fi Adapter:** - A USB Wi-Fi adapter is a small external device that adds wireless connectivity to a computer or other devices that lack built-in Wi-Fi capabilities. It enables devices to connect to Wi-Fi networks, providing wireless internet access.

i. **IOT server:** - An IoT server is responsible for receiving and ingesting data from IoT devices. This data can include sensor readings, status updates, and other information generated by connected devices.

j. **Cloud:** - Smartphones, laptops, tablets, and other smart devices are used for internet cloud service facilities.

13.5 FLOWCHART AND ALGORITHM

The flowchart and algorithm of the proposed system have been depicted below.

Figure 13.3 depicts the steps of the proposed system. Designing a wireless video surveillance robot using the Raspberry Pi involves several steps, including hardware setup, software development, and the integration of various components. Here's a high-level algorithmic overview of the process:

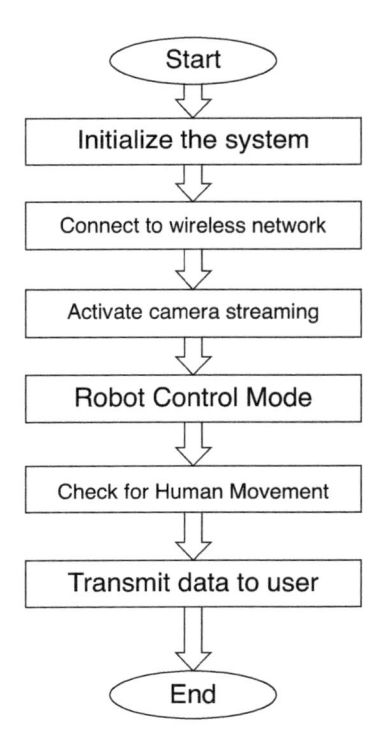

Figure 13.3 Flowchart of the system.

Hardware setup:

    a. Assemble the robot:
        - Connect motors and wheels for movement.
        - Attach a camera module for video capture.
        - Add any additional sensors (e.g., ultrasonic sensors for obstacle avoidance).
    b. Install Raspberry Pi:
        - Connect the Raspberry Pi board to the robot chassis.
        - Attach necessary peripherals (Wi-Fi dongle, camera module, etc.).
    c. Power Supply:
        - Connect a suitable power source for the Raspberry Pi and motors.

Software development:

    a. Operating system:
        - Install and configure the Raspbian OS on the Raspberry Pi.
    b. Wireless setup:
        - Configure the Wi-Fi settings on the Raspberry Pi for wireless communication.
    c. Camera interface:
        - Set up the camera module on the Raspberry Pi.
        - Use the Picamera library or OpenCV to interface with the camera for video capture.
    d. Motor control:
        - Implement code to control the motors for movement.
        - Use GPIO pins or motor driver modules for motor control.
    e. Communication:
        - Implement a communication protocol for wireless control.
        - Use sockets or a messaging protocol to send commands from a remote device.
    f. Streaming video:
        - Set up a video streaming server on the Raspberry Pi.
        - Use tools like Motion or implement a custom solution for video streaming.
    g. Obstacle avoidance (optional):
        - If using additional sensors, implement obstacle avoidance algorithms.
        - Adjust robot movement based on sensor inputs.

Remote control:

    a. Remote interface:
        - Develop a remote-control interface (web-based, mobile app, etc.).
        - Implement controls for movement, camera adjustments, and other functionalities.

b. Wireless connection:
- Connect the remote-control device to the robot over a wireless network.

Integration:

a. Combine modules:
- Integrate camera, motor control, and communication modules into a cohesive program.
b. Error handling:
- Implement error handling mechanisms for communication and hardware failures.

Testing:

a. Functional testing:
- Test individual components (cameras, motors, and sensors) for proper functionality.
- Conduct integrated testing to ensure seamless operation.
b. Performance optimization:
- Optimize code for performance, considering the limited resources of the Raspberry Pi.

Deployment:

a. Environmental considerations:
- Consider the deployment environment for factors like lighting, obstacles, and wireless signal strength.
b. Security measures:
- Implement security measures to prevent unauthorized access to the robot.

Maintenance and upgrades:

a. Remote update mechanism:
- Include mechanisms for remote software updates.
- Monitor and address any issues that arise during operation.

Keep in mind that the specific details of each step will depend on the exact hardware components, sensors, and software libraries you choose to use in your project. Additionally, documentation and proper version control can be essential for maintaining and upgrading the system over time.

13.6 ADVANTAGES AND LIMITATIONS

A wireless video surveillance robot using a Raspberry Pi can have several advantages:

1. **Live streaming:** The internet can be used for visualizing the camera feed from anywhere.
2. **HD streaming:** The live feed of the camera can provide HD streaming.
3. **Face recognition:** The robot can identify unknown humans using a face recognition algorithm.
4. **Bluetooth control:** The robot can be controlled using Bluetooth.
5. **Surveillance:** The robot can surveil areas and protect them from adversaries.
6. **Cost effective:** This video surveillance is a cost-effective system.

A Raspberry Pi-based surveillance system can be beneficial for:

- Determining crime
- Gathering evidence
- Detecting thefts
- Controlling limited areas using Wi-Fi
- Providing surveillance of its location

A wireless video surveillance robot using a Raspberry Pi can be a versatile and cost-effective solution, but it also comes with certain limitations. Here are some potential limitations to consider:

1. Limited processing power.
2. Limited memory.
3. Limited wireless range.

To overcome some of these limitations, one might consider using external peripherals, optimizing software for performance, and integrating additional sensors for better navigation and environmental awareness. It's essential to carefully plan the deployment based on the specific requirements and constraints of the surveillance environment.

13.7 COST CALCULATION

Table 13.1 shows the approximated calculation of cost for this system.

Table 13.1 Approximated cost calculation

Sl. no.	Component name	Quantity	Approx. price (Rs.)
1.	Raspberry Pi 2	1	4000
2.	Power Bank (5V)	1	500
3.	SD Card (8GB)	1	250
4.	RF-Module (Tx-Rx Pair)	1	250
5.	HT12E	1	120
6.	HT12D	1	120
7.	Battery (12V-1.2A)	1	400
8.	Robot Chassis	1	300
9.	Wheels	4	200
10.	Motor (12V-200RPM)	2	750
11.	L293D	1	280
12.	Switches On-Off	4	100
13.	Cell Battery (1.5V)	2	250
14.	USB Wi-Fi Adapter (Edimax)	1	1700
15.	IR Sensor (Pair)	2	200
16.	Motion Sensor	1	350
17.	Raspberry Pi Camera	1	380
18.	Connectors	Required as per implementation	100
Total			10,250

13.8 FUTURE SCOPES

The future scope of this project is at its peak because, in low cost, one can make a small four-wheeler robot using massive working features. This robot can take pictures from any surface, i.e., soil. It is used as a space rover in space research projects in the moon or on the military battle field. This can be used as a spy robot and help the militants. Some uses of the project are discussed below:

a. During a conflict, it can be utilized to gather intelligence from the enemy's terrain, monitor that intelligence from a safe distance, and securely organize a counterattack.
b. Monitoring terrorist organizations' locations and scheduling an attack for a convenient moment.
c. Installing video surveillance in disaster-affected areas is inaccessible to people.

A complete program for a wireless video surveillance robot using the Raspberry Pi involves a substantial amount of code and would typically be spread across multiple files. In this example, the program sets up a server socket for communication and listens for a single incoming connection. It captures video frames from the camera, converts them to bytes, and sends them to the connected client (e.g., a remote-control interface). The robot's movement is controlled based on the received commands (e.g., forward, backward, left, right, and stop). Note that one needs to adapt the GPIO pin numbers, motor control logic, and camera settings based on their specific hardware configuration. Additionally, one needs a remote-control interface (e.g., a web-based interface or a mobile app) to send control commands to the robot.

13.9 CONCLUSIONS AND FUTURE SCOPE

More investigation into the body's or legs' mechanical architecture would enable the finished project to more successfully accomplish our list of goals. More advancements in control system technology would enable the robot to respond to its environment and make human controls simpler and more understandable. This project lowers risks and can be expanded to replace humans. The robotic system can be remotely controlled with a smart smartphone thanks to the Raspberry Pi module. It takes very little time to process the remote commands and react appropriately. The demonstrations indicate that a clear, fast, and up to 15 frames per second video may be fetched, which is a noteworthy factor.

REFERENCES

1. Liu, J. N. K., Wang, M., and Feng, B. iBotGuard: An internet-based intelligent robot security system using invariant face recognition against intruder. *IEEE Transactions on Systems, Man, and Cybernetics, Part C (Applications and Reviews)*, vol. 35, no. 1, pp. 97–105, 2005. doi: 10.1109/TSMCC.2004.840051.
2. Harshitha, R. and Hussain, M. H. S. Surveillance robot using raspberry Pi and IoT. In: *2018 International Conference on Design Innovations for 3Cs Compute Communicate Control (ICDI3C)*, Bangalore, India, 2018, pp. 46–51, doi: 10.1109/ICDI3C.2018.00018.
3. Kumar, S. and Solanki, S. S. Remote home surveillance system. In: *2016 International Conference on Advances in Computing Communication & Automation (ICACCA)*, 29 September, 2016, Dehradun, India, IEEE, doi: 10.1109/ICACCA.2016.7578890.
4. Maity, A., Paul, A., Goswami, P., and Bhattacharya, A. Android application based bluetooth controlled robotic car. *International Journal of Intelligent Information Systems*, vol. 6, no. 5, pp. 62–66, 2017. doi: 10.11648/j.ijiis.20170605.12.

5. https://www.taylorfrancis.com/books/edit/10.1201/9781003217398/internet-things-data-mining-modern-engineering-healthcare-applications-ankan-bhattacharya-bappadittya-roy-samarendra-nath-sur-saurav-mallik-subhasis-dasgupta

6. Bhattacharya, A., Roy, B., Chowdhury, S. K., and Bhattacharjee, A. K. A compact fractal monopole antenna with defected ground structure for wideband communication. *ACES Journal*, vol. 33, no. 03, pp. 347–350, 2021.

7. Bhattacharya, A., Roy, B., Chowdhury, S. K., and Bhattacharjee, A. K. Design and analysis of a koch snowflake fractal monopole antenna for wideband communication. *ACES Journal*, vol. 32, no. 06, pp. 548–554, 2021.

8. Bhattacharya, A., Roy, B., Chowdhury, S. K., and Bhattacharjee, A. K. Compact slotted UWB monopole antenna with tuneable band-notch characteristics. *Microwave and Optical Technology Letters*, vol. 59, pp. 2358–2365, 2017. doi: 10.1002/mop.30730.

9. Roy, B., Bhattacharya, A., Bhattacharjee, A. K., and Chowdhury, S. K. UWB monopole antenna design in a different substrate using Sierpinski Carpet Fractal Geometry. In: *2015 2nd International Conference on Electronics and Communication Systems (ICECS)*, Coimbatore, India, 2015, pp. 382–385. doi: 10.1109/ECS.2015.7124930.

10. De, A., Roy, B., Bhattacharya, A., and Bhattacharjee, A. K. Investigations on a circular UWB antenna with Archimedean spiral slot for WLAN/Wi-MAX and satellite X-band filtering feature. *International Journal of Microwave and Wireless Technologies*, vol. 14, no. 6, pp. 781–789, 2022. doi: 10.1017/S1759078721000891.

11. Bhattacharya, A. and Roy, B. Investigations on an extremely compact MIMO antenna with enhanced isolation and bandwidth. *Microwave and Optical Technology Letters*, vol. 62, pp. 845–851, 2020. doi: 10.1002/mop.32084.

12. Bhattacharya, A., Roy, B., Chowdhury, S. K., and Bhattacharjee, A. K. Compact printed hexagonal ultra wideband monopole antenna with band-notch characteristics. *Indian Journal of Pure & Applied Physics (IJPAP)*, vol. 57, no. 4, pp. 272–277, 2019.

13. Bhattacharya, A. An analytical approach to study the behavior of defected patch structures. In: Sengupta, S., Das, K., and Khan, G. (eds) *Emerging Trends in Computing and Communication. Lecture Notes in Electrical Engineering*, vol. 298, 2014, Springer, New Delhi. doi: 10.1007/978-81-322-1817-3_43.

14. Bhattacharya, A., De, A., Roy, B., and Bhattacharjee, A. K. Investigations on a low-profile, filter backed, printed monopole antenna for UWB communication. *Indian Journal of Pure & Applied Physics (IJPAP)*, vol. 58, no. 2, pp. 106–112, 2020.

15. https://www.taylorfrancis.com/chapters/edit/10.1201/9781003217398-17/extremely-compact-low-cost-antenna-sensor-designed-iot-integrated-biomedical-applications-ankan-bhattacharya-souvik-pal

Chapter 14

AI-enabled road traffic flow control and mobility prediction using YOLO

Shiladitya Pujari and Ankit Kumar

14.1 INTRODUCTION

Road traffic monitoring is a crucial area of study. Current traffic situation may be assessed, and actionable information may be given to traffic control management organizations by assessing different types of vehicles and traffic on the route. These data can assist these organizations in making choices that enhance the quality of life of the general population. For instance, during holidays, information on the volume of traffic on the roads may be utilized to advise drivers of other routes to redirect traffic from crowded places. Additionally, roadside signs can be erected to notify drivers and decrease traffic accidents on routes that are often used by big trucks. Additionally, it is possible to identify and monitor the automobiles of criminals using the make, model, and color of a particular vehicle. All of the aforementioned programs analyze data gathered by a traffic monitoring system. Therefore, numerous researchers have employed various techniques to accomplish vehicle detection and categorization in order to gather information about passing cars.

There are primarily two categories of traditional vehicle detecting techniques: (1) static-based techniques [1–7] that produce and validate vehicle prediction frames using sliding windows techniques; and (2) techniques [8–12] that leverage dynamic characteristics of moving objects to extract the contour of the item from the picture. Mohamed et al. [1] prescribed a moving vehicle detection system which inputs those features that is extracted into an artificial neural network (ANN) to achieve vehicle classification. The system employs some characteristics (Haar-like) to extract data about the shape of the vehicles. In order to extract the edge properties of cars and feed them into AdaBoost to filter crucial features, Wen et al. [2] also employed Haar-like characteristics. SVM, i.e., support vector machine, was then used to categorize the filtered features which increased the recognition accuracy. In order to identify whether a vehicle is visible in a picture, Sun [3] and David & Athira [4] employed a filter called Gabor filter for gathering attributes of vehicles. A two-step vehicle-detection approach was created by Wei et al. [5]. They first obtained the region of interest with automobiles using Haar-like features and AdaBoost, and then they reverified the region using the HOG known as histogram of oriented gradients [6] with the

DOI: 10.1201/9781003517689-14

help of SVM. Their technology demonstrated increased vehicle-detecting capabilities, according to the findings of their experiments. Yan et al.'s [7] vehicle-detection system selected the borders of cars using vehicle shadows and extracted features using the HOG. Then, for verification, these characteristics were fed into classifiers named AdaBoost and SVM. That strategy lowers the detection impact because when cars obstruct one another, they are treated as one vehicle since the shadows are interconnected.

Dynamicswise, Seenouvong et al.'s approach for detecting and counting vehicles is based on dynamic characteristics. A difference map was generated from a specific current image using background subtraction in order to segment the related foreground image. Additionally, a number of morphological procedures were utilized to identify moving cars, count the number of vehicles traveling through a certain region, and determine the bounding box and contour of a moving item. In order to address the issue of background subtraction brought on by background pictures, several studies have employed Gaussian Mixture Models [9,10] for simulating adaptive background [11–13]. Gradual fluctuations in brightness are the root cause of poor foreground segmentation. The aforementioned static and dynamic approaches can only do so much to solve this issue. For instance, classic feature extraction methods are labor-intensive since they must be hand-crafted by professionals using their knowledge and expertise. The dynamic feature technique not only performs poorly in circumstances of large backdrop changes, but it also makes subsequent image processing processes more difficult. These traditional approaches have increasingly been supplanted by deep learning (DL) techniques.

14.2 LITERATURE REVIEW

DL has been successfully applied and used in various sectors in the recent era, producing accurate prediction outcomes. The convolutional neural network (CNN) approach significantly enhances the accuracy of picture identification as compared to conventional methods that call for artificial feature determination. The LeNet model was initially put up by Lecun et al. [14] to address the issue of identifying handwritten numbers in the banks. AlexNet has been introduced by Krizhevsky [15] to enhance the conventional CNN. GoogLeNet, which Szegedy et al. [16] proposed, employs several filters having various sizes for feature extraction that augment feature information. Two models, known as VGG-16 and VGG-19, were proposed by Simonyan and Zisserman [17]. They demonstrated that increasing a model's depth may increase accuracy by substituting gradually employing many tiny convolution kernels for the main convolution kernel. The ResNet model was suggested by He et al. To address the issue of gradient disappearance and convergence impossibility brought on by an excessive network depth, they employed residential blocks. MoblieNet, which employs deep separation convolution for extracting fewer and meaningful features that lower the redundancy of number of parameters in a CNN model, was proposed by Howard et al. [18].

The aforementioned works have concentrated on enhancing a CNN's feature description skills so that CNNs may be applied to more challenging issues, such as object identification. To address the vehicle-detection issue, some papers [18–23] already employed region-based CNN (R-CNN) models. R-CNN extracts an object's location using the Region Proposal Network (RPN) [24] before classifying it with a conventional CNN. The most recent network design for R-CNN models is called RetinaNet [25]. The R-CNN architecture employs a multilayer NN for classification and has a two-stage methodology [26,27]. This design is inadequate for real-time detection since it significantly increases the amount of parameters needed and slows down performance. One-stage mechanism solutions like the YOLO framework [27–30] and SSD, i.e., single-shot multibox detector [31] framework, have been developed to address this issue. Single-stage algorithms can identify real-time objects quickly, but they have a lower classification accuracy than R-CNN approaches [32,33].

These issues with the object detection techniques mentioned so far are as follows: (1) Although two-stage detection methods offer good accuracy in classification, the pace of detection is slowed down by the huge number of network parameters. (2) One-stage object detection techniques are faster in real time but less accurate than two-stage techniques. (3) The network as a whole needs to be retrained in order to expand the number of object categories, which takes time and limits the method's capacity to scale. Fuzzy neural networks (FNNs), which resemble human fuzzy inference mechanisms and have the powerful learning capabilities of NN, have gained a bunch of popularity in an array of domains which consists of classification, control, and forecasting. Classification issues were addressed by Asim et al. using adaptive network-based fuzzy inference approach [34]. This approach produced greater classification accuracy when compared to conventional NN. When Lin et al. [35] employed a tool chip and a type-2 FNN to forecast, their technique produced better prediction outcomes.

A few researchers have successfully solved system identification and prediction issues using FNN networks [36–38]. In order to optimize classification performance and minimize network parameter use, an FNN was integrated into deep neural network (DNN) in this study. For feature fusion, traditional CNNs employ the global pooling and channel pooling [39,40] techniques. Global pooling techniques, which can be further broken down into global average pooling and global max pooling [41,42], add up the spatial data. Global pooling techniques avoid overfitting and are therefore more resistant to spatial translations. Feature fusion is accomplished using channel pooling techniques, such as channel average pooling (CAP) [43] and channel max pooling (CMP) [44], which compute average or maximum values of those pixels occupying the same places in the feature map. These techniques produce subpar classification outcomes since they just compress information and do not include learnable weights. To increase the usage of fusion of different features and find the efficacy of various fusion approaches, a novel feature fusion technique called network mapping was presented in this study.

14.3 YOLO ALGORITHM DEVELOPMENT

YOLO target identification technique is most popular for the compactness and quick computational speed of the model. It can determine the category and bounding box position instantly with the implementation of NN. As YOLO possesses the ability of generalization and can achieve vastly widespread characteristics, it can be the trump card for various applications.

There are altogether twenty four layers which lead two layers in the original YOLO design which are fully coupled. YOLO predicts multiple bounding boxes for each individual grid cell; however, those bounding boxes have the greatest intersection over union (IOU) [13]. YOLO has got two flaws known as imprecise location and a lower recall rate. YOLO V2 generally becomes better. Additionally, YOLO V2 just simplifies the network rather than deepening or expanding it.

YOLO V2 has two improvements: (1) better and (2) faster.

When the input is (416 416), (13 13), (26 26), and (52 52), YOLO V3 adopts feature graphs of three scales. K-means is used for getting 9 previous boxes which are then divided into feature maps having three scales; as for each position, YOLO V3 employs three prior boxes. Feature maps of larger scale employ smaller preceding boxes. The YOLO V3 feature extraction network, alternatively, uses the residual model, also known as Darknet-53 since it included 53 convolution layers in comparison to the 19 utilized by YOLO V2. Table 14.1 shows the versionwise improvement of YOLO. The following figure (Figure 14.1) depicts the YOLO algorithm evolution timeline below.

14.4 REAL-TIME OBJECT TRACKING
IN VIDEO WITH YOLO

YOLO, which stands for "You Only Look Once", is a quick and accurate moving object detection algorithm. It works by applying its algorithm to a picture first. In our illustration, the picture is separated into 3×3 matrix grids. After the picture has been segmented, each grid is classified and subjected to item localization. Each grid's objectness or confidence score is determined. In the next section, the bounding box prediction is described. Anchor boxes are also employed in order to improve and enhance the accuracy of object detection capabilities, as it is further detailed below.

Table 14.1 Improvement journey of YOLO from VI to V8

YOLO	Detection of objects is done using grid divisions.
YOLO V2	Anchoring is done with K-means added. Training is done in two different stages. Complete convolutional network (CN) is employed.
YOLO V3	In this version, detection is multi-scaling in nature. Feature Pyramid Network (FPN) is used.
YOLO V4	MISH activation function Spatial Pyramid Pooling (SPP)Loss function is used Data enhancement mosaic/mixup
YOLO V5	Flexibility in model size controlling, data enhancement, and hardswish activation function
YOLO V6	Separation of heads; additional layers are employed in the network for separating these features from the final head.
YOLO V7	This one is the fastest; this version is more accurate in real-time object detection.
YOLO V8	This one is the most cutting-edge, state-of-the-art model in terms of performance and versatility.

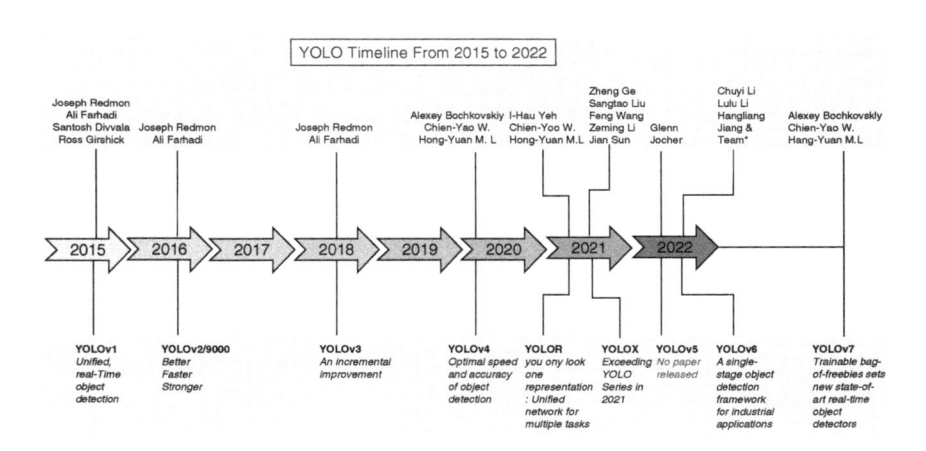

Figure 14.1 YOLO algorithm evolution timeline.

The workings of the module employing the cloud, server, and embedded system are shown in Figure 14.2. The YOLO framework is used by the backdrop detector to assist with object detection. We have modified YOLO in our system to only identify vehicles belonging to various categories, such as trucks, cars, bikes, etc. YOLO is installed on a server that connects to an embedded system through the Amazon Web Services (AWS) cloud. The relay is connected to the signals, and the Raspberry Pi board is connected to the on-site camera. We will have to use the Wi-Fi module that is linked to the Raspberry Pi to connect to the internet since we need to

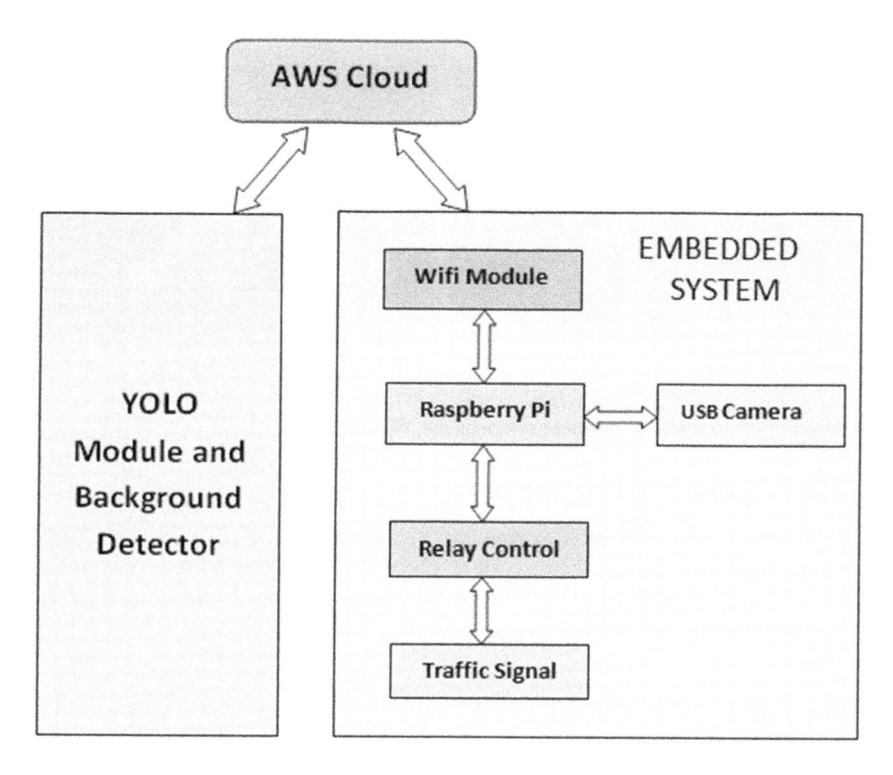

Figure 14.2 Architecture modules.

connect it to the cloud. Raspberry Pi facilitates the upload of footage from a camera to the cloud. To determine lane density, the backdrop detector uses an algorithm with a YOLO module. It is a clever method that uses a cutting-edge YOLO architecture to govern traffic flow.

14.4.1 How does YOLO object detection work

Let's have a high-level overview of how the YOLO algorithm performs object detection using a simple use case. Imagine we built a YOLO application that detects players and soccer balls from a given image. It can be understood how YOLO works and how to get image (b) from image (a) as shown in Figure 14.3.

The algorithm works on the following four approaches:

- Residual blocks approach,
- Bounding box regression,
- IOU, and
- Non-maximum suppression.

Figure 14.3 (a) Input and (b) output images for YOLO.

14.4.1.1 Residual blocks approach

It primary divides the actual image (A) into a grid of $N \times N$ cells having equal shapes as shown in Figure 14.4. Here, in the given image, N is 4. Each individual cell is accountable for predicting the class of the object covered by it, along with the probability/confidence value.

14.4.1.2 Bounding box regression

The bounding boxes that match the rectangles that highlight each object in the picture are determined in the second stage. Bounding boxes can be added to a picture in an equal number as the number of items it contains. For each of the bounding boxes, Y denotes the ultimate vector representation.

$$Y = \left[p_c, b_x, b_y, b_h, b_w, c_1, c_2 \right]$$

Figure 14.4 4×4 Grid cells of Input image.

- p_c is called the probability score related to the grid containing object, such as grids having a probability score higher than zero will be depicted as red.
- b_x, b_y are the x and y coordinates of the center of the bounding box for enveloping cells.
- b_h, b_w denotes the height and the width of the bounding box with respect to the enveloping grid cell.
- c_1 and c_2 corresponds to two classes (i.e., the player and the ball), respectively.

For better understanding, pay closer attention to the player on the bottom right in Figures 14.5 and 14.6.

Figure 14.5 Significant grid.

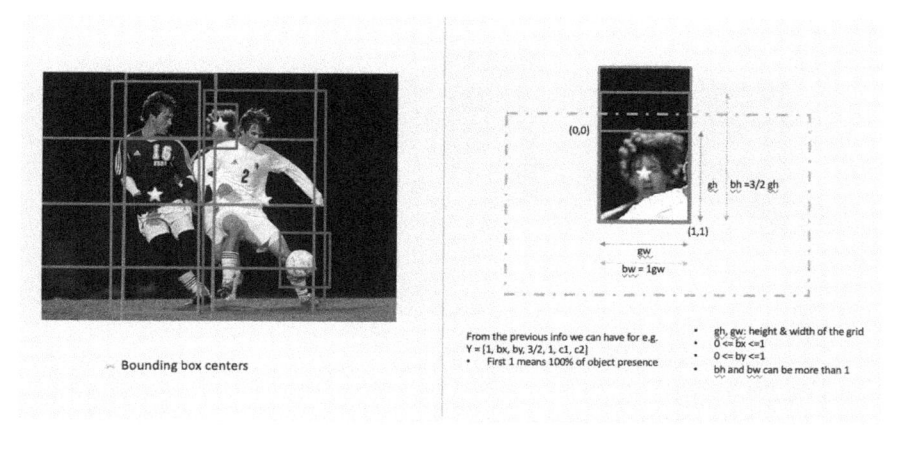

Figure 14.6 Bounding box center and its measurements.

14.4.1.3 IOU

Even while not all of the grid box candidates for a given item in a picture are meaningful, most of the time there might be many of them. By eliminating such grid boxes and keeping only the necessary ones, the IOU (a value between 0 and 1) seeks to achieve this. That makes sense as follows:

- IOU selection threshold is user defined, for instance, it is 0.5.
- IOU of individual grid cells are computed which is the intersection area divided by the union area.
- Next, it ignores the grid cells having an IOU less than or equal to the threshold and only considers those having IOU greater than the threshold.

Following example shows the application of grid selection process as shown in Figure 14.7.

14.5 YOLO ARCHITECTURE

YOLO can be considered a smart CNN used to recognize real-time moving objects. It can operate in real-time and attain more accuracy; YOLO is also considered well liked in the sense that it only has to perform one forward propagation run through the NN to produce predictions; the algorithm "only looks once" at the picture.

In our network, as many as 24 convolutional layers precede 2 fully linked layers. We only utilize 1×1 reduction layers followed by 3×3 convolutional layers in place of the inception modules used by Google neural network as depicted in Figure 14.8.

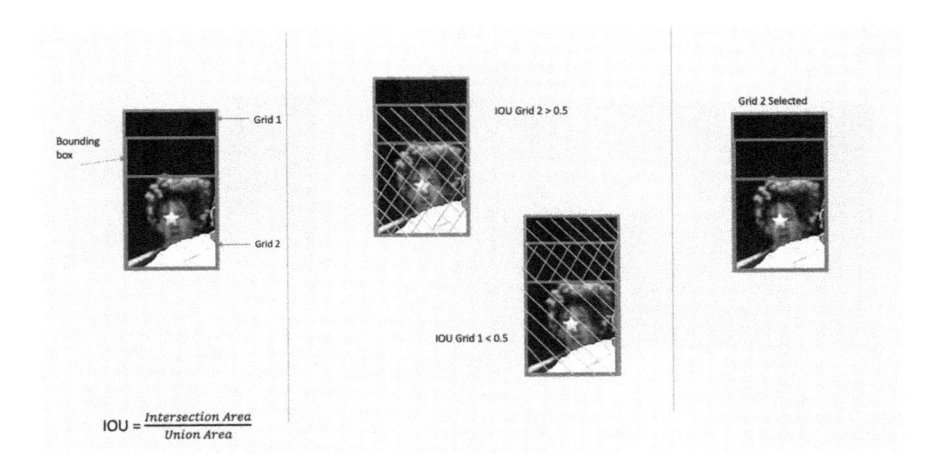

Figure 14.7 IOU grid selection.

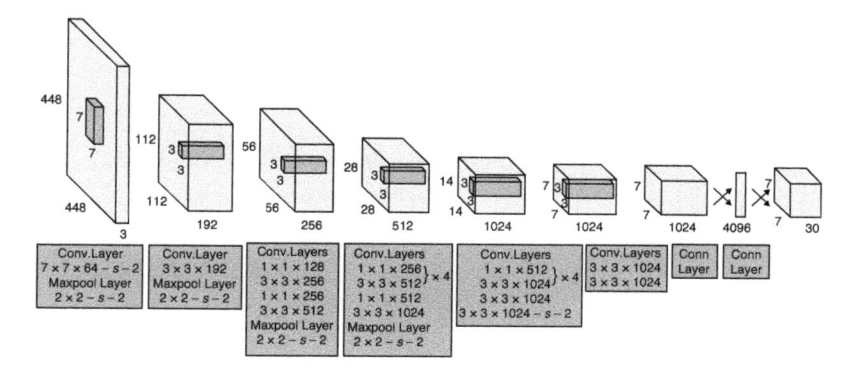

Figure 14.8 YOLO architecture layers.

14.5.1 DNN

DNN may be considered as an ANN having more than two layers in between the input and output layers. It chooses the mathematically optimal operation to transform the input into the output, regardless of whether there exists any kind of linear or non-linear relationship between the input and the output. Each output's probability is computed as the network moves through the levels. Upon examining the results, the user has the option to select those probabilities (those over a threshold, for example) the network displays and provides the recommended label.

Generally speaking, feedforward networks, or DNNs, transfer data directly from the input to the output layer. The DNN starts by creating a map of virtual neurons and assigning "weights," or random integer values, to the connections between them [45].

14.5.2 CNN

A classical CNN consists of an input layer, an output layer, and several other layers which are hidden. The intermediate hidden layers of a CNN are composed of convolutional layers. There are further convolutions after the function known as activation function, that is usually a rectified linear unit (ReLU) layer [46], including pooling layers, normalizing layers, and completely connected layers. These are called hidden layers of CNN as the input and output of these layers are concealed by the activation function and final convolution layer.

14.6 YOLO-BASED INTELLIGENT TRAFFIC MONITORING SYSTEM

An intelligent traffic monitoring system is described in this section. Three tasks are performed by the suggested system: vehicle detection, counting, and categorization. In Figure 14.9, the system architecture is depicted.

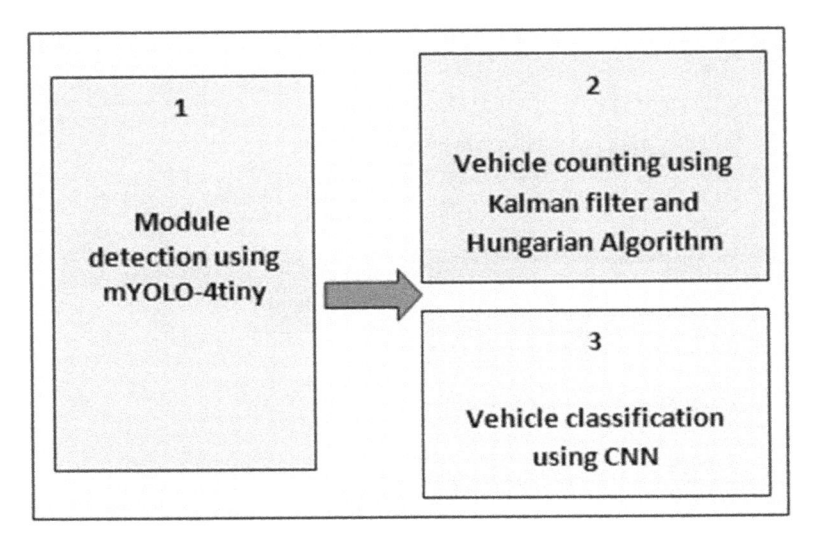

Figure 14.9 Functions involved in the proposed intelligent traffic monitoring system.

Figure 14.10 displays the flowchart of the suggested system. Traffic cameras are used to first acquire real-time photos of the roads. Next, the suggested mYOLOv4-tiny model is employed to determine a vehicle's location.

A counting method is implemented to track vehicles to address the issue of the redundant recording of the same vehicle in many different frames. To put it another way, a vehicle receives the same identification (ID) throughout all frames. To lessen the computing load, the target vehicle's virtual detection region is screened before the counting technique is run.

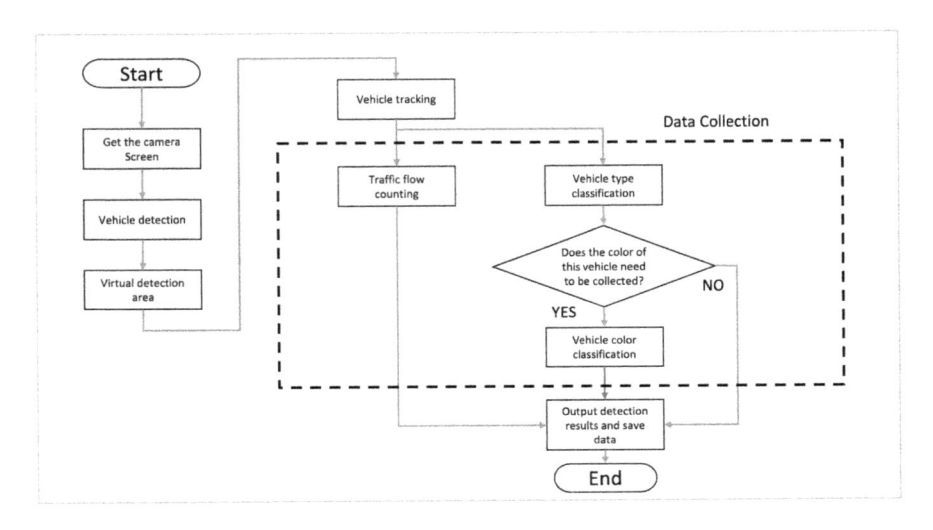

Figure 14.10 Flowchart of the proposed system.

14.6.1 Detection of vehicle using YOLOv4-tiny

Using YOLO, the traditional YOLOv4-tiny is a lighter version of the network. To extract object characteristics, max-pooling layers as well as convolutional layers are used. For admixing the features and extending the amount of feature data, YOLOv4-tiny also employs UpSampling and Concat layers. YOLOv4-tiny offers a quicker detection speed when compared to previous YOLO and SSD approaches. However, because of its much-reduced network design, YOLOV4-tiny's detection accuracy is poorer than that of YOLO and SSD techniques. Traditionally, the YOLOv4-tiny employs two outputs to detect objects. Three max pooling layers and 24 convolutional layers altogether were used. Finally, three scales were employed for prediction: 25×15, 50×30, and 100×60. The mYOLOv4-tiny model is exclusively utilized for vehicle object detection in this system.

14.6.1.1 Vehicle counting method

A car may be recognized and its position determined from a single image using the YOLO object detection approach previously discussed. However, a continuous visual frame is given as the input in practical traffic applications. Vehicles found in several picture frames are separate from one another. As a result, the same car gets counted more than once, and the data gathered on vehicles is inaccurate. The ID of the identified vehicle is set up to avoid multiple counting in order to fix this issue.

The multiobject counting approach [47] is used in this work to pre-detect the accurate location of the moving vehicle in the current frame by using vehicle position data.

14.6.1.2 Vehicle classification using CNN

The multiobject counting approach [47] is used in this work to forecast the vehicle position in the current frame from the previous one as collected by the detection procedure. IOU is then calculated and the vehicle's current position is anticipated using the Kalman filter. For tracking the vehicles, the Hungarian algorithm is then used to match vehicles.

The convolutional layer in the CNN model extracts features from the captured picture and the maximum pooling layers are used.

14.7 SYSTEM CONTROL FLOW

The system uses camera sensors on traffic lights which will detect vehicles as objects. After detecting the vehicles from all the passing sides of roads, the most important task is to make decision, "Which Road will get pass first?" For that, our first approach depends upon the number of vehicles.

The side which has more vehicles will get pass first and the traffic light on that side will turn into green. At the same time, other lights will be on hold, that is red (if very less vehicles) and orange (if slightly fewer vehicles). In this way, the system will work. We will also use ML for this project which will help the system to make decisions in some uncommon situations like sometimes if a very big vehicle comes, then it needs to pass first because it can cause massive jam due to its size.

Initially, all lanes of the traffic circle will transmit the footage. In the live video of these routes, the values will be read frame by frame. The cloud-based processing engine receives all of the input videos that the camera has recorded. As soon as the videos are received by the cloud, a background program filters them to determine how many vehicles are in each lane. Each lane is then prioritized depending on the threshold values, and the timeframe is set properly. YOLO works using a background algorithm to count vehicles in real time at a pace of 24 frames per second. For transport vehicles, density is calculated based on vehicle category; thus, containers will have a higher density than vehicles. Every lane in this case has its overall density recorded. There is a threshold value that is determined by two factors: the lane's priority and its density count.

Consider the following lanes: L1, L2, L3, L4,..., Ln in Figure 14.11. Each lane will have an equal duration T_{max}, which indicates the maximum amount of time that a given lane's green light will be lit. To prevent lengthier wait times in any lane with lower density, the lights will turn green one lane at a time. No green light will switch on for that lane until its density count meets the threshold if there is extremely little density, i.e., if the lane is empty. After being converted to a resolution of $S \times S$ size, the video of the lanes with red signals would be uploaded to the cloud. The object is subsequently detected by YOLO's background detector, after which the density count is determined and the lanes' priorities are set. This will make employing a Raspberry Pi with DL for automated traffic control effective.

There are many challenges that are to meet by the system. Particular situations are the case studies that must be taken into consideration, such as crossing of pedestrians during rush hour; passing of emergency priority vehicles, if needed, if any kind of road blockage happened due to a road accident or other reasons; etc.

14.7.1 Handling emergency or priority vehicles

For handling emergency vehicles case, we will add a note case algorithm which will be able to detect emergency vehicles like ambulance, police car, etc. If there is any emergency vehicle, then our whole traffic module will give access to that lane first. When that emergency vehicle passes away the whole module will work as it was working earlier. The illustrative visuals are given below.

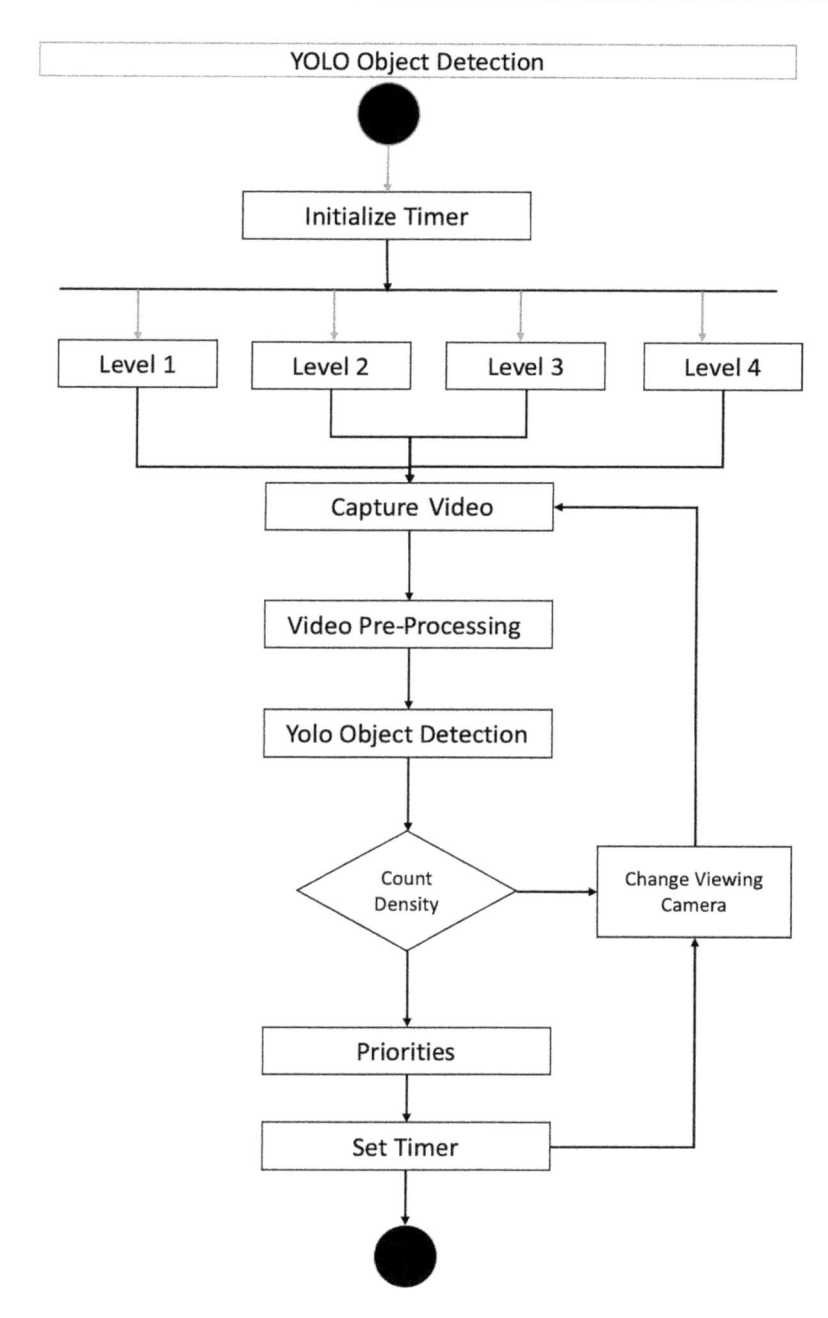

Figure 14.11 System control flow.

14.7.2 Pedestrian case in automated traffic light signal

Automated traffic light signals are engineered to control the flow of transportation and pedestrian traffic at junctions. In the case of pedestrian traffic, automated traffic lights incorporate various features and systems to ensure the safe and efficient movement of pedestrians across roadways. Here are some key aspects related to pedestrian cases in automated traffic light signals:

1. **Pedestrian crosswalk:** Automated traffic light signal system includes designated crosswalk for pedestrian, which are typically marked with painted lines on the road surface. These crosswalks provide a designated area for pedestrians to safely cross the road.
2. **Pedestrian signal phase:** Traffic light signals will have specific phases dedicated to pedestrian. These phases include the "Walk" signal which indicates that the pedestrian can cross the road and "Wait" signal which indicates that the pedestrian should not cross or should finish crossing if already in progress.
3. **Signal timing:** The timing of pedestrian signal phases is determined based on factors, such as the width of the road, the expected numbers of pedestrians, and the traffic patterns in the area.
4. **Pedestrian detectors:** Many automated traffic light signals use sensors and detectors to detect the pedestrians at crosswalk. These detectors can be in the form of pressure-sensitive mats, infrared sensors, or cameras.

14.7.3 Working flow

STEP 1: Firstly, the data are taken by the sensors.

STEP 2: The camera will look in the particular direction and determine the number of people in each lane among the four.

STEP 3: Camera will detect pedestrian crossing or intending to cross the road.

STEP 4: After detecting, the person who needs the most or who is in a hurry the signal will notify that particular person to proceed first.

STEP 5: And on the road, markers are activated from where a pedestrian can proceed and it will not allow any vehicle to move until the needy person will not cross the road.

STEP 6: After allowing that person to proceed, the signal will allow all the pedestrians to proceed and clear that road.

STEP 7: Again, the same thing will happen to the next lane.

Figure 14.12 shows the pedestrian detection method as described above.

Figure 14.12 Pedestrian detection method.

14.7.4 Objectives

The objectives of pedestrian cases in automated traffic light signals primarily focuses on ensuring the safety and efficient movement of pedestrians at intersections. Here are some common objectives:

A. **Pedestrian safety:** The foremost objective is to protect the pedestrian and minimize the risk of accidents or collisions involving the vehicles.
B. **Efficient pedestrian flow:** Automated traffic light signals are designed to allocate appropriate time intervals for pedestrians to cross the road.
C. **Accessibility:** The traffic signals should cater to the needs of all pedestrians, including those with a disability or reduce mobility.

14.7.5 Limitations and considerations

1. Automated traffic light signals operate based on pre-defined timing patterns or fixed algorithms. They may not be able to respond effectively to changing traffic conditions or unexpected events.
2. In some cases, individual traffic signals may not be adequately coordinated with each other, leading to suboptimal traffic flow.
3. Automated traffic light signals may not provide enough flexibility to accommodate specific traffic needs or patterns.
4. As cities strive to promote sustainable transportation options like public transit, cycling, and walking, automated traffic light signals may not adequately prioritize these modes.
5. Automated traffic light signals lack human judgment and intuition. It may struggle to handle complex or exceptional scenarios that require subjective judgment.

The future of automated traffic light signals holds significant potential for advancements and improvements in traffic management. Here are some of the key areas where we can expect to see developments.

- Pedestrian detection and priority,
- Pedestrian-friendly signal and timings.
- Improved accessibility,
- Smart pedestrian crossing system, and
- Integration with mobile apps and devices.

14.7.6 Challenges to handle

Uses of object detection technology span various sectors, encompassing autonomous vehicles, recognition of objects in aerial views, text recognition, security surveillance, search and rescue missions, robotic vision, identifying faces, tracking pedestrians, enhancing visual search engines, measuring items of interest, recognizing brands, and numerous other applications.

Majority of ML- and DL-based object detector algorithms fail to handle the following typical challenges:

- **Training across different scales:** Commonly, object detection models are optimized for images of a specific resolution. When faced with images of different scales or resolutions, these models often exhibit reduced accuracy.
- **Foreground–background class imbalance:** An imbalance between the number of examples from different classes, particularly foreground and background categories, can notably influence the effectiveness of a model.
- **Detection of smaller objects:** All object detection techniques will work well on larger objects if the model is trained for objects larger than others. Hence, these models perform poorly on smaller objects.
- **Requirement for computing power and huge datasets:** DL object tracking and detection techniques need large datasets for computation. They also need powerful computational resources for processing [48]. Due to the rise in produced data from multiple sources exponentially, annotating each and every item in the visual contents has become a tiresome and labor-intensive operation [49].
- **Smaller datasets:** While DL models perform better than classic ML algorithms by a wide margin, they perform poorly when tested on smaller datasets.
- **Incorrect localization and wrong predictions:** Background pixels are typically incorporated during predictions, which impact the algorithm's accuracy. The majority of localization mistakes are caused by background occupancy in forecasts and identifying comparable objects [48].

14.8 CONCLUSION

Automated traffic light signals in the case of pedestrians offer significant benefits in terms of safety and traffic efficiency. By utilizing advanced technologies, such as pedestrian sensors and real-time data analysis, this system can enhance pedestrian safety, optimize traffic flow, and improve the overall transportation experience. However, careful implementation, reliable sensors, and public awareness campaigns are crucial to ensure the effectiveness and successful integration of these automated systems.

REFERENCES

1. A. Mohamed, A. Issam, B. Mohamed, and B. Abdellatif, "Real-time detection of vehicles using the Haar-like features and artificial neuron networks," *Proc. Comput. Sci.*, vol. 73, pp. 24–31, Jan. 2015.
2. X. Wen, L. Shao, W. Fang, and Y. Xue, "Efficient feature selection and classification for vehicle detection," *IEEE Trans. Circuits Syst. Video Technol.*, vol. 25, no. 3, pp. 508–517, Mar. 2015.
3. Z. Sun, G. Bebis, and R. Miller, "On-road vehicle detection using Gabor filters and support vector machines," In *Proc. 14th Int. Conf. Digit. Signal Process. (DSP)*, Jul. 2002, pp. 1019–1022. https://ieeexplore.ieee.org/document/1028263
4. H. David and T. A. Athira, "Improving the performance of vehicle detection and verification by log Gabor filter optimization," In *Proc. 4th Int. Conf. Adv. Comput. Commun.*, Aug. 2014, pp. 50–55. https://ieeexplore.ieee.org/document/6905987
5. Y. Wei, Q. Tian, J. Guo, W. Huang, and J. Cao, "Multi-vehicle detection algorithm through combining Harr and HOG features," *Math. Comput. Simul.*, vol. 155, pp. 130–145, Jan. 2018.
6. S. Bougharriou, F. Hamdaoui, and A. Mtibaa, "Linear SVM classifier based HOG car detection," In *Proc. 18th Int. Conf. Sci. Techn. Autom. Control Comput. Eng. (STA)*, Dec. 2017, pp. 241–245. https://ieeexplore.ieee.org/document/8314922
7. G. Yan, M. Yu, Y. Yu, and L. Fan, "Real-time vehicle detection using histograms of oriented gradients and AdaBoost classification," *Optik*, vol. 127, no. 19, pp. 7941–7951, 2016.
8. N. Seenouvong, U. Watchareeruetai, C. Nuthong, K. Khongsomboon, and N. Ohnishi, "A computer vision based vehicle detection and counting system," In *Proc. 8th Int. Conf. Knowl. Smart Technol. (KST)*, Feb. 2016, pp. 224–227. https://ieeexplore.ieee.org/document/7440510/
9. P. K. Bhaskar and S.-P. Yong, "Image processing based vehicle detection and tracking method," In *Proc. Int. Conf. Comput. Inf. Sci. (ICCOINS)*, Jun. 2014, pp. 1–5. https://ieeexplore.ieee.org/document/6868357

10. N. Seenouvong, U. Watchareeruetai, C. Nuthong, K. Khongsomboon, and N. Ohnishi, "Vehicle detection and classification system based on virtual detection zone," In *Proc. 13th Int. Joint Conf. Comput. Sci. Softw. Eng. (JCSSE)*, Jul. 2016, pp. 1–5. https://ieeexplore.ieee.org/document/7748886

11. M. Anandhalli and V. P. Baligar, "Improvised approach using background subtraction for vehicle detection," In *Proc. IEEE Int. Advance Comput. Conf. (IACC)*, Jun. 2015, pp. 303–308. https://ieeexplore.ieee.org/document/7154719

12. N. S. Sakpal and M. Sabnis, "Adaptive background subtraction in images," In *Proc. Int. Conf. Adv. Commun. Comput. Technol. (ICACCT)*, Feb. 2018, pp. 439–444. https://ieeexplore.ieee.org/document/8529323

13. N. Shah, A. Pingale, V. Patel, and N. V. George, "An adaptive background subtraction scheme for video surveillance systems," In *Proc. IEEE Int. Symp. Signal Process. Inf. Technol. (ISSPIT)*, Dec. 2017, pp. 13–17. https://ieeexplore.ieee.org/document/8388311

14. Y. LeCun, L. Bottou, Y. Bengio, and P. Haffner, "Gradient-based learning applied to document recognition," *Proc. IEEE*, vol. 86, no. 11, pp. 2278–2324, Nov. 1998.

15. A. Krizhevsky, I. Sutskever, and G. Hinton, "ImageNet classification with deep Convolutional Neural Networks," In *Proc. 25th Int. Conf. Neural Inf. Process. Syst. (NIPS)*, vol. 1, Dec. 2012, pp. 1097–1105. https://papers.nips.cc/paper_files/paper/2012/hash/c399862d3b9d6b76c8436e924a68c45b-Abstract.html

16. C. Szegedy, W. Liu, Y. Jia, P. Sermanet, S. Reed, D. Anguelov, D. Erhan, V. Vanhoucke, and A. Rabinovich, "Going deeper with convolutions," In *Proc. IEEE Conf. Comput. Vis. Pattern Recognit. (CVPR)*, Jun. 2015, pp. 1–9. https://arxiv.org/abs/1409.4842

17. K. Simonyan and A. Zisserman, "Very deep convolutional networks for large-scale image recognition," 2014, arXiv:1409.1556.

18. A. G. Howard, M. Zhu, B. Chen, D. Kalenichenko, W. Wang, T. Weyand, M. Andreetto, and H. Adam, "MobileNets: Efficient Convolutional Neural Networks for mobile vision applications," 2017, arXiv:1704.04861.

19. K. Shi, H. Bao, and N. Ma, "Forward vehicle detection based on incremental learning and fast R-CNN," In *Proc. 13th Int. Conf. Comput. Intell. Secur. (CIS)*, Dec. 2017, pp. 73–76. https://ieeexplore.ieee.org/document/8288446

20. S.-C. Hsu, C.-L. Huang, and C.-H. Chuang, "Vehicle detection using simplified fast R-CNN," In *Proc. Int. Workshop Adv. Image Technol. (IWAIT)*, Jan. 2018, pp. 1–3. https://ieeexplore.ieee.org/document/8369767

21. S. Rujikietgumjorn and N. Watcharapinchai, "Vehicle detection with sub-class training using R-CNN for the UA-DETRAC benchmark," In *Proc. 14th IEEE Int. Conf. Adv. Video Signal Based Surveill. (AVSS)*, Aug. 2017, pp. 1–5. https://ieeexplore.ieee.org/document/8078520

22. W. Zhang, Y. Zheng, Q. Gao, and Z. Mi, "Part-aware region proposal for vehicle detection in high occlusion environment," *IEEE Access*, vol. 7, pp. 100383–100393, 2019.

23. L. Wang, Y. Lu, H. Wang, Y. Zheng, H. Ye, and X. Xue, "Evolving boxes for fast vehicle detection," In *Proc. IEEE Int. Conf. Multimedia Expo. (ICME)*, Jul. 2017, pp. 1135–1140. https://ieeexplore.ieee.org/document/8019461

24. S. Ren, K. He, R. Girshick, and J. Sun, "Faster R-CNN: Towards real-time object detection with region proposal networks," *IEEE Trans. Pattern Anal. Mach. Intell.*, vol. 39, no. 6, pp. 1137–1149, Jun. 2017.

25. T.-Y. Lin, P. Goyal, R. Girshick, K. He, and P. Dollar, "Focal loss for dense object detection," *IEEE Trans. Pattern Anal. Mach. Intell.*, vol. 42, no. 2, pp. 318–327, Feb. 2020.

26. N. A. Al-Sammarraie, Y. M. H. Al-Mayali, and Y. A. Baker El-Ebiary, "Classification and diagnosis using back propagation artificial neural networks (ANN)," In *Proc. Int. Conf. Smart Comput. Electron. Enterprise (ICSCEE)*, Jul. 2018, pp. 1–5. https://ieeexplore.ieee.org/document/8538383

27. O. I. Abiodun, A. Jantan, A. E. Omolara, K. V. Dada, N. A. Mohamed, and H. Arshad, "State-of-the-art in artificial neural network applications: A survey," *Heliyon*, vol. 4, no. 11, Nov. 2018, Article no. e00938.

28. J. Redmon and A. Farhadi, "YOLOv3: An incremental improvement," 2018, arXiv:1804.02767.

29. A. Bochkovskiy, C.-Y. Wang, and H.-Y. M. Liao, "YOLOv4: Optimal speed and accuracy of object detection," 2020, arXiv:2004.10934.

30. Z. Jiang, L. Zhao, S. Li, and Y. Jia, "Real-time object detection method based on improved YOLOv4-tiny," 2020, arXiv:2011.04244.

31. W. Liu, D. Anguelov, D. Erhan, C. Szegedy, S. Reed, C. Fu, and A. C. Berg, "SSD: Single shot multibox detector," In *Proc. Eur. Conf. Comput. Vis.*, Amsterdam, The Netherlands, Oct. 2016, pp. 21–37. https://link.springer.com/chapter/10.1007/978-3-319-46448-0_2

32. J. Huang, V. Rathod, C. Sun, M. Zhu, A. Korattikara, A. Fathi, I. Fischer, Z. Wojna, Y. Song, S. Guadarrama, and K. Murphy, "Speed/accuracy trade-offs for modern convolutional object detectors," In *Proc. IEEE Conf. Comput. Vis. Pattern Recognit. (CVPR)*, Jul. 2017, pp. 3296–3297. https://arxiv.org/abs/1611.10012

33. P. Soviany and R. T. Ionescu, "Optimizing the trade-off between single-stage and two-stage deep object detectors using image difficulty prediction," In *Proc. 20th Int. Symp. Symbolic Numeric Algorithms Sci. Comput. (SYNASC)*, Sep. 2018, pp. 209–214. https://arxiv.org/abs/1803.08707

34. Y. Asim, B. Raza, A. K. Malik, A. R. Shahid, M. Faheem, and Y. J. Kumar, "A hybrid adaptive neuro-fuzzy inference system (ANFIS) approach for professional bloggers classification," In *Proc. 22nd Int. Multitopic Conf. (INMIC)*, Nov. 2019, pp. 1–6. https://ieeexplore.ieee.org/document/9022776

35. C.-J. Lin, J.-Y. Jhang, S.-H. Chen, and K.-Y. Young, "Using an interval type-2 fuzzy neural network and tool chips for flank wear prediction," *IEEE Access*, vol. 8, pp. 122626–122640, 2020.

36. D. K. Bebarta, R. Bisoi, and P. K. Dash, "Locally recurrent functional link fuzzy neural network and unscented H-infinity filter for shortterm prediction of load time series in energy markets," In *Proc. IEEE Power, Commun. Inf. Technol. Conf. (PCITC)*, Oct. 2015, pp. 663–670. https://ieeexplore.ieee.org/document/7438080

37. J.-W. Yeh and S.-F. Su, "Efficient approach for RLS type learning in TSK neural fuzzy systems," *IEEE Trans. Cybern.*, vol. 47, no. 9, pp. 2343–2352, Sep. 2017.

38. C.-J. Lin, C.-H. Lin, and J.-Y. Jhang, "Dynamic system identification and prediction using a self-evolving Takagi-Sugeno-Kang-type fuzzy CMAC network," *Electronics*, vol. 9, no. 4, p. 631, Apr. 2020.

39. M. Lin, Q. Chen, and S. Yan, "Network in network," 2013, arXiv:1312.4400.

40. Z. Ma, D. Chang, J. Xie, Y. Ding, S. Wen, X. Li, Z. Si, and J. Guo, "Fine-grained vehicle classification with channel max pooling modified CNNs," *IEEE Trans. Veh. Technol.*, vol. 68, no. 4, pp. 3224–3233, Apr. 2019.

41. V. Christlein, L. Spranger, M. Seuret, A. Nicolaou, P. Kral, and A. Maier, "Deep generalized max pooling," In *Proc. Int. Conf. Document Anal. Recognit. (ICDAR)*, Sep. 2019, pp. 1090–1096. https://arxiv.org/abs/1908.05040

42. Z. Li, S.-H. Wang, R.-R. Fan, G. Cao, Y.-D. Zhang, and T. Guo, "Teeth category classification via seven-layer deep Convolutional Neural Network with max pooling and global average pooling," *Int. J. Imag. Syst. Technol.*, vol. 29, no. 4, pp. 577–583, May 2019.

43. Z. Gao, Y. Li, Y. Yang, N. Dong, X. Yang, and C. Grebogi, "A coincidence-filtering-based approach for CNNs in EEG-based recognition," *IEEE Trans. Ind. Informat.*, vol. 16, no. 11, pp. 7159–7167, Nov. 2020.

44. L. Cheng, D. Chang, J. Xie, R. Ma, C. Wu, and Z. Ma, "Channel max pooling for image classification," In *Intelligence Science and Big Data Engineering. Visual Data Engineering*, Z. Cui, J. Pan, S. Zhang, L. Xiao, and J. Yang, Eds. Cham, Switzerland: Springer, 2019, pp. 273–284.

45. A. Zhou, A. Yao, Y. Guo, L. Xu, and Y. Chen, "Incremental network quantization: Towards lossless CNNs with lowprecision weights," 2017, arXiv:1702.03044.

46. A. F. Agarap, "Deep learning using rectified linear units (ReLU)," 2018, arXiv:1803.08375.

47. A. Bewley, Z. Ge, L. Ott, F. Ramos, and B. Upcroft, "Simple online and realtime tracking," In *Proc. IEEE Int. Conf. Image Process. (ICIP)*, Sep. 2016, p. 346.

48. L. Liu, W. Ouyang, X. Wang, P. Fieguth, J. Chen, X. Liu, and M. Pietikäinen Deep learning for generic object detection: A survey. *Int. J. Comput. Vis.*, vol. 128, no. 2, pp. 261–318, 2020. doi:10.48550/arXiv.1809.02165

49. C.Y. Wang, H. Y. Mark Liao, Y. H. Wu, P. Y. Chen, J.W. Hsieh, and I. H. Yeh CSPNet: A new backbone that can enhance learning capability of CNN. In: *Proc. IEEE/CVF Conf. Comput. Vis. Pattern Recog. Workshops*, pp. 390–391, 2020. https://arxiv.org/abs/1911.11929

Index

www.ingramcontent.com/pod-product-compliance
Ingram Content Group UK Ltd.
Pitfield, Milton Keynes, MK11 3LW, UK
UKHW021243190125
453847UK00005B/23